Electron Transport in Nanosystems

NATO Science for Peace and Security Series

This Series presents the results of scientific meetings supported under the NATO Programme: Science for Peace and Security (SPS).

The NATO SPS Programme supports meetings in the following Key Priority areas: (1) Defence Against Terrorism; (2) Countering other Threats to Security and (3) NATO, Partner and Mediterranean Dialogue Country Priorities. The types of meeting supported are generally "Advanced Study Institutes" and "Advanced Research Workshops". The NATO SPS Series collects together the results of these meetings. The meetings are co-organized by scientists from NATO countries and scientists from NATO's "Partner" or "Mediterranean Dialogue" countries. The observations and recommendations made at the meetings, as well as the contents of the volumes in the Series, reflect those of parti-cipants and contributors only; they should not necessarily be regarded as reflecting NATO views or policy.

Advanced Study Institutes (ASI) are high-level tutorial courses intended to convey the latest developments in a subject to an advanced-level audience

Advanced Research Workshops (ARW) are expert meetings where an intense but informal exchange of views at the frontiers of a subject aims at identifying directions for future action

Following a transformation of the programme in 2006 the Series has been re-named and re-organised. Recent volumes on topics not related to security, which result from meetings supported under the programme earlier, may be found in the NATO Science Series.

The Series is published by IOS Press, Amsterdam, and Springer, Dordrecht, in conjunction with the NATO Public Diplomacy Division.

Sub-Series

A.	Chemistry and Biology	Springer
B.	Physics and Biophysics	Springer
C.	Environmental Security	Springer
D.	Information and Communication Security	IOS Press
E.	Human and Societal Dynamics	IOS Press

http://www.nato.int/science
http://www.springer.com
http://www.iospress.nl

Series B: Physics and Biophysics

Electron Transport in Nanosystems

Edited by

Janez Bonča
J. Stefan Institute,
Ljubljana, Slovenia

and

Sergei Kruchinin
Bogolyubov Institute for Theoretical Physics,
Kiev, Ukraine

 Springer

Published in cooperation with NATO Public Diplomacy Division

Proceedings of the NATO Advanced Research Workshop on
Electron Transport in Nanosystems
Yalta, Ukraine
17–21 September 2007

Library of Congress Control Number: 2008936556

ISBN 978-1-4020-9145-2 (PB)
ISBN 978-1-4020-9144-5 (HB)
ISBN 978-1-4020-9146-9 (e-book)

Published by Springer,
P.O. Box 17, 3300 AA Dordrecht, The Netherlands.

www.springer.com

Printed on acid-free paper

PREFACE

These proceedings of the NATO-ARW "Electron transport in nanosystems" held at the "Russia" Hotel, Yalta, Ukraine from 17–21 September 2007 resulted in many discussions between various speakers.

The wide range of topics discussed at the Yalta NATO meeting included the new nanodevice applications, novel materials, superconductivity and sensors. There have been many significant advances in the past 2 years and some entirely new directions of research in these fields are just opening up. Recent advances in nanoscience have demonstrated that fundamentally new physical phenomena are found when systems are reduced in size with dimensions, comparable to the fundamental microscopic length scales of the investigated material. Late developments in nanotechnology and measurement techniques now allow experimental investigation of transport properties of nanodevices. Great interest in this research is focused on development of spintronics, molecular electronics and quantum information processing and graphene. At the workshop, important open problems concerning cuprate superconductity, mesoscopic superconductors and novel superconductors such MgB_2, $CeCoIn_5$ where considered. There was much discussion of the mechanism and symmetry of pairing for cuprate superconductors as well as the nature of the pseudogap. In the session on novel superconductors, the physical properties of MgB_2 were discussed. There were also lively debates about two-gap superconductivity in MgB_2.

We would like to thank the NATO Science Committee for the essential financial support, without which the meeting could not have taken place. We also acknowledge the National Academy of Science of Ukraine, Ministry of Ukraine for Education and Science, J. Stefan Institute, Ljubljana, Slovenia and Faculty of Mathematics and Physics, University of Ljubljana, Slovenia for their generous support.

Ljubljana, Kiev,
June 2008

Janez Bonča
Sergei Kruchinin

CONTENTS

Part IV Sensors

25 Rapid methods for multiply determining potent xenobiotics based on the optoelectronic imaging

26 Electronic nanosensors based on nanotransistor with bistability behaviour

27 A bio-inspired electromechanical system: Artificial hair cell

LIST OF CONTRIBUTORS

Fujiwara A. School of Materials Science, Japan Advanced Institute of Science and Technology, 1-1 Asahidai, Nomi, Ishikawa 923-1292, Japan. fujiwara@jaist.ac.jp

Konishi A. School of Materials Science, Japan Advanced Institute of Science and Technology, 1-1 Asahidai, Nomi, Ishikawa 923-1292, Japan

Shikoh E. School of Materials Science, Japan Advanced Institute of Science and Technology, 1-1 Asahidai, Nomi, Ishikawa 923-1292, Japan

Nawrocki W. Faculty of Electronics and Telecommunications, Poznan University of Technology, ul. Piotrowo 3, 60-965 Poznan, Poland. nawrocki@et.put.poznan.pl

Wade T.L. Laboratoire des Solides Irradies ECOLE Polytechnique Route de Saclay, 91128 Palaiseau Cedex France. travis.wade@polytechnique.edu

Abdulla A.A. Laboratoire des Solides Irradies ECOLE Polytechnique Route de Saclay, 91128 Palaiseau Cedex France

Ciornei M.C. Laboratoire des Solides Irradies ECOLE Polytechnique Route de Saclay, 91128 Palaiseau Cedex France

Pribat D. Laboratoire de Physique des Interfaces et des Couches Minces ECOLE Polytechnique, Route de Saclay, Palaiseau France

Cojocaru C. Laboratoire des Solides Irradie ECOLE Polytechnique Route de Saclay, 91128 Palaiseau Cedex France

Wegrowe J.-E. Laboratoire des Solides Irradies ECOLE Polytechnique Route de Saclay, 91128 Palaiseau Cedex France

Gorczyca A. Department of Theoretical Physics, Institute of Physics, University of Silesia, 40-007 Katowice, Poland. marcin@phys.us.edu.pl

Maśka M. Department of Theoretical Physics, Institute of Physics, University of Silesia, 40-007 Katowice, Poland

Mierzejewski M. Department of Theoretical Physics, Institute of Physics, University of Silesia, 40-007 Katowice, Poland

Koong Chee Wen Department of Physics, National University of Singapore, Singapore 117542 and IMRE, 3 Research Link, Singapore 117602

Chandrasekhar N. Department of Physics, National University of Singapore, Singapore 117542 and IMRE, 3 Research Link, Singapore 117602. n-chandra@imre.a-star.edu.sg

Miniatura C. Department of Physics, National University of Singapore, Singapore 117542 and INLN, UNS, CNRS, 1361 route des Lucioles, F-06560 Valbonne, France

Berthold-Georg Englert Department of Physics, National University of Singapore, Singapore 117542 and Centre for Quantum Technologies, National University of Singapore, Singapore 117543

Burmistrov I.S. L.D. Landau Institute for Theoretical Physics, Kosygina street 2, 117940 Moscow, Russia. burmi@itp.ac.ru

Pruisken A.M.M. Institute for Theoretical Physics, University of Amsterdam, Valcke- nierstraat 65, 1018XE Amsterdam, The Netherlands. pruisken@science.uva.nl

Knight J.M. Department of Physics and Astronomy University of South Carolina, Columbia, SC 29208, USA. knight@physics.sc.edu

Kunchur M.N. Department of Physics and Astronomy University of South Carolina, Columbia, SC 29208, USA. kunchur@sc.edu

Shanenko A.A. TGM, Departe- ment Fysica, Universiteit Antwerpen, Groenenborgerlaan 171, B-2020 Antwerpen, Belgium. arkady.shanenko@ua.ac.be and Bogoliubov Laboratory of Theo- retical Physics, Joint Institute for Nuclear Research, 141980 Dubna, Russia

Croitoru M.D. EMAT, Departement Fysica, Universiteit Antwerpen, Groenenborgerlaan 171, B-2020 Antwerpen, Belgium

Peeters F.M. TGM, Departement Fysica, Universiteit Antwerpen, Groenenborgerlaan 171, B-2020 Antwerpen, Belgium

Nagao H. Division of Mathematical and Physical Science, Graduate School of Natural Science and Technology, Kanazawa University, Kakuma, Kanazawa 920-1192, Japan. nagao@wriron1.s.kanazawa-u. ac.jp

Kruchinin S.P. Bogolyubov Institute for Theoretical Physics, The Ukrainian National Academy of Science, Kiev 03143, Ukraine. skruchin@i.com.ua

Rydh A. Department of Physics, Stockholm University, AlbaNova Uni- versity Center, SE-106 91 Stockholm, Sweden. arydh@physto.se

Tagliati S. Department of Physics, Stockholm University, AlbaNova University Center, SE-106 91 Stockholm, Sweden

Nilsson R.A. Department of Physics, Stockholm University, AlbaNova University Center, SE-106 91 Stockholm, Sweden

Xie R. Materials Science Division, Argonne National Laboratory, 9700 South Cass Avenue, Argonne, IL 60439, USA

Pearson J.E. Materials Science Division, Argonne National Laboratory, 9700 South Cass Avenue, Argonne, IL 60439, USA

Welp U. Materials Science Division, Argonne National Laboratory, 9700 South Cass Avenue, Argonne, IL 60439, USA

Kwok W.-K. Materials Science Division, Argonne National Laboratory, 9700 South Cass Avenue, Argonne, IL 60439, USA

Divan R. Center for Nanoscale Materials, Argonne National Laboratory, 9700 South Cass Avenue, Argonne, IL 60439, USA

Movshovich R. Los Alamos National Laboratory, MS K764, Los Alamos, NM 87545 USA. roman@lanl.gov

Tokiwa Y. Los Alamos National Laboratory, MS K764, Los Alamos, NM 87545 USA

Ronning F. Los Alamos National Laboratory, MS K764, Los Alamos, NM 87545 USA

Bianchi A. Los Alamos National Laboratory, MS K764, Los Alamos, NM 87545 USA

Capan C. Los Alamos National Laboratory, MS K764, Los Alamos, NM 87545 USA

Young B.L. Los Alamos National Laboratory, MS K764, Los Alamos, NM 87545 USA

Urbano R.R. Los Alamos National Laboratory, MS K764, Los Alamos, NM 87545 USA

Curro N.J. Los Alamos National Laboratory, MS K764, Los Alamos, NM 87545 USA

Park T. Los Alamos National Laboratory, MS K764, Los Alamos, NM 87545 USA

Thompson J.D. Los Alamos National Laboratory, MS K764, Los Alamos, NM 87545 USA

Bauer E. Los Alamos National Laboratory, MS K764, Los Alamos, NM 87545 USA

Sarrao J.L. Los Alamos National Laboratory, MS K764, Los Alamos, NM 87545 USA

Alexandrov A.S. Department of Physics, Loughborough University, Loughborough, United Kingdom. a.s.alexandrov@lboro.ac.uk

Latyshev Yu.I. Institute of Radio-Egineering and Electronics, Russian Academy of Sciences, Mokhovaya 11-7, 125009 Moscow, Russia. lat@mail.cplire.ru

Kristoffel N. Institute of Theoretical Physics, University of Tartu, Tähe 4,51010 Tartu, Estonia. kolja@fi.tartu.ee

Rubin P. Institute of Physics, University of Tartu, Riia 142, 51014 Tartu, Estonia

Baňacký P. Faculty of Natural Science, Institute of Chemistry, Chemical Physics division, Comenius University, Mlynská dolina CH2, 84215 Bratislava, Slovakia and S-Tech a.s., Dubravská cesta 9, 84105 Bratislava, Slovakia. banacky@fns.uniba.sk

Dow J.D. Department of Physics, Arizona State University, Tempe, AZ 85287-1504 U.S.A. cats@dancris.com

Harshman D.R. Physikon Research Corporation, Lynden, WA 98264 U.S.A. and Department of Physics, Arizona State University, Tempe, AZ 85287-1504 U.S.A.

Freericks J.K. Department of Physics, Georgetown University, Washington, DC 20057, U.S.A. freericks@physics.georgetown.edu

Joura A.V. Department of Physics, Georgetown University, Washington, DC 20057, U.S.A.

Bennemann K.H. Institut für Theoretische Physik, Freie Universität Berlin, Germany. khb@physik.fu-berlin.de

Pruschke T. Institute for Theoretical Physics, University of Göttingen, Friedrich-Hund-Platz 1, D-37077 Göttingen, Germany. pruschke@theorie.physik.uni-goettingen.de

Gezzi R. Institute for Theoretical Physics, University of Göttingen, Friedrich-Hund-Platz 1, D-37077 Göttingen, Germany

Dirks A. Institute for Theoretical Physics, University of Göttingen, Friedrich-Hund-Platz 1, D-37077 Göttingen, Germany

Prischepa S.L. Belarus State University of Informatics and Radioelectronics, P. Brovka str.6, Minsk 220013, Belarus. prischepa@bsuir.by

Kushnir V.N. Belarus State University of Informatics and Radioelectronics, P. Brovka str.6, Minsk 220013, Belarus

Ilyina E.A. Belarus State University of Informatics and Radioelectronics, P. Brovka str.6, Minsk 220013, Belarus

Attanasio C. Dipartimento di Fisica "E.R. Caianiello" and Laboratorio Regionale SuperMat CNR/INFM-Salerno, Università degli Studi di Salerno, Baronissi (Sa), I-84081, Italy

Cirillo C. Dipartimento di Fisica "E.R. Caianiello" and Laboratorio Regionale SuperMat CNR/INFM-Salerno, Università degli Studi di Salerno, Baronissi (Sa), I-84081, Italy

Aarts J. Kamerling Onnes Laboratory, Leiden University, P.O. Box 9504, 2300 RA Leiden, The Netherlands

Chtchelkatchev N.M. L.D. Landau Institute for Theoretical Physics, Russian Academy of Sciences, 117940 Moscow, Russia. `nms@itp.ac.ru`

Burmistrov I.S. L.D. Landau Institute for Theoretical Physics, Russian Academy of Sciences, 117940 Moscow, Russia. `burmi@itp.ac.ru`

Yudson V. Institute of Spectroscopy, Russian Academy of Sciences, Troitsk, Moscow region, 142190, Russia. `yudson@isan.troitsk.ru`

Maslov D. Department of Physics, University of Florida, P.O. Box 118440, Gainesville, FL 32611-8440 USA. `maslov@phys.ufl.edu`

Syzranov S. Theoretische Physik III, Ruhr-Universität Bochum, 44801 Bochum, Germany. `sergey.syzranov@ruhr-uni-bochum.de`

Makhlin Y. Landau Institute for Theoretical Physics, Kosygin st. 2, 119334 Moscow, Russia. `makhlin@itp.ac.ru` and Moscow Institute of Physics and Technology, Institutskii per. 9, Dolgoprudny, Russia

Guskos N. Solid State Physics, Department of Physics, University of Athens, Panepistimiopolis, 15 784 Zografos, Athens, Greece. `ngouskos@phys.uoa.gr` and Institute of Physics, Szczecin University of Technology, Al. Piastow 17, 70-310 Szczecin, Poland

Anagnostakis E.A. Solid State Physics, Department of Physics, University of Athens, Panepistimiopolis, 15 784 Zografos, Athens, Greece

Karkas K.A. Solid State Physics, Department of Physics, University of Athens, Panepistimiopolis, 15 784 Zografos, Athens, Greece

Guskos A. Institute of Physics, Szczecin University of Technology, Al. Piastow 17, 70-310 Szczecin, Poland

Biedunkiewicz A. Mechanical Department, Szczecin University of Technology, Poland

Figiel P. Mechanical Department, Szczecin University of Technology, Poland

Snopok B.V. Lashkaryov Institute of Semiconductor Physics, National Academy of Sciences, 41 Prospect Nauki, 03028 Kyiv, Ukraine. `snopok@isp.kiev.ua`

Ermakov V.N. Bogolyubov Institute for Theoretical Physics, Metrologichna str. 14 b, 03143, Kiev-143 Kyiv, Ukraine. `vlerm@bitp.kiev.ua`

Ahn Kang-Hun Department of Physics, Chungnam National University, Daejeon 305-764, Republic of Korea. `ahnkh@cnu.ac.kr`

Litovchenko V.G. Institute of Semiconductor Physics, National Academy of Sciences of Ukraine, 41 prospect Nauki, 03028, Kiev, Ukraine. `Lvg@isp.kiev.ua`

Solntsev V.S. Institute of Semiconductor Physics, National Academy of Sciences of Ukraine, 41 prospect Nauki, 03028, Kiev, Ukraine

Abid A. El Group of Condensed Matter Physics, Physics Department, Faculty of Sciences, B.P 8106 Hay Dakhla, University Ibn Zohr, 80000 Agadir, Morocco

Nafidi A. Group of Condensed Matter Physics, Physics Department, Faculty of Sciences, B.P 8106 Hay Dakhla, University Ibn Zohr, 80000 Agadir, Morocco. nafidi21@yahoo.fr

Kaaouachi A. El Group of Condensed Matter Physics, Physics Department, Faculty of Sciences, B.P 8106 Hay Dakhla, University Ibn Zohr, 80000 Agadir, Morocco

Chaib H. Group of Condensed Matter Physics, Physics Department, Faculty of Sciences, B.P 8106 Hay Dakhla, University Ibn Zohr, 80000 Agadir, Morocco

Part I

Electron transport in nanodevices

1

OPTICAL PROPERTIES AND ELECTRONIC STRUCTURE OF ORGANIC-INORGANIC NANO-INTERFACE

A. FUJIWARA, A. KONISHI, AND E. SHIKOH

School of Materials Science, Japan Advanced Institute of Science and Technology, 1-1 Asahidai, Nomi, Ishikawa 923-1292, Japan.
fujiwara@jaist.ac.jp

Abstract. Interfaces between C_{60} and electrode materials (gold (Au) or indium tin oxide ($In_2O_3 + SnO_2$ 10 wt%, ITO)) have been investigated by means of optical absorption. For the bi-layer of C_{60} and Au, a clear shift in onset of absorption at long wave length (λ) region originated from Au absorption was observed. This can be attributed to the red shift of plasma frequency and decrease in electron density of Au. On the other hand, for the bi-layer of C_{60} and ITO, clear enhancement of absorption was observed for $\lambda > 300$ nm. The origin is suggested to be low energy charge transfer from ITO to C_{60}, which is consistent with low Schottky barrier height at the interface. Our results suggest that electronic structures at and around the interface between an organic semiconductor and an inorganic metal are modified by charge transfer from a metal to an organic material.

Keywords: organic-inorganic interface, nano-interface

1.1 Introduction

In last decades, semiconductor devices have been developed to hold the Moore's law [1]. Researches and developments in this trend, namely, miniaturization of devices, have contributed to the improvement of the device performance (higher processing speed and higher memory capacity). This progress in electronics has stimulated an innovation of information technology. In forthcoming ubiquitous computing society, however, not only high performance but also many other functions, such as high shock-resistance, flexibility, low cost, light weight, ability of large-area fabrication and low environmental load will be required. Organic electronic devices are one of the promising ones, because they can meet the demand. Examples of emerging applications are organic

J. Bonča, S. Kruchinin (eds.), *Electron Transport in Nanosystems*.

light emitting diode (OLED), organic field effect transistor (OFET) and organic solar cell. In late 2007, an OLED TV system is already on the market.

1.1.1 ORGANIC FIELD EFFECT TRANSISTOR

Field effect transistor (FET) is the most widely used in semiconductor devices and one of the most important elements. Development of OFET is, therefore, one of the most important issues in the research field of organic electronics. Since the first report on OFETs in 1984 [2] they have attracted considerable interest both for basic research and application. Recently, the field effect mobility μ of OFETs which is one of the indexes of device performance has increased rapidly and has become comparable to that of amorphous Si ($1 \ cm^2/Vs$).

Development of OFETs has been mainly owing to the material search and improvement of material quality so far. In the development, OFETs with p-type operation have been intensively investigated after the first report [2]. On the other hand, OFETs with n-type operation was behind in development because of instability of materials in air. For the device action of n-type OFET, operation in vacuum or passivation is required [3,4]. N-type operation of fullerene FET was reported in 1993 [5], and rather high performance in C_{60} FET, on-off ratio of 10^6 and mobility μ of 0.08 cm^2/Vs, was reported by Haddon et al. in 1995 [6]. Now, fabrication method of C_{60} FETs with μ higher than 0.1 cm^2/Vs has been established, [7,8] and reproducibility was confirmed by researchers. FETs using other fullerenes, such as higher fullerenes, [9–12] endohedral metallofullerenes [13–17] and fullerene derivatives, [18–22] also show n-type operation.

The increase of μ of OFETs has slowed in last years. For further development of OFETs, detailed understand of device operation and clear guideline of device design are required. Device structure of OFET which has been intensively studied is similar to that of Si-based metal-oxide-semiconductor FET (MOSFET). In spite of the similarity of device structure, mechanism of device operation of OFETs is different from that of MOSFETs. Because doping technique used for inorganic semiconductors is not available for organic semiconductors: organic semiconductors are used not as n-type or p-type semiconductor but as intrinsic semiconductors. This results in the qualitative difference in device operation. One is a carrier injection, and another is a channel formation in organic semiconductors.

In the case of Si-based MOSFET, carrier is smoothly injected from p-type (n-type) source electrode to an inversion channel formed in n-type (p-type) semiconductors. For carrier injection, there is no potential barrier in principle. On the other hand, carrier is injected from inorganic-metal electrode to intrinsic semiconductor in OFETs. In most cases, metal-semiconductor interface forms a Schottky contact which acts as a potential barrier for carrier injection.

For a channel formation, minority carrier in the inversion layer is used in Si-based MOSFETs. In this case, operation type, n-type or p-type, depends on the type of semiconductor used in channel region. In the case of OFETs, however, an intrinsic semiconductor is used as a channel material. Carrier is produced in accumulation layer as major carrier, and both electron and hole can be accumulated in channel in principle. Operation type is not determined by channel material itself but depend on the charier injection efficiency [23–29].

1.1.2 INTERFACE EFFECTS ON ORGANIC FIELD EFFECT TRANSISTOR

Both issues of channel formation and carrier injection can be considered from the viewpoint of effects of organic-inorganic interfaces (Fig. 1.1). In the former issuer, it is well known that a quality of interface between organic semiconductor and inorganic insulator substrate affects the device performance. Because, organic materials are hardly grown on an inorganic substrate epitaxially, the crystal structure or molecular alignment at channel region near the interface tends to be disordered. The disorder results in the trapping site for the carrier. For reduction of the interface effect, detailed analysis for understanding this effect has been studied [22, 30–33]. Improvements of device performance with channel region by inserting buffer layer or by using single crystal of organic semiconductor were reported [34–41].

Carrier injection from inorganic metal electrode to organic semiconductor, the latter issue, is related to the Schottky barrier formation at the metal-semiconductor interface. Microscopic electronic structure of organic-inorganic interfaces, however, has not been fully explored so far. Schottky barrier height

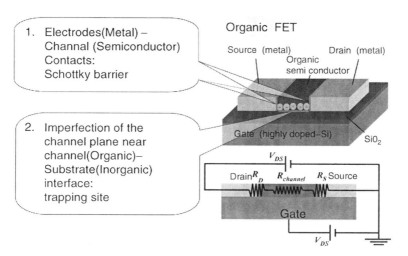

Fig. 1.1. Two important issues of the organic field-effect transistor.

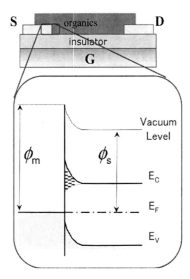

Fig. 1.2. Device structure of bottom-contact, back-gate organic field effect transistor (upper), and expected electronic structure at the interface between an electrode and an organic semiconductor (lower).

between metal and semiconductor is, in principle, estimated to be difference between Fermi level of metal and lowest unoccupied molecular orbital (LUMO) for electron, and between Fermi level of metal and highest occupied molecular orbital (HOMO) for hole. In most cases, Schottky barrier heights for widely used materials are expected to be about or more than 1 eV by this estimation method. On the other hand, their values estimated by transport measurements for contact between inorganic metal electrode and organic semiconductor are about 0.1 eV [29]. Although the origin of the discrepancy has not yet clarified, it has been suggested that the interface states reduce the effective Schottky barrier height (Fig. 1.2). Understanding and controlling the Schottky barrier height at the contact is very important for the improvement of device performance of OFETs.

In this study, bi-layer samples of inorganic metal and organic semiconductor have been fabricated and investigated in order to clarify local electronic structure at the interface. C_{60} which shows one of the highest values of electron mobility was used as an organic semiconductor. Au and ITO were used as inorganic "metal" materials. Although ITO is heavily doped semiconductor in the strict sense, it is widely accepted that it is a transparent electrode. Because C_{60} FETs with ITO electrodes show high device performance [28], it is worthy to investigate properties of contact between C_{60} and ITO. In addition, ITO is very useful for the observation of change in absorption spectrum at low energy (long wave length) region. Finally results in this study will be discussed with device performance of C_{60} FETs, especially with carrier injection efficiency.

1.2 Experimental details

1.2.1 SAMPLE STRUCTURE

Bi-layer samples of an organic semiconductor and an inorganic metal used in this study was designed so that the interface effect became maximum. A roughness of single-layer thin films (C_{60}, Au or ITO) grown by conventional evaporation methods in this study is less than 10 nm. However, roughness of bi-layer in which an inorganic material deposited on an organic material is about 10 nm or more. This can be attributed to fact that the deposited inorganic materials lodged in the soft organic material, resulting in the rough interface. In order to enhance the interface effect in optical absorption spectra, namely, to increase the relative numbers of metal atoms and organic molecules which face to the interface relative to total numbers of atoms and molecules, the thickness of molecules was fixed to be 20 nm (comparable to the roughness of bi-layer).

Three types of bi-layer samples have been investigated: Quartz/C_{60} (20 nm)/Au (10 nm), Quartz/Au (10 nm)/C_{60} (20 nm), and Quartz/ITO (10 nm)/C_{60} (20 nm). For reference, single-layer samples of Quartz/C_{60} (20 nm), Quartz/Au (10 nm), and Quartz/ITO (10 nm) have been also investigated. Here, values of thickness refer the nominal ones monitored by the quartz crystal resonator during deposition process. Schematic sample structures are shown in Fig. 1.3.

Structures of Quartz/C_{60}/Au and Quartz/Au/C_{60} samples can be correlated with the top-contact configuration and bottom-contact configuration

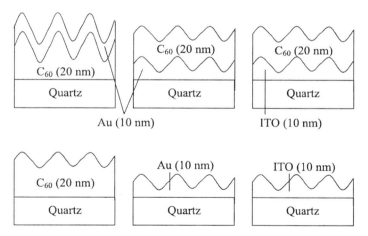

Fig. 1.3. Schematic sample structures investigated in this study. Thickness values of C60, Au and ITO are nominal ones estimated from the monitor. Thickness of quarts (1 mm) is reduced in the figure. Roughness of interface is expected to be about 10 nm, except for Quarts/C60/Au sample with roughness of about 20 nm.

of planer-type FETs, respectively. Therefore, comparison between electronic structures of these two types of sample structure will give information about difference in device performance between of top-contact and bottom-contact FETs. Because RF magnetron sputtering of ITO on C_{60} (detailed experimental procedure will be shown in the next sub-section) will result in serious damage of the surface of C_{60}, we did not investigate the bi-layer of Quartz/C_{60}/ITO.

1.2.2 EXPERIMENTAL PROCEDURE

Thin-films of C_{60} and/or electrode materials (Au, ITO) for optical absorption measurements were fabricated on quartz substrate (1 mm thickness). Commercially available C_{60} (99.98%) was used for the formation of the thin films. A thin film of C_{60} with a thickness of 20 nm was formed using vacuum ($<10^{-5}$ Pa) vapor deposition at the deposition rate of 0.1 nm/s. A thin film of ITO with a thickness of 10 nm was deposited using RF magnetron sputtering system (Tokuda CFS-4ES) at a nominal deposition rate of 0.4 nm/s under argon flow at the pressure of 0.50 Pa. A thin film of Au with a thickness of 10 nm was deposited using electron beam evaporation system at a deposition rate of about 0.06 nm/s.

Optical absorption measurements were performed at room temperature under dry air flow by UV-VIS Spectrophotometer (Shimadzu UV-2500PC) with the wave length range from 190 to 900 nm. Contribution of quartz substrate was eliminated by subtracting a spectrum of substrate from observed spectra.

1.3 Results and discussion

1.3.1 INTERFACE BETWEEN C_{60} AND AU

Figure 1.4 shows optical absorption spectra for samples of Quartz/Au/C_{60}, Quartz/C_{60}/Au, Quartz/Au, and Quartz/C_{60}. Clear three peaks, observed at wave length λ of about 220, 270, and 350 nm for samples of Quartz/Au/C_{60}, Quartz/C_{60}/Au, and Quartz/C_{60}, originate from the dipole-allowed electronic transition of C_{60} [42]. High optical density observed at $\lambda > 500$ nm for the sample of Quartz/Au can be attributed to the optical absorption by Au: the optical absorption edge of Au corresponds to the wave length of about 500 nm. Similar enhancement of optical density was also observed for samples of Quartz/Au/C_{60} and Quartz/C_{60}/Au. An interface effect can be obtained by comparison between spectra of bi-layer samples (Quartz/Au/C_{60} and Quartz/C_{60}/Au) and a sum of spectra of single-layer samples (Quartz/Au and Quartz/C_{60}). Spectra of Quartz/Au/C_{60} and Quartz/C_{60}/Au samples can be roughly explained by a sum of spectrum of Quartz/Au sample and that of Quartz/C_{60} ones.

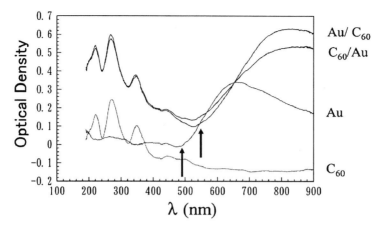

Fig. 1.4. Optical absorption spectra for samples of Quartz/Au/C$_{60}$, Quartz/C$_{60}$/Au, Quartz/Au, and Quartz/C$_{60}$. Arrows show the onset of optical density for Quartz/Au/C$_{60}$ and Quartz/C$_{60}$/Au(right), and Quartz/Au(left).

One of the most important results is similarity of spectra between Quartz/Au/C$_{60}$ and Quartz/C$_{60}$/Au samples. This result shows that the C$_{60}$ molecular structure in Quartz/C$_{60}$/Au sample remains intact, showing that C$_{60}$ molecules are not disintegrated during the deposition process of Au atoms on C$_{60}$ layer. If they are disintegrated during the deposition, intensity of three peaks originating from C$_{60}$ should decrease. In addition, local electronic structures for Quartz/Au/C$_{60}$ and Quartz/C$_{60}$/Au samples are similar each other, showing that they do not depend on the order of deposition. This result shows that the local electronic structures at the interface between metal electrodes and active layer of C$_{60}$ in top-contact and bottom contact FETs are similar each other, and difference in device performance of two types of FETs does not originate from difference in carrier injection efficiency due to the different electronic structure. In contrast to the pentacene FETs, difference in device performance can not be attributed to difference in crystalinity, because of C$_{60}$ is spherical molecule, and disorder effect is not so large to transport properties. Another important result is a red shift of the onset of optical absorption edge originating from Au in Quartz/Au/C$_{60}$ and Quartz/C$_{60}$/Au samples, as shown by right arrow in Fig. 1.4: this can be attributed to the interface effect. It has been reported that charge transfer occurs from noble metal to C$_{60}$ at the interface by means of UPS measurements of C$_{60}$ [43,44], transport measurements [45], Kelvin probe methods [46], and density functional theory [47,48]. A red shift of optical absorption edge is consistent with decrease in carrier number of Au, and then, consistent with charge transfer from Au to C$_{60}$ [43–48]. It should be mentioned that the direct observation of electronic-structure change of Au at the interface between Au and organic materials has not been reported so far. This is the first observation, to our knowledge.

It should be also mentioned that the free-electron like electronic states are observed in the organic monolayer on a metal substrate [49]. It is interesting to consider the role of this "metallic like" behavior in device operation.

1.3.2 INTERFACE BETWEEN C_{60} AND ITO

Figure 1.5 shows optical absorption spectra for samples of Quartz/ITO/C_{60}, Quartz/ITO, and Quartz/C_{60}. Here, intensity of spectrum of the Quartz/C_{60} sample was multiplied by 2.985 for comparison. As in the case of spectra shown in Fig. 1.4, clear three peaks originate from the dipole-allowed electronic transition of C_{60} were observed in samples containing a C_{60} layer. In contrast to spectra of samples containing an Au layer, optical density of Quartz/ITO sample is almost zero for $\lambda > 400$ nm. As mentioned at the end of introduction, ITO is heavily doped semiconductor and does not show metallic reflection at long λ (low energy) region, although ITO is used as electrodes: ITO is transparent material for $\lambda > 400$ nm. Consequently, the sample by use of an electrode material of ITO is very useful for the investigation on the behavior of low energy excitation at the interface between electrode and organic semiconductor.

In order to extract interface effect, we obtained a differential spectrum by subtracting spectra of Quartz/ITO, and Quartz/C_{60} samples from a spectrum of Quartz/ITO/C_{60}, using a trial and error method. The factor of 2.985 for the spectrum of the Quartz/C_{60} was used so that the intensity of three peaks becomes smallest. Extracted differential spectrum is shown in Fig. 1.6. The differential spectrum shows two features. One is additional absorption above

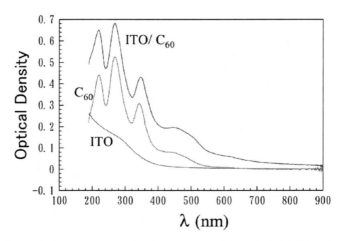

Fig. 1.5. Optical absorption spectra for samples of Quartz/ITO/C_{60}, Quartz/ITO, and Quartz/C_{60}. Intensity of spectrum of the Quartz/C_{60} sample was multiplied by 2.985 for comparison: detailed is discussed in the text.

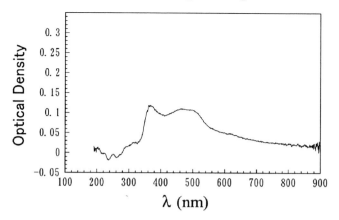

Fig. 1.6. Differential spectrum by subtracting spectra of the spectra of Quartz/ITO and Quartz/C_{60} samples from the spectrum of Quartz/ITO/C_{60}, using a trial and error method.

350 nm whose intensity monotonically decreases with increasing λ. Another is enhancement of broad peak observed in the region of 420–550 nm.

The former feature can be attributed to the low energy excitation due to the interface effect. From the investigation of transport properties in C_{60} FETs, Schottky barrier height is as low as 0.1 eV, corresponding to about λ of 10 μm. Therefore, photo-assisted charge transfer from electrode to C_{60} through Schottky barrier can occur up to λ of 10 μm. Because appearance of new absorption above 350 nm should originate from interface effect, it is plausible that the origin is charge transfer from ITO to C_{60}. In order to verify this argument, optical absorption measurements and/or photoconductivity measurements in infrared region are effective tool.

The latter feature is also worthy to be mentioned. A broad peak in the region of 420–550 nm is hardly observed for C_{60} in solution, whereas it is observed for C_{60} in thin film. It is suggested that this peak is closely related to delocalized excitons [50,51]. According to the argument, C_{60} at the interface is more favorable situation for delocalized excitons. Relation between the existence of delocalized exciton and transport properties is interesting issue to be clarified.

1.4 Conclusion

We have investigated bi-layer samples of an organic semiconductor (C_{60}) and an inorganic metal (Au or ITO) in order to clarify local electronic structure at the interface. From the experiments on interface between C_{60} and Au, a change in electronic structure of Au originating from charge transfer from Au to C_{60} has been observed. On the other hand, from the experiments on

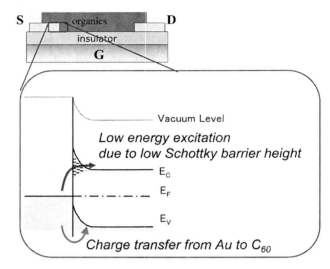

Fig. 1.7. Proposed model of electronic structure at the interface between an electrode and an organic semiconductor.

interface between ITO and C_{60}, low energy excitation which can be attributed to photo-assisted charge transfer from an electrode material (ITO) to organic semiconductor (C_{60}) has been observed. From the results in this study and previous reports on transport properties of organic FETs, we propose a simple model of local electronic structure at interface between an electrode and organic active layer as shown in Fig. 1.7. At the interface between an organic semiconductor and a metal electrode material, charge transfer occurs from metal to organic semiconductor, resulting in charge redistribution and decrease in effective Schottky barrier height. The model is consistent with low Schottky barrier height estimated from transport properties of organic FETs and metallization of organic layer facing to metals.

Finally, it should be noticed that the local electronic structures for Quartz/Au/C_{60} and Quartz/C_{60}/Au samples are similar each other, showing that they do not depend on the order of deposition. It is suggested that difference in device performance between top-contact and bottom-contact C_{60} FETs can not be attributed to the difference in local electronic structure or difference in local crystallinity. For an improvement of device performance by increase of carrier injection efficiency, therefore, electronic structures at not only the interface but also all current paths should be considered, and precise electronic structure control of these entire regions should be performed. This development is next generation research in the field of organic electronics.

Acknowledgements

Contributions by members in magnetic materials laboratory, Japan Advanced Institute of Science and Technology are gratefully acknowledged. This work is supported in part by the Grant-in-Aid for Scientific Research (Grant No. 17540322) from the Ministry of Education, Culture, Sports, Science and Technology (MEXT) of Japan, and the NEDO Grant (Grant No. 04IT5) form the New Energy and Industrial Technology Development Organization (NEDO).

References

1. G. Moore, Cramming more components onto integrated circuits, Electronics **38**, April 19 (1965).
2. K. Kudo, M. Yamashina and T. Moriizumi, Field effect measurement of organic dye films, Jpn. J. Appl. Phys. **23**, 130 (1984).
3. K. Horiuchi, K. Nakada, S. Uchino, S. Hashii, A. Hashimoto, N. Aoki, Y. Ochiai and M. Shimizu, Passivation effects of alumina insulating layer on C_{60} thin-film field-effect transistors, Appl. Phys. Lett. **81**, 1911 (2002).
4. M. Chikamatsu, A. Itakura, Y. Yoshida, R. Azumi, K. Kikuchi and K. Yase, Correlation of molecular structure, packing motif and thin-film transistor characteristics of solution-processed n-type organic semiconductors based on dodecyl-substituted C_{60} derivatives, J. Photochem. Photobiol. A **182**, 245 (2006).
5. J. Paloheimo, H. Isotalo, J. Kastner and H. Kuzmany, Conduction mechanisms in undoped thin films of C_{60} and $C_{60}/70$, Synth. Met. **56**, 3185 (1993).
6. R. C. Haddon, A. S. Perel, R. C. Morris, T. T. M. Palstra, A. F. Hebard and R. M. Fleming, C_{60} thin film transistors, Appl. Phys. Lett. **67**, 121 (1995).
7. S. Kobayashi, T. Takenobu, S. Mori, A. Fujiwara and Y. Iwasa, Fabrication and characterization of C_{60} thin-film transistors with high field-effect mobility, Appl. Phys. Lett. **82**, 4581 (2003).
8. K. Itaka, M. Yamashiro, J. Yamaguchi, M. Haemori, S. Yaginuma, Y. Matsumoto, M. Kondo and H. Koinuma, High-mobility C_{60} field-effect transistors fabricated on molecular-wetting controlled substrates, Adv. Mater. **18**, 1713 (2006).
9. R. C. Haddon, C70 thin film transistors, J. Am. Chem. Soc. **118**, 3041 (1996).
10. H. Sugiyama, T. Nagano, R. Nouchi, N. Kawasaki, Y. Ohta, K. Imai, M. Tsutsui, Y. Kubozono and A. Fujiwara, Transport properties of field-effect transistors with thin films of C76 and its electronic structure, Chem. Phys. Lett. **449**, 160 (2007).
11. Y. Kubozono, Y. Rikiishi, K. Shibata, T. Hosokawa, S. Fujiki and H. Kitagawa, Structure and transport properties of isomer-separated C82, Phys. Rev. B **69**, 165412 (2004).
12. K. Shibata, Y. Kubozono, T. Kanbara, T. Hosokawa, A. Fujiwara, Y. Ito, H. Shinohara, Fabrication and characteristics of C84 fullerene field-effect transistors, Appl. Phys. Lett. **84**, 2572 (2004).
13. T. Nagano, H. Sugiyama, E. Kuwahara, R. Watanabe, H. Kusai, Y. Kashino and Y. Kubozono, Fabrication of field-effect transistor device with higher fullerene, C88, Appl. Phys. Lett. **87**, 023501 (2005).

14   A. Fujiwara et al.

14. S. Kobayashi, S. Mori, S. Iida, H. Ando, T. Takenobu, Y. Taguchi, A. Fujiwara, A. Taninaka, H. Shinohara and Y. Iwasa, Conductivity and field effect transistor of La2@C80 metallofullerene, J. Am. Chem. Soc. **125**, 8116 (2003).
15. Y. Rikiishi, Y. Kubozono, T. Hosokawa, K. Shibata, Y. Haruyama, Y. Takabayashi, A. Fujiwara, S. Kobayashi, S. Mori and Y. Iwasa, Structural and electronic characterizations of two isomers of Ce@C82, J. Phys. Chem. B **108**, 7580 (2004).
16. T. Nagano, E. Kuwahara, T. Takayanagi, Y. Kubozono and A. Fujiwara, Fabrication and characterization of field-effect transistor device with C2v isomer of Pr@C82, Chem. Phys. Lett. **409**, 187 (2005).
17. T. Kanbara, K. Shibata, S. Fujiki, Y. Kubozono, S. Kashino, T. Urisu, M. Sakai, A. Fujiwara, R. Kumashiro and K. Tanigaki, N-channel field effect transistors with fullerene thin films and their application to a logic gate circuit, Chem. Phys. Lett. **379**, 223 (2003).
18. C. Waldauf, P. Schilinsky, M. Perisutti, J. Hauch, C. J. Brabec, Solution-processed organic n-type thin-film transistors, Adv. Mater. **15**, 2084 (2003).
19. T. -W. Lee, Y. Byun, B. -W. Koo, I. -N. Kang, Y. -Y. Lyu, C. H. Lee, L. Pu and S. Y. Lee, All-solution-processed n-type organic transistors using a spinning metal process, Adv. Mater. **17**, 2180 (2005).
20. M. Chikamatsu, S. Nagamatsu, Y. Yoshida, K. Saito, K. Yase and K. Kikuchi, Solution-processed n-type organic thin-film transistors with high field-effect mobility, Appl. Phys. Lett. **87**, 203504 (2005).
21. T. Nagano, H. Kusai, K. Ochi, T. Ohta, K. Imai, Y. Kubozono and A. Fujiwara, Fabrication of field-effect transistor devices with fullerene related materials, Phys. Stat. Sol. (b) **243**, 3021 (2006).
22. N. Kawasaki, T. Nagano, Y. Kubozono, Y. Sako, Y. Morimoto, Y. Takaguchi, A. Fujiwara, C. -C. Chu, T. Imae, Transport properties of field-effect transistor with Langmuir-Blodgett films of C_{60} dendrimer and estimation of impurity levels, Appl. Phys. Lett. **91**, 243515 (2007).
23. I. Yagi, K. Tsukagoshi and Y. Aoyagi, Direct observation of contact and channel resistance in pentacene four-terminal thin-film transistor patterned by laser ablation method, Appl. Phys. Lett. **84**, 813 (2004).
24. M. Chikamatsu, S. Nagamatsu, T. Taima, Y. Yoshida, N. Sakai, H. Yokokawa, K. Saito and K. Yase, C_{60} thin-film transistors with low work-function metal electrodes, Appl. Phys. Lett. **85**, 2396 (2004).
25. S. Kobayashi, T. Nishikawa, T. Takenobu, S. Mori, T. Shimoda, T. Mitani, H. Shimotani, N. Yoshimoto, S. Ogawa, Y. Iwasa, Control of carrier density by self-assembled monolayers in organic field-effect transistors, Nat. Mater. **3**, 317 (2004).
26. T. Nishikawa, S. Kobayashi, T. Nakanowatari, T. Mitani, T. Shimoda, Y. Kubozono, G. Yamamoto, H. Ishii, M. Niwano and Y. Iwasa, Ambipolar operation of fullerene field-effect transistors by semiconductor/metal interface modification, J. Appl. Phys. **97**, 104509 (2005).
27. Y. Matsuoka K. Uno, N. Takahashi, A. Maeda, N. Inami, E. Shikoh, Y. Yamamoto, H. Hori and A. Fujiwara, Intrinsic transport and contact resistance effect in C_{60} field-effect transistors, Appl. Phys. Lett. **89**, 173510 (2006).
28. N. Takahashi, A. Maeda, K. Uno, E. Shikoh, Y. Yamamoto, H. Hori, Y. Kubozono and A. Fujiwara, Output properties of C_{60} field-effect transistors with different source/drain electrodes, Appl. Phys. Lett. **90**, 083503 (2007).

29. T. Nagano, M. Tsutsui, R. Nouchi, N. Kawasaki, Y. Ohta, Y. Kubozono, N. Takahashi and A. Fujiwara, Output Properties of C_{60} Field-effect transistors with Au electrodes modified by 1-alkanethiols, J. Phys. Chem. C **111**, 7211 (2007).

30. G. Horowitz, M. E. Hajlaoui and R. Hajlaoui, Temperature and gate voltage dependence of hole mobility in polycrystalline oligothiophene thin film transistors, J. Appl. Phys. **87**, 4456 (2000).

31. A. R. Volkel, R. A. Street and D. Knipp, Carrier transport and density of state distributions in pentacene transistors , Phys. Rev. B **66**, 195336 (2002).

32. D. V. Lang, X. Chi, T. Siegrist, A. M. Sergent, and A. P. Ramirez, Amorphous-like density of gap states in single-crystal pentacene, Phys. Rev. Lett. **93**, 086802 (2004).

33. N. Kawasaki, Y. Ohta, Y. Kubozono, A. Konishi, A. Fujiwara, An investigation of correlation between transport characteristics and trap states in n-channel organic field-effect transistors, Appl. Phys. Lett. **92**, 163307 (2008).

34. V. Podzorov, V. M. Pudalov and M. E. Gershenson, Field-effect transistors on rubrene single crystals with parylene gate insulator, Appl. Phys. Lett. **82**, 1739 (2003).

35. R. W. I. de Boer, T. M. Klapwijk and A. F. Morpurgo, Field-effect transistors on tetracene single crystals, Appl. Phys. Lett. **83**, 4345 (2003).

36. J. Takeya, C. Goldmann, S. Haas, K. P. Pernstich, B. Ketterer and B. Batlogg, Field-induced charge transport at the surface of pentacene single crystals: A method to study charge dynamics of two-dimensional electron systems in organic crystals, J. Appl. Phys. **94**, 5800 (2003).

37. E. Menard, V. Podzorov, S. -H. Hur, A. Gaur, M. E. Gershenson and J. A. Rogers, High-performance n- and p-type single-crystal organic transistors with free-space gate dielectrics, Adv. Mater. **16**, 2097 (2004).

38. J. Takeya, T. Nishikawa, T. Takenobu, S. Kobayashi, Y. Iwasa, T. Mitani, C. Goldmann, C. Krellner and B. Batlogg, Effects of polarized organosilane self-assembled monolayers on organic single-crystal field-effect transistors, Appl. Phys. Lett. **85**, 5078 (2004).

39. V. Podzorov, E. Menard, A. Borissov, V. Kiryukhin, J. A. Rogers and M. E. Gershenson, Intrinsic charge transport on the surface of organic semiconductors, Phys. Rev. Lett. **93**, 086602 (2004).

40. V. C. Sundar, J. Zaumseil, V. Podzorov, E. Menard, R. L. Willett, T. Someya, M. E. Gershenson and J. A. Rogers, Elastomeric transistor stamps: reversible probing of charge transport in organic crystals, Science **303**, 1644 (2004).

41. J. Takeya, K. Tsukagoshi, Y. Aoyagi, T. Takenobu, and Y. Iwasa, Hall effect of quasi-hole gas in organic single-crystal transistors, Jpn. J. Appl. Phys. **44**, L1393 (2005).

42. Y. Wang, J. M. Holden, A. M. Rao, W. -T. Lee, X. S. Bi, S. L. Ren, G. W. Lehman, G. T. Hager and P. C. Eklund, Interband dielectric function of C_{60} and $M6C_{60}$ (M=K, Rb, Cs), Phys. Rev. B **45**, 14396 (1992).

43. B. W. Hoogenboom, R. Hesper, L. H. Tjeng and G. A. Sawatzky, Charge transfer and doping-dependent hybridization of C_{60} on noble metals, Phys. Rev. B **57**, 11939 (1998).

44. S. C. Veenstra, A. Heeres, G. Hadziioannou, G. A. Sawatzky and H. T. Jonkman, On interface dipole layers between C_{60} and Ag or Au, Appl. Phys. A **75**, 661 (2002).

16   A. Fujiwara et al.

45. R. Nouchi, and I. Kanno, Charge transfer and formation of conducting C_{60} monolayers at C_{60}/noble-metal interface, J. Appl. Phys. **97**, 103716 (2005).
46. N. Hayashi, H. Ishii, Y. Ouchi and K. Seki, Examination of band bending at buckminsterfullerene (C_{60})/metal interfaces by the Kelvin prove method, J. Appl. Phys. **92**, 3784 (2002).
47. L. -L. Wang and H. -P. Cheng, Rotation, translation, charge transfer, and electronic structure of C_{60} on Cu(111) surface, Phys. Rev. B **69**, 045404 (2004).
48. L. -L. Wang, and H. -P. Cheng, Density functional study of the adsorption of a C_{60} monolayer on Ag(111) and Au(111) surfaces, Phys. Rev. B **69**, 165417 (2004).
49. R. Temirov, S. Soubatch, A. Luican and F. S. Tautz, Free-electron-like dispersion in an organic monolayer film on a metal substrate, Nature **444**, 350 (2006).
50. K. Harigaya and S. Abe, Optical-absorption spectra in fullerenes C_{60} and C_{70}: Effects of Coulomb interactions, lattice fluctuations, and anisotropy, Phys. Rev. B **49**, 16746 (1994).
51. K. Harigaya and S. Abe, Exciton and Lattice-fluctuation effects in optical spectra of C_{60}, Mol Cryst. Liq. Cryst. **256**, 825 (1994).

2

ELECTRON TRANSPORT IN NANOWIRES – AN ENGINEER'S VIEW

W. NAWROCKI

Faculty of Electronics and Telecommunications, Poznan University of Technology, ul. Piotrowo 3, 60-965 Poznan, Poland.
nawrocki@et.put.poznan.pl

Abstract. In the paper technological problems connected to electron transport in mesoscopic- and nanostructures are considered. The electrical conductance of nanowires formed by metallic contacts in an experimental setup proposed by Costa-Kramer et al. The investigation has been performed in air at room temperature measuring the conductance between two vibrating metal wires with standard oscilloscope. Conductance quantization in units of $G_0 = 2e^2/h = (12.9\,k\Omega)^{-1}$ up to five quanta of conductance has been observed for nanowires formed in many metals. The explanation of this universal phenomena is the formation of a nanometer-sized wire (nanowire) between macroscopic metallic contacts which induced, due to theory proposed by Landauer, the quantization of conductance. Thermal problems in nanowirese are also discussed in the paper.

Keywords: electron transport, nanowires, conductance

2.1 Introduction

Prognoses of the development of the semiconductor industry (ITRS) foresee that sizes of electronic components in integrated circuits will be smaller than 10 nm in the course several years, in the year 2020 will even amount 6 nm (Table 2.1). From this reason and many others it is need to study of electric and thermal proprieties of nanostructures. Electric and thermal proprieties of electronic components about nanometric sizes are not more described by the classical theory of conductance and by the Boltzmann transport equation, but by quantum theories. Classical theories of electrical and thermal conductance assume a huge number of atoms and free electrons. However number of atoms and free electrons in nanosize structure is not sufficient for a statistical processing of their behaviour. Let's assume a silicon cube with one side dimension

Table 2.1. Data of integrated circuits (IC) according to The International Technology Roadmap for Semiconductors (Edition 2006)

Year		2006	2010	2013	2016	2020
Clock frequency (on chip)	GHz	6.8	15	23	40	72
Functionality of IC	Gbit	8.6	34,3	68,7	137	275
Supply voltage	V	1.1	1.0	0.9	0.8	0.7
Dissipeted power (cooling on)	W	180	198	198	198	198
Gate length in FET transistors in an integrated circuit	nm	28	18	13	9	6

of a and with common doping of 10^{16}cm^{-3}. In a n-doped silicon cube with the size $(100\,\text{nm})^3$ there are 5×10^7 atoms and 50 free electrons, but in the Si cube with the size $(10\,\text{nm})^3$ there are 5×10^4 atoms and 5% chance only to find **one** free electron.

In last 20 years considerable attention has been focused on the quantization of both electrical and thermal conductance in nanostructures. It is to underline that the electrical and the thermal conductance of a nanostructure describe the same process: the electron transport in the nanostructure. Therefore there are several analogues between the two physical quantities. The theoretical quantum unit of electrical conductance $G_0 = 2e^2/h$ was predicted by Landauer [1] in his new theory of electrical conductance. In 1987 Gimzewski and Moller [2] published results on measurements of the quantization of conductance in metals at room temperature observed with a scanning tunnelling microscope. In 1988 two groups [3, 4] reported the discovery of the conductance quantization in controllable two-dimensional electron gas (2DEG) in the $GaAs$ constriction at the temperatures less than several kelvin. Formation of nanowires in the process of breaking contact between ordinary metallic wires was published by Costa-Kramer [5] in 1995.

For integrated circuits heat exchange in nanostructers are very important as well. It is generally known that limits for speed-up the digital circuits, especially microprocessors, are determined by thermal problems. Therefore many groups investigate a heat exchange and a thermal conductance in nanostructures. First theoretical analyses of thermal conductance in structures in the ballistic regime were made by P. Streda [6], last papers come from several groups, e.g. [7, 8].

2.2 Ballistic electron transport

Transport of electrons can be described classically by the Boltzmann transport equation (Drude model) which introduces mean free path. At relatively

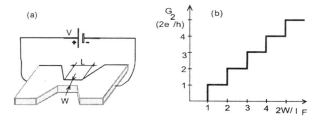

Fig. 2.1. Conductance quantization in a nanowire (conductor with length $L < \Lambda$ and width W comparable with the length of Fermi wave λ_F): (a) nanowire outline (the third dimension is not considered); (b) conductance quantization G versus width W.

low temperatures the considerable contribution to conductivity is given by electrons with energy close to Fermi surface. Hence, conductivity is given by:

$$\sigma = \frac{ne^2\tau}{m^*} \tag{2.1}$$

where n is the concentration of the carriers, m^* – the effective mass of the electron, τ – relaxation time.

Another parameter characterizing the system is Fermi wavelength $\lambda_F = 2\pi/k_F$, where k_F is the Fermi wavevector. For metals like copper or gold $\lambda_F \approx 0,5\,\text{nm}$ is much less then free electron path $\Lambda(\Lambda_{Au} = 14\,\text{nm})$. If the dimensions of the system is less than free electron path, the impurity scattering is negligible, so the electrons transport can be regarded as ballistic. If a metal wire has outside diameter of W, comparable with Fermi wavelength λ_F, and the length L is less than Λ, the system can be regarded as one-dimensional (1D), the electron – as a wave, and one can expect quantum effects. Let's consider perfect conductor with diameter W and the length L (Fig. 2.1) connecting two wide contacts (reservoirs of the electrons) between which the conductivity is measured. Assuming that the wide contacts are infinitely large, the electrons moves are in the thermodynamic equilibrium described by Fermi-Dirac statistic.

When the electrons enter 1D conductor nonequilibrium states occur with negative and positive velocities. If there is a resultant current, the states with positive velocities correspond to higher energies [2]. According to the Buttiker [9] model the hamiltonian of the perfect conductor can be expressed as follows:

$$H = \frac{1}{2m^*}\left(\hbar^2 k_x^2 + \hbar k_y^2\right) + V(x) \tag{2.2}$$

where y is along the wire, x is in the transverse direction and m^* is the effective mass. $V(x)$ denotes the potential well of the width W. The wavefunction of the electron can be separated:

$$\Psi_{j,k}(x,y) = \exp\left(ik_y y\right) f_j(x) \tag{2.3}$$

where k_y is a wavevector along y, $f_j(x)$ is a wavefunction along x with energy eigenvalue of E_{Tj}.

Because of the narrowness of the potential well $V(x)$ the energy for the transverse propagation is quantized:

$$E_{Tj} = \frac{\hbar^2 k_x^2}{2m^*} = \frac{\hbar^2}{2m^*} \left(\frac{j\pi}{W}\right)^2 \qquad (2.4)$$

Total energy equals $E_j = E_{Tj} + E_{Lj}$. For the Fermi level $E_F = E_j$ there is a number $N \sim 2W/\lambda_F$ of states E_{Tj} below Fermi surface. Let us assume that thermal energy $k_B T$ is much smaller then the energy gap between levels, and that the wide contacts are characterized by chemical potentials μ_1 and μ_2 with $(\mu_1 > \mu_2)$. Then current of electrons in jth state equals to:

$$I_j = ev_j \left(\frac{dn}{dE}\right)_j \Delta\mu, \qquad (2.5)$$

where v_j is the velocity along y and $(dn/dE)_j$ is the density of states at the Fermi level for jth state. For 1D conductor the density of states is

$$\frac{dn}{dk} = \frac{1}{2\pi} \ and \ \left(\frac{dn}{dE}\right)_j = \left(\frac{dn}{dk}\frac{dk}{dE}\right)_j = \frac{2}{hv_j} \qquad (2.6)$$

The factor of 2 results from spin degeneracy. Hence, the current for jth state $I_j = \frac{2e^2}{h}V$ does not depend on j (where the voltage difference $V = \Delta\mu/e$). Total current $I = \sum\limits_{j=1}^{N} I_j$, hence conductivity is expressed as

$$G = \frac{2e^2}{h}N \qquad (2.7)$$

where N depends on the width of the wire (Fig. 2.1).

However, defects, impurities and irregularities of the shape of the conductor can induce scattering, then conductivity is given by the Landauer equation:

$$G = \frac{2e^2}{h}\sum\limits_{i,j=1}^{N} t_{ij} \qquad (2.8)$$

where t_{ij} denotes probability of the transition from jth to ith state. In the absence of scattering $t_{ij} = \delta_{ij}$ thus Eq. (2.8) is reduced to Eq. (2.7).

Figure 2.2 presents a resistance of a gold cube sample versus the dimension W (W is a cube side). Resistance has been calculated according to the Drude theory (W from 14 nm to 10 μm) and to the Landauer theory (W from 0.56 to 14 nm). For the size of the cube W equal to Λ_{Au} (Λ_{Au} is the mean free path for gold at 295 K, Λ_{Au}=14 nm) one can compare results from this two theories: $R_D = 1.6\,\Omega$ (Drude theory) and $R_L = 240\,\Omega$ (Landauer theory). Measurements of electrical resistance (or conductance) of a sample of the size about Λ mesoscopic range) show that the Landauer theory better describes real parameters of the sample.

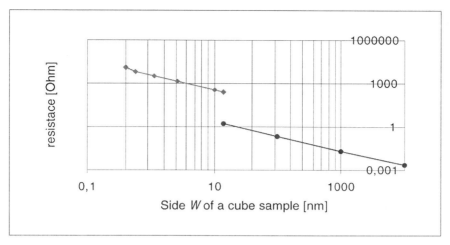

Fig. 2.2. Resistance of a gold cube with a side W: from the Drude theory (right line, up 10 μm) and the Landauer theory (link curve, from 0.56 to 14 nm). $W =$ 14 nm $= \Lambda_{Au}$ is a length of the mean free path for Au at 295 K.

2.3 Electrical conductance measurements

The experimental setup consisted of a pair of metallic wires (they formed a nanowire), a digital oscilloscope, a motion control system (doesn't show on the pocyure) and a PC (Fig. 2.3). Instruments are connected in one system using the IEEE-488 interface. There was the resistor $R_1 = 1\,k\Omega$ in series to the connected wires. The circuit was fed by the constant voltage V_{sup} and measurements of current I have been performed. Conductance was determined by current I accordingly to:

$$G = I\frac{1}{V - IR_1} \tag{2.9}$$

Transient effects of making contact or breaking the contact give time dependent current. The voltage on the resistor R was measured with computer controlled oscilloscope. The piezoelectric device is used to control the backward and forward movement of the macroscopic wires between which nanowires occur. A high voltage amplifier controlled by a digital function generator supplies the piezoelectric device. Both electrodes (macroscopic wires) are made of wire 0.5 mm in diameter. The conductance was measured between two metallic electrodes, moved to contact by the piezoelectric tube actuator. The oscilloscope was triggered by a single pulse. All experiments were performed at room temperature and at ambient pressure.

In order to compare our results with those published before by other groups the first experiment was performed for gold wires. Even if quantization of conductivity by $G_0 = 2e^2/h$ does not depend on the metal and on

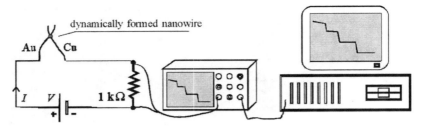

Fig. 2.3. A system for measurements of conductance quantization in nanowires formed between two macroscopic wires (e.g. Au and Cu wires).

temperature, the purpose of studying quantization for different metals was to see how properties of the metal affect the contacts between wires. Therefore, we have investigated the conductance quantization of nanowires for three non-magnetic metals (gold, copper and tungsten) and for magnetic metals (cobalt and nickel). All measurements were carried out at room temperature.

2.4 Measurement results on electrical conductance in metallic nanowires

The quantization of electric conductance depends neither on the kind of metal nor on temperature. However, the purpose of studying the quantization for different metals was to observe how the metal properties affect the contacts between wires. For nonmagnetic metals, the conductance quantization in units of $G_0 = 2e^2/h = 7.75 \times 10^{-5}[A/V] = (12.9\,\Omega k)^{-1}$ was previously observed for the following nanowires: Au-Au, Cu-Cu, Au-Cu, W-W, W-Au, W-Cu. The quantization of conductance in our experiment was evident. All characteristics showed the same steps equal to $2e^2/h$. We observed two phenomena: quantization occurred when breaking the contact between two wires, and quantization occurred when establishing the contact between the wires. The characteristics are only partially reproducible; they differ in number and height of steps, and in the time length. The steps can correspond to 1, 2, 3 or 4 quanta. It should be emphasised that quantum effects were observed only for some of the characteristics recorded. The conductance quantization has been so far more pronouncedly observable for gold contacts. Figures 2.4 and 2.5 show example plots of conductance vs. time during the process of drawing a gold and copper nanowire, respectively, for the bias voltage $V_{sup} = 0.420\,V$ (the measurements were carried out by Dr. M. Wawrzyniak from PUT, Poznan). Figures 2.4 and 2.5 show the conductance histogram obtained from 6000 consecutive characteristics in the conductance range from $0.5G_0$ to $4G_0$.

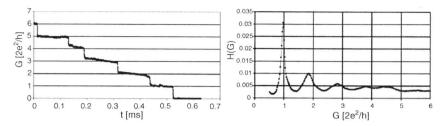

Fig. 2.4. Conductance quantization in gold nanowires: a time plot (left) and a histogram from 6,000 consecutive formations of a nanowire.

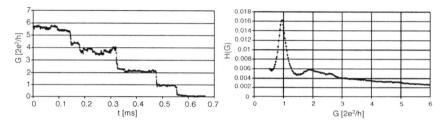

Fig. 2.5. Conductance quantization in copper nanowires: a time plot (left) and a histogram from 6,000 consecutive formations of a nanowire.

2.5 Thermal problems in nanowires

Both electrical G_E and thermal G_T conductance of a nanostructure describe the same process: electron transport in nanostructures. Therefore there are several analogues between the two physical quantities.

Beside observations of electrical conductance quantization in nanowires one can expect the thermal conductance quantization as well. Electron transport in a nanowire does two effects: an electrical current I and a heat flux density Q_D.

Electrical current: $I = G_E \times \Delta V$,

Heat flux density: $Q_D = G_T \times \Delta T$,

where G_E – electrical conductance of a sample, ΔV – difference of electrical potentials, G_T – thermal conductance of a sample, ΔT – temperature difference.

$$G_E = \sigma \times A/l, \ G_T = \lambda \times A/l \tag{2.10}$$

where σ – electrical conductivity, λ – thermal conductivity, l – length of a sample (e.g. nanowire), A – area of a cross-section of a sample.

Quantized thermal conductance in one-dimensional systems (e.g. nanowires) was predicted theoretically by Rego [10] by using the Landauer theory. The thermal conductance is considered in a similar way like the electrical conductance. In one-dimension systems are formed conductive channels. Each channel contributes to a total thermal conductance with the quantum of thermal conductance G_{T0}. Quntized thermal conductance and its quantum (unit)

G_{T0} was confirmed experimentally by Schwab [7]. The quantum of thermal conductance

$$G_{T0}[W/K] = (\pi^2 k_B^2 / 3h)T = 9.5 \times 10^{-13}T \tag{2.11}$$

depends on the temperature (2.11). At $T = 300\,\text{K}$ value of $G_{T0} = 2.8 \times 10^{-10}[W/K]$. This value is determined for an ideal ballistic transport (without scattering) in a nanowire, with the transmission coefficient $t_{ij} = 100\%$. It means that in all practical cases (for $t_{ij} < 100\%$) the thermal conductance is below the limit given by formula (2.11). One can obtain a similar (but not the same) value of the quantum of thermal conductance G_{T0} using the Wiedemann-Franz law. The law describes the relation between the thermal conductivity λ, and electrical conductivity σ of a sample (2.12) for macroscopic objects.

$$\frac{\lambda}{\sigma} = \frac{1}{3}\left(\frac{\pi k_B}{e}\right)^2 T = L \times T = 2,35 \times 10^{-8}T \tag{2.12}$$

where e – electron charge, k_B – Boltzmann constant, L – Lorenz number, T – temperature.

The Wiedemann-Franz law is valid for the relation between conductances G_T/G_E as well. The value of G_{T0} can be obtained directly from (2.13).

$$G_{T0} = L \times T \times G_{E0} = \frac{1}{3}\left(\frac{\pi k_B}{e}\right)^2 T \times \frac{2e^2}{h} = \frac{2}{3}\frac{(\pi k_B)^2}{h} \times T = 1,89 \times 10^{-12}T \tag{2.13}$$

The value of G_{T0} obtained from the Wiedemann-Franz law is twice larger than this from (2.11) and it is not correct. A single nanowire should be consider together with its terminals (Fig. 2.6). They are colled reservoirs of electrons. Electron transport in the nanowire itself is ballistic, it means the transport without scattering of electrons and without energy dissipation. The energy dissipation takes part in terminals. Because of the energy dissipation the local temperature T_{term} in terminals is higher then the temperature T_{wire} of nanowires itself (Fig. 2.6). A heat distribution in terminals of a nanostructe should be analyzed.

In small structures a dissipated energy is quite large. For the first step of conductance quantization, $G_E = G_{E0} = 7.75 \times 10^{-5}[A/V]$, and at the

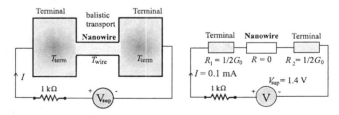

Fig. 2.6. Conductance quantization in a nanowire (conductor with length $L < \Lambda$ and width W comparable with the length of Fermi wave λ_F): (a) nanowire outline (the third dimension is not considered); (b) conductance quantization G versus width W.

supply voltage $V_{sup} = 1.4V$ the current in the circuit $I = 100\,\mu A(I = 190\,\mu A$ for the second step of quantization). The power dissipation in terminals of nanowires is $P = I^2/G_{E0} = 130\,\mu W$ for the first step and $P = 230\,\mu W$ for the second step. One ought to notice that the density of electric current in nanowires is extremely high. The diameter of the gold nanowire on the first step of quantization can be estimated to $D = 0.4\,\text{nm}$, so for $I = 100\,\mu A$ the current density $J \approx 8 \times 10^{10}[A/\text{cm}^2]$.

Conclusions

Conductance quantization has proved to be observable in a simple experimental setup, giving opportunity to investigate subtle quantum effects in electrical conductivity. The energy dissipation in nanowires takes part in their terminals. Because of the energy dissipation the local temperature in terminals is higher then the temperature of nanowires itself.

References

1. R. Landauer, Conductance determined by transmission: probes and quantised constriction resistance, J. Phys.: Cond. Matter **1**, 8099–8120 (1989).
2. J. K. Gimzewski, R. Moller, Transition from the tunneling regime to point contact studied using scanning tunneling microscopy, Phys. Rev. **B 36**, 1284 (1987).
3. D. A. Wharam, T. J. Thornton, R. Newbury, M. Pepper, H. Ahmed, J. E. F. Frost, D. G. Hasko D. C. Peacock, D. A. Richie, G. A. C. Jones, One-dimensional transport and the quantization of the ballistic resistance, J. Phys. **C 21**, L209–L214 (1988).
4. B. J. van Wees, H. van Houten, C. W. J. Beenakker, J. G. Williamson, L. P. Kouwenhoven, D. van der Marel, C. T. Foxon, Quantized conductance of point contacts in a two-dimensional electron gas, Phys. Rev. Lett. **60**, 848–850 (1988).
5. J. L. Costa-Kramer, N. Garcia, P. Garcia-Mochales, P. A. Serena, Nanowire formation in macroscopic metallic contacts: quantum mechanical conductance tapping a table top, Surf. Sci. **342**, S. L1144 (1995).
6. P. Streda, Quantised thermopower of a channel in the ballistic regime, J. Phys.: Cond. Matter 1, 1025–1027 (1989).
7. K. Schwab, E. A. Henriksen, J. M. Worlock, M. L. Roukes, Measurement of the quantum of thermal conductance, Nature **404**, 974–977 (2000).
8. Y. Tanaka, F. Yoshida, S. Tamura, Lattice thermal conductance In nanowires at low temperatures: Breakdown and recovery of quantization, Phys. Rev. B **71**, 205308 (2005).
9. M. Buttiker, Absence of backscattering in the quantum Hall effect in multiprobe conductors, Phys. Rev. B 38, 9375–9380 (1988).
10. L. G. C. Rego, G. Kirczenow, Qunatized thermal conductance of dielectric quantum wires, Phys. Rev. Lett. **81**, 232–235 (1998).

3

NANOPOROUS ANODIC ALUMINA WIRE
TEMPLATES FOR NANOWIRE DEVICES

T.L. WADE[1], A.A. ABDULLA[1], M.C. CIORNEI[1],
D. PRIBAT[2], C. COJOCARU[2],
AND J.-E. WEGROWE[1]

[1] *Laboratoire des Solides Irradies ECOLE Polytechnique Route de Saclay, 91128 Palaiseau Cedex FRANCE.*
travis.wade@polytechnique.edu
[2] *Laboratoire de Physique des Interfaces et des Couches Minces ECOLE Polytechnique, Route de Saclay, Palaiseau France*

Abstract. Aluminum wires are electrochemically etched into bi-directional anodic alumina (AAO) templates for the growth and contacting of nanowires as three terminal devices. The use of this nanostructured template is shown by a ZnO nanowire surrounding gate field-effect transistor. Fabrication procedures and preliminary device characteristics of this bottom-up approach to nanowire transistors are shown.

Keywords: aluminum, anodization, electrodeposition, nanowire, semiconductor, template synthesis, transistor, ZnO

3.1 Introduction

Electrodeposition, sol-gel synthesis, and CVD are some of the many techniques that can produce nanostructures with nanometer control [1–5]. The quality and reproducibility of these nanostructures are well developed. The difficulty is, however, is to organize and contact these nano-objects [6, 7]. The role of the template is two fold. First, it allows the production of the structure with the best possible reproducibility and it plays the role of a skeleton in order to organize the different functions of a device, the active components and the different interfaces (building blocks, electric contacts, gate voltage, bias fields, optical sensors) on a rigid body. Second, this nanoscaffold is used to link the structure to the macroscopic world, i.e. the contacts. In the scheme of template synthesis it is possible to identify three different steps: (1) the creation of the

J. Bonča, S. Kruchinin (eds.), *Electron Transport in Nanosystems.*
© Springer Science + Business Media B.V. 2008

building blocks, the nanowires or nanodots, (2) the assembly of the nano building blocks into a functional architecture within the template, and (3) the fabrication and control of the contacts to the macroscopic world. The first and second steps coincide for metallic nanowires or semiconductors that are made by electrodeposition. For carbon nanotubes and silicon nanowires a catalytic layer is made by electrodeposition followed by CVD for the carbon nanotubes or solid-liquid-vapor growth for the silicon nanowires. The final step, and perhaps most important, is the contact of the nanoscale objects to the macroscopic world.

Nanoporous anodic alumina is commonly used as a template for nanowire synthesis [8]. It is made by anodization of aluminum in acidic solutions, which forms a self-assembled, hexagonal network of nanometer diameter pores [9–13]. These are interesting templates because the pore diameter, distribution, and pore length can be tailored to suit the needs of the user by varying the anodization conditions: electrolyte, voltage, time, and temperature. The chosen pore size determines the resulting nanowire dimensions. Another advantage is that once the nanowires have been made in the alumina template, they can be electrically contacted at the top and bottom of the membrane for physical measurements without the need for lithography. Thus, the template acts as a mould for the nanowires.

3.2 Background

Anodization of a metal is the controlled anodic growth of a metal-oxide film on a metal surface, mainly aluminum, in an electrolytic bath [10–19]. There are two types of metal-oxide films formed by anodization, amorphous barrier films and porous films. These oxide films can be a micron thick for barrier films to many tens of microns thick for porous films, as opposed to the 2–3 nm thick metal-oxide films that exist on many metals as a result of ambient atmospheric oxidation. The barrier layer films are formed in pH neutral aqueous electrolytes, such as ammonium borate, in which aluminum is insoluble. Porous films are formed in acidic aqueous electrolytes such as dilute, 1 molar, sulfuric acid, in which the oxide layer is formed but also dissolves at the same time, enhanced by the local electric field. For these oxide films, the metal to be anodized is the positive electrode, anode, in an aqueous electrolyte solution with an inert negative electrode, cathode, to complete the circuit. When a potential of several volts is applied to the cell, current flows from the anode to the cathode, the electrode/solution interfaces are polarized, and electrochemical reactions occur at these interfaces. The important reactions for us occur at the anode (Fig. 3.1a). Here a metal oxide film is growing due to the field-induced migration of cations, Al^{3+}, from the electrode and anions, O^{2-}, from the solution. The cations react with water at the oxide/solution interface, equation 3.1, and the anions react with the metal/oxide interface, equation 3.2. The oxide growing at the metal/oxide interface is pure oxide and

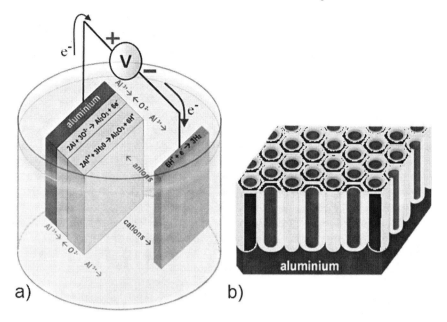

Fig. 3.1. (a) Chemical reactions during the anodic formation of alumina. (b) Schematic of nanoporous anodized aluminum.

the oxide growing at the oxide/solution interface incorporates anions from the electrolyte. The reaction occurring at the cathode results in the production of hydrogen, equation 3.3. The overall reaction is thus oxide film growth and hydrogen gas formation, equation 3.4.

$$2Al^{3+} + 3H_2O \rightarrow Al_2O_3 + 6H^+ \tag{3.1}$$

$$2Al + 3O^{2-} \rightarrow Al_2O_3 + 6e^- \tag{3.2}$$

$$6H^+ + 6e^- \rightarrow 3H_2 \tag{3.3}$$

$$2Al + 3H_2O \rightarrow Al_2O_3 + 3H_2 \tag{3.4}$$

In the case of neutral electrolytes, i.e. Al^{3+} insoluble, the barrier layer will grow to less than 1 μm with applied voltages of 500–700 V. After this thickness is obtained, the film will undergo dielectric breakdown. Anodization in an acidic electrolyte such as sulphuric acid, however, changes the structure of the oxide layer, Fig. 3.1b. This is caused by two phenomena: First, the Al^{3+} cations are now soluble in the solution so the oxide layer dissolves at the same time it is forming. The thickness of the oxide layer is proportional to the voltage and inversely proportional to the solubility of the aluminium in the solution or the competition between oxide growth and dissolution; the more acidic the solution the more soluble the aluminium and the thinner the oxide layer and the smaller the pores. Second, the mechanical stress between the aluminium and the oxide layer volume expansion of 1.2 [12] results in heterogeneous dissolution of the oxide layer or i.e. the formation of pores. The dissolution of the aluminium is field assisted so the oxide dissolves in areas where the oxide layer is thinner. The pores become deeper and other areas become isolated from dissolution. The competition between oxide dissolution and growth modulated by the film stress can result in an ordered porous structure, in stabilizing regimes of the Laplace pressure of the pores and the elastic stress [20,21], Fig. 3.1b. The heterogeneous nature of the oxide film is the same as the case of the barrier film with the pure oxide is light blue, Figs. 3.1a and b, and the anion contaminated oxide is light green, Figs. 3.1a and 3.1b, while the pores are depicted as black, Fig. 3.1b. The dimensions and interpore spacing of the pores are proportional to the anodization voltage [12]. Also, as stated earlier, the more soluble the aluminium the smaller the diameter of the pores, 5–30 nm for sulphuric acid, 40–60 nm for oxalic acid, and 80–130 nm for phosphoric acid. These values coincide with the increasing pH's of the solutions. For all electrolytes the oxide layer is thinnest at the pore bottom where it is dissolving. The pore diameter, Φ_p, and the interpore distance, D_int, are proportional to the anodisation voltage U. [1] In other words, $\Phi_p = k_1 U$ and $D_{int} = k_2 U$, where $k_1 \sim 1.29$ nm/V and $k_2 \sim 2.5$ nm/V [12,22].

Electrodeposition of materials in porous structures is performed by connecting the aluminum part of the template to the working electrode lead of a potentiostat. This sample is then placed in a metal salt electrolyte such as $ZnNO_3$. Potentiostatic electrodeposition of the material (M) is then performed in the pores by means of the working electrode, where M (Co, Cu, Ni, Zn etc.) is a metal and n is the number of electrons, equation 3.5.

$$M^{n+} + ne^- \rightarrow M^0 \qquad (3.5)$$

During the electrodeposition the metal ions diffuse into the pores to the bottoms and are deposited by applying a potential between the working and the reference electrode, which is kept at a constant value for potentiostatic control. The counter anode serves as the current source for the system.

The standard template approach facilitates two contacts for two terminal devices. [7,23–25] Another contact or electrode is needed for the realization of three terminal devices necessary for electronic applications. We have developed a new 3D alumina template that allows placement of a third electrode close enough to the nanowires for an electric-field effect, Fig. 3.2 [7,26] Gray colored area of the diagram is an aluminum wire, the end of which has been electrochemically etched to about 3 μm or less in diameter, the bottom 10 mm is anodized about 500 nm deep perpendicular to its axis to form an isolatated layer, green. The bottom 5 mm of the layer is sputtered coated by gold as a gate electrode, yellow. The sputtered gold layer does not cover all of the oxidized area and is thus isolated from the aluminum. A small section of the bottom is cut off to expose the interior aluminum. This is then anodized parallel to the wire to form a porous alumina template about one micron

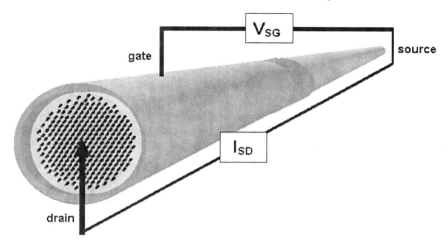

Fig. 3.2. Scheme of a bi-directionally anodized wire transistor. The gray area is the aluminum wire and the yellow area is the gold gate separated from the aluminum wire by the external oxide, green. The nanowires are the black dots in the light blue internal porous alumina template.

deep, light blue. The nanowires, black, are electrodeposited into the parallel template by connecting the aluminum wire as the cathode of an electrochemical cell. The nanowires are connected as the drain, the aluminum wire as the source, and the gold layer as the gate for transport measurements. This design is a vertical surrounding gate design [27–31] and the cylindrical geometry also provides an enhancement of the gate field.

The idea of anodizing Al wires and tubes has been tried for applications such as extracting fibers or chromatography columns [32, 33], however, using anodized wires as templates and multidirectional anodization are new concepts. These templates could be used for transistors, microelectrodes, diodes, micro/nano fluidics, and moulds for MEMs etc.

3.3 Experimental

120-micron diameter aluminum wire is the starting material for the transistor template. This was annealed at 500°C for 24 h in a vacuum tube furnace and cut into 3 cm long pieces. The wire was then electrochemically etched in a 25% $HClO_4$ 75% ethanol solution at +10 V to a few microns in diameter. The etch rate is approximately 1.5 μm per second. Next, the wire is anodized at +40 V in 0.3 M oxalic acid for 2–4 min to produce an external insolent layer on the exterior of the wire. At this voltage the growth rate of the oxide layer is 200 nm per minute. The tip of the wire is immersed in a 1 M NaOH solution to dissolve the oxide layer. Now, a gold layer is sputtered on the exterior oxide layer of the wire. The gold layer, which is electrically isolated from the interior aluminum, will function as a gate electrode. The wires are anodized internally at +40 V in a 0.3 M oxalic acid for 5 min to form an internal nanoporous template. The pores are 40–50 nm in diameter and should be about 1 μm deep. This will serve as the template for the electrodeposition of the nanowires. The nanowires can then be grown and contacted with the gate electrode already in place.

ZnO and other transparent conducting oxides are interesting as materials for UV lasers, light-emitting diodes, photo detectors, and for applications in flat panel displays and solar cells. It has a band gap of 3.35 eV and is normally a n-type direct gap material [34–37].

Once the porous template is made the ZnO nanowires can be electrodeposited into it. This is done by potentiostatic electrodeposition of ZnO from a 0.001M $Zn(NO_3)_2$ solution (pH 6.8) at −1.500 V vs. SCE in the pores for 1,000 s. Since this is an unbuffered solution the polarization causes an increase of the pH at the sample surface, due to a loss of H^+ by H_2 gas formation, which results in the precipitation of ZnO [38, 39]. Optimization of the solution and an increase in the solution deposition temperature could produce single crystal ZnO nanowires [39, 40]. When the ZnO arrives at the surface and extends beyond the template it can be electrically contacted. Then the aluminum wire base of the template and the gold gate electrode can be contacted by silver paste. Chemical analysis by backscattering of the electrodeposited ZnO on Au substrates revealed stoichiometric ZnO, however, the XRD analysis showed many phases of polycrystalline ZnO and Zn metal.

3.4 Results

Figure 3.3 shows the mean current of six IV scans, three scans with the gate bias starting at −30 V and stepped to zero and three scans starting with the gate bias at 0 V and stepping it to −30 V. It seems in Fig. 3.3 that the noise

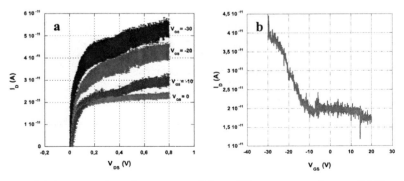

Fig. 3.3. (a) Average of six drain-source IVs at different gate voltages. (b) Transfere characteristics at a drain-source voltage of 1.0.

in the individual scans are slightly dependent on the gate bias. Figure 3.3b shows the transfer characteristics at a drain-source bias of 1.0 V and shows that the transistor operates as a p-channel depletion mode device. From Fig. 3.3b the on/off ratio is only two.

3.5 Conclusion

The gate dependent drain currents and transfer characteristics demonstrate that it is possible to make nanowire transistor devices in a beaker without clean rooms or lithography. We are in the process of trying this template with carbon nanotubes and silicon nanorods. The electrodeposition of the ZnO needs to be optimized as well as removing the barrier layer before electrodeposition of the ZnO nanowires. Doping of the ZnO to lower its resistance is also being explored. Standardizing the fabrication and combining the techniques with lithography should greatly improve the device performance.

References

1. C. N. R. Rao, F. L. Deepak, G. Gundiah, A. Govindaraj, Inorganic nanowires. Progress in Solid State Chemistry **31**, 5–147 (2003).
2. B. Doudin, A. Blondel, J. P. Ansermet, Arrays of multilayered nanowires. Journal of Applied Physics **79**, (8), 6090–6094 (1996).
3. S. Dubois, J. L. Duvall, A. Fert, J. M. George, J. L. Maurice, L. Piraux, Perpendicular giant magnetoresistance in Co/Cu and permalloy/Cu multilayered nanowires. Journal of Applied Physics **81**, (8), 4569–4569 (1997).
4. W. Schwarzacher, O. I. Kasyutich, P. R. Evans, M. G. Darbyshire, G. Yi, V. M. Fedosyuk, F. Rousseaux, E. Cambril, D. Decanini, Metal nanostructures prepared by template electrodeposition. **199**, 185–190 (1999).
5. P. Levy, A. G. Leyva, H. E. Troiani, R. D. Sanchez, Nanotubes of rare-earth oxide. Applied Physics Letters **83**, (25), 5247–5249 (2003).
6. L. Piraux, A. Encinas, L. Vila, S. Màtéfi-Tempfli, M. Màtéfi-Tempfli, M. Darques, F. Elhoussine, S. Michotte, Magnetic and Superconducting Nanowires. Journal of Nanoscience and Nanotechnology **5**, 372–389 (2005).
7. T. L. Wade, J.-E. Wegrowe, Template synthesis of nanomaterials. The European Physical Journal Applied Physics **29**, 3–22 (2005).
8. K. Nielsch, F. Müller, A.-P. Li, U. Gösele, Uniform nickel deposition into ordered alumina pores by pulsed electrodeposition. Advanced Materials **12**, (8), 582–586 (2000).
9. F. Keller, M. S. Hunter, D. L. Robinson, Journal of the Electrochemical Scoiety 100(9), 411 (1953).
10. J. P. O'sullivan, G. C. Wood, Morphology and Mechanism of Formation of Porous Anodic Films on Aluminium. Proceedings of the Royal Society of London Series a-Mathematical and Physical Sciences **317**(1531), 511–520 (1970).
11. H. Masuda, K. Fukuda, Ordered metal nanohole arrays made by a 2-Step replication of honeycomb structures of anodic alumina. Science **268**(5216), 1466–1468 (1995).

12. K. Nielsch, J. Choi, K. Schwirn, R. B. Wehrspohn, U. Gosele, Self-ordering regimes of porous alumina: The 10% porosity rule. Nano Letters 2(7), 677–680 (2002).
13. O. Jessensky, F. Muller, U. Gosele, Self-organized formation of hexagonal pore arrays in anodic alumina. Applied Physics Letters 72(10), 1173–1175 (1998).
14. G. E. Thompson, R. C. Furneaux, G. C. Wood, J. A. Richardson, J. S. Goode, Nucleation and Growth of Porous Anodic Films on Aluminum. Nature 272(5652), 433–435 (1978).
15. A. P. Li, F. Muller, A. Birner, K. Nielsch, U. Gosele, Hexagonal pore arrays with a 50–420 nm interpore distance formed by self-organization in anodic alumina. Journal of Applied Physics 84(11), 6023–6026 (1998).
16. K. Nielsch, F. Muller, A. P. Li, U. Gosele, Uniform nickel deposition into ordered alumina pores by pulsed electrodeposition. Advanced Materials 12(8), 582–586 (2000).
17. K. Nielsch, R. Hertel, R. B. Wehrspohn, J. Barthel, J. Kirschner, U. Gosele, S. F. Fischer, H. Kronmuller, Switching behavior of single nanowires inside dense nickel nanowire arrays. Ieee Transactions on Magnetics 38(5), 2571–2573 (2002).
18. Y. T. Pang, G. W. Meng, L. D. Zhang, W. J. Shan, X. Y. Gao, A. W. Zhao, Y. Q. Mao, Arrays of ordered Pb nanowires with different diameters in different areas embedded in one piece of anodic alumina membrane. Journal of Physics-Condensed Matter 14(45), 11729–11736 (2002).
19. Y. H. Wang, Y. Q. Xu, W. L. Cai, J. M. Mo, New method to prepare CdS nanowire arrays. Acta Physico-Chimica Sinica 18(10), 943–946 (2002).
20. G. K. Singh, A. A. Golovin, I. S. Aranson, V. M. Vinokur, Formation of nanoscale pore arrays during anodization of aluminum. Europhysics Letters 70(6), 836–842 (2005).
21. G. K. Singh, A. A. Golovin, I. S. Aranson, Formation of self-organized nanoscale porous structures in anodic aluminum oxide. Physical Review B 73(20), 205422 (2006).
22. D. Y. H. Lo, A. E. Miller, M. Crouse, Self-ordered pore structure on anodized aluminum on silicon and patteren transfer. Applied Physics Letters 76(1), 49–51 (2000).
23. X. Hoffer, C. Klinke, J. -M. Bonard, L. Gravier, J. -E. Wegrowe, Spin-dependent magnetoresistance in multiwall carbon nanotubes. Europhysics Letters 67(1), 103–109 (2004).
24. J. -E. Wegrowe, S. E. Gilbert, D. Kelly, B. Doudin, J. -P. Ansermet, Anisotropic Magnetoresistance as a Probe of Magnetization Reversal in Individual Nano-Sized Nickel Wires. IEEE Transactions on Magnetics 34(4), 903–905 (1998).
25. J. F. Dayen, T. L. Wade, M. Konczykowski, J. E. Wegrowe, X. Hoffer, Conductance in multiwall carbon nanotubes and semiconductor nanowires. Physical Review B 27(7), 073402 (2005).
26. T. Wade, J. -E. Wegrowe Procede de fabrication de composants electroniques et composants electroniques obtenus par ce procede. 2003.
27. T. Bryllert, L. E. Wernersson, L. E. Froberg, L. Samuelson, Vertical high-mobility wrap-gated InAs nanowire transistor. Ieee Electron Device Letters 27(5), 323–325 (2006).
28. J. Goldberger, A. I. Hochbaum, R. Fan, P. Yang, Silicon vertically integrated nanowire field effect transistors. Nano Letters 6(5), 973–977 (2006).

29. T. Schulz, W. R?sner, E. Landgraf, L. Risch, U. Langmann, Planar and vertical double gate concepts. Solic-State Electronics **46**, 985–989 (2002).
30. J. Chen, R. Konenkamp, Vertical nanowire transistor in flexible polymer foil. Applied Physics Letters **82**(26), 4782–4784 (2003).
31. A. K. Sharma, S. H. Zaidi, S. Lucero, S. R. J. Brueck, N. E. Islam, Mobility and transverse electric field effects in channel conduction of wrap-around-gate nanowire MOSFETs. IEE Proc.-Circuits Devices Syst. **151**(5), 422–430 (2004).
32. D. Djozan, Y. Assadi, S. H. Haddadi, Anodic aluminium wire as a solid-phase microextraction fiber. Analytical Chemistry **73**(16), 4054–4058 (2001).
33. D. Djozan, A.-Z.M., Anodizing of inner surface of long and small-bore aluminum tube. Surface Coatings Technology **173**(2–3), 185–191 (2003).
34. Z. Fan, D. Wang, P. -C. Chang, W. -Y. Tseng, J. G. Lu, ZnO nanowire field-effect transistor and oxygen sensing property. Applied Physics Letters **85**(24), 5923 (2004).
35. Y. W. Heo, D. P. Norton, L. C. Tien, Y. Kwon, B. S. Kang, F. Ren, S. J. Pearton, Zno nanowire growth and devices. Materials Science and Engineering R **47**, 1–47 (2004).
36. Y. W. Heo, L. C. Tien, Y. Kwon, D. P. Norton, S. J. Pearton, Depletion-mode ZnO nanowire field-effect transistor. Applied Physics Letters **85**(12), 2274–2276 (2004).
37. H. T. Ng, J. Han, T. Yamada, P. Nguyen, Y. P. Chen, M. Meyyappan, Single Crystal Nanowire Vertical Surrounding-Gate Field-Effect Transistor. Nano Letters **4**(7), 1247–1252 (2004).
38. M. Pourbaix, Atlas d'Equilibres Electrochimiques. Gauthier-Villars: Paris, 1963, Vol. Chapter **15** Zinc.
39. S. Peulon, D. Lincot, Mechanistic Study of cathodic electrodeposition of zinc oxide and hydroxychloride films from oxygenated aqueous zinc chloride solutions. Journal of the Electrochemical Society **145**(3), 864–874 (1998).
40. Y. Leprince-Wang, G. Y. Wang, X. Z. Zhang, D. P. Yu, Study on the microstructure and growth mechanism of electrochemical deposited ZnO nanowires. Journal of Crystal growth **287**, 89–93 (2006).

4

FRIEDEL OSCILLATIONS IN NANOWIRES AT FINITE BIAS VOLTAGE

A. GORCZYCA, M. MAŚKA,
AND M. MIERZEJEWSKI
*Department of Theoretical Physics, Institute of Physics, University
of Silesia, 40-007 Katowice, Poland. marcin@phys.us.edu.pl*

Abstract. We investigate the charge density oscillations in a nanowire coupled asymmetrically to two leads. Depending on this asymmetry, the Friedel oscillations can either be characterized by a single wave-vector or become a superposition of oscillations with different wave-vectors. Using the formalism of nonequilibrium Keldysh Green functions, we derive a simple equation that determines bias voltage dependence of the wave-length of the oscillations. Finally, we discuss limitations of the commonly used formula that describes the spatial character of the Friedel oscillations.

Keywords: nanowire, Friedel oscillations, nonequilibrium Green functions

4.1 Introduction

The conductance of nanosystems has recently been the subject of intensive experimental and theoretical investigations, because of their possible application in the electronic devices. This ongoing research concerns the transport properties of various systems, e.g., nanowires, single molecules and quantum dots [2,3,4,8,18]. The properties of nanosystems differ significantly from the transport properties of macroscopic materials, what shows up in nonlinear or even irregular current–voltage characteristics.

In many cases, the charge carriers are distributed inhomogeneously in the nanosystem. On one hand, such inhomogeneity may be caused by the mechanisms, which are well known from the analysis of macroscopic materials. In particular, it may originate from impurities [7] or/and correlations which lead to charge density waves [12, 15, 17, 19, 23]. On the other hand, inhomogeneous distribution of the charge carriers may originate from mechanisms, which are typical for nanosystems. In analogy to the case of a quantum well, one may expect that electrons are inhomogeneously distributed just due to the spatial confinement at the nanoscale [13]. The standing waves have recently been observed at the end of a carbon nanotube [14]. Inhomogeneous charge distribution may originate also from the applied bias voltage [2, 6, 11].

Recent theoretical analysis of the Friedel oscillations in nanosytems has been carried out for zero or vanishingly small bias voltage V. The most of these investigations concerned the electron correlations [1, 5, 10, 21, 22] which, as has been shown, suppress the decay of oscillations [3, 20]. In this paper we neglect the electron–electron interaction. Instead, we focus on the nonequilibrium case that occurs for finite V. We investigate a nanowire that is connected to two macroscopic particle reservoirs with different Fermi energies. Since this difference increases with V, the meaning of the Fermi momentum, k_F, becomes ambiguous. Therefore, the well known $2k_F x$ spatial dependence of the density oscillations (that holds also in the presence of correlations) may be modified for $V \neq 0$. We have recently considered a nanowire that is symmetrically coupled to the leads. We have shown that the wave-vector of the oscillations decreases monotonically with bias voltage and vanishes when V is equal to the band width. These results hold both for the charge density waves [16] and the Friedel oscillations [9]. Here, we extend our previous analysis of the Friedel oscillations and discuss the case of asymmetric coupling. We account also for a linear potential drop in the nanowire. We demonstrate that these effects seriously affect the spatial dependence of the density oscillations.

4.2 Model

We investigate a nanowire coupled to two macroscopic leads. Therefore, we consider a Hamiltonian, that consists of three terms H_{el}, H_{nano}, and $H_{nano-el}$, which describe the electrodes, nanowire and the coupling between the electrodes and the nanowire, respectively

$$H = H_{el} + H_{nano} + H_{nano-el}. \tag{4.1}$$

The electrodes are modeled by the electron gas:

$$H_{el} = \sum_{k,\sigma,\alpha} (\varepsilon_{k,\alpha} - \mu_\alpha) c^\dagger_{k\sigma\alpha} c_{k\sigma\alpha}, \tag{4.2}$$

where μ_α is the chemical potential, $\alpha \in \{L,R\}$ indicates the left or right electrode. $c^\dagger_{k\sigma\alpha}$ creates an electron with momentum k and spin σ in the electrode α. We assume that the nanowire can be described by the tight-binding Hamiltonian:

$$\mathcal{H}_{nano} = -t \sum_{\langle ij \rangle \sigma} d^\dagger_{i\sigma} d_{j\sigma} + \sum_{i\sigma} \varepsilon_i n_{i\sigma} + U \sum_\sigma n_{l\sigma}, \tag{4.3}$$

where $d_{i\sigma}^{\dagger}$ is the creation operator of an electron with spin σ at site i of the nanosystem, $n_{i\sigma}$ – occupation number operator, ε_i – atomic energy level. In order to investigate the Friedel oscillations we have introduced a single impurity with the potential U. This impurity is located in the site l. The coupling between the nanowire and the leads is given by:

$$H_{\text{nano-el}} = \sum_{\mathbf{k},i,\alpha,\sigma} \left(g_{\mathbf{k},i,\alpha} c_{\mathbf{k}\sigma\alpha}^{\dagger} d_{i\sigma} + \text{H.c.} \right), \tag{4.4}$$

where the matrix elements $g_{\mathbf{k},i,\alpha}$ are nonzero only for the edge atoms of the nanowire.

4.3 Results and discussion

The electron density fluctuations have been determined by means of the nonequilibrium Keldysh Green functions. First, we have calculated the local carrier density, that is expressed by the lesser Green function,

$$\langle d_{i\sigma}^{\dagger} d_{i\sigma} \rangle = \frac{1}{2\pi i} \int d\omega \, G_{i\sigma,i\sigma}^{<}(\omega). \tag{4.5}$$

$G_{i\sigma,i\sigma}^{<}$ is, in turn, determined by the retarded and advanced Green functions:

$$\hat{G}^{<}(\omega) = i \sum_{\alpha \in \{L,R\}} \hat{G}^{r}(\omega) \hat{\Gamma}_{\alpha}(\omega) \hat{G}^{a}(\omega) f_{\alpha}(\omega), \tag{4.6}$$

where

$$\left[\hat{\Gamma}_{\alpha}(\omega) \right]_{ij} = 2\pi \sum_{\mathbf{k}} g_{\mathbf{k},i,\alpha}^{*} g_{\mathbf{k},j,\alpha} \delta(\omega - \varepsilon_{\mathbf{k},\alpha}), \tag{4.7}$$

and $f_{\alpha}(\omega)$ stands for the Fermi distribution function of the electrode α. We have assumed that the energy bands of the leads are wide enough, and one can neglect the frequency dependence of $\hat{\Gamma}_{\alpha}$. Finally, the retarded self-energy is determined by the coupling between the nanowire and the leads

$$\hat{\Sigma}^{r}(\omega) = \frac{1}{2} \sum_{\alpha \in \{L,R\}} \left[\frac{1}{\pi} P \int d\Omega \frac{\hat{\Gamma}_{\alpha}(\Omega)}{\omega - \Omega} - i\hat{\Gamma}_{\alpha}(\omega) \right]. \tag{4.8}$$

Numerical analysis of the above equations allows one to investigate spatial distribution of charge carriers and the properties of the Friedel oscillations. In the equilibrium case, these oscillations asymptotically decay with the distance x from the impurity as $\cos(Qx + \phi_0) x^{-\delta}$, where $Q = 2k_F$ and η parametrizes the interaction. As we have pointed out in the introduction, the wave-vector of the oscillations, Q, may nontrivially depend on the bias voltage. This dependence has been obtained from the fast Fourier transform of the the spatial electron distribution. On one hand, this approach enables accurate investigations of the Friedel oscillations for arbitrary model parameters. On the other hand, such numerical analysis does not explain the underlying physical mechanisms.

The bias voltage dependence of the wave-vector Q obtained from the above numerical analysis is exactly the same as previously reported for the charge density waves (CDW) [16]. Therefore, one may expect that there is a common physical mechanism that determines $Q(V)$ in both cases. In the following, we investigate a correlation function, that in the equilibrium case is related to the CDW instability. On the one hand, it allows us to explain $Q(V)$ obtained for the charge density waves. On the other hand, it provides a simple (even if only phenomenological) description of $Q(V)$ in the case of Friedel oscillation. The discussed above numerical analysis allowed us to confirm the applicability of this description.

In the equilibrium case the wave-length of the oscillations can be analyzed with the help of an appropriate correlation function. Usually one assumes *periodic boundary* conditions and investigates

$$\chi(Q,\omega) = -\langle\langle\hat{\rho}(Q)|\hat{\rho}^{\dagger}(Q)\rangle\rangle, \tag{4.9}$$

where

$$\hat{\rho}(Q) = \sum_{i,\sigma} \exp(iQR_i)d_{i\sigma}^{\dagger}d_{i\sigma}. \tag{4.10}$$

Since the nanowire is connected to macroscopic leads, application of periodic boundary conditions is unjustified and the choice of open boundary conditions seems to be much more appropriate. At this stage of our analysis we assume that $U = 0$ and $\varepsilon_i = \varepsilon_0 = const$. The second assumption implies that there is no potential drop inside the nanowire. Then, one can use the following unitary transformation to diagonalize H_{nano} (4.3):

$$d_{i\sigma}^{\dagger} = \sqrt{\frac{2}{N+1}} \sum_{k} \sin(kR_i)d_{k\sigma}^{\dagger} \tag{4.11}$$

where the values of k are given by: $k = \frac{\pi}{N+1}, \frac{2\pi}{N+1}, \ldots, \frac{N\pi}{N+1}$ and N denotes the number of sites of the nanowire. It is straightforward to check that

$$\mathcal{H}_{\text{nano}} = \sum_{k\sigma} \varepsilon_k d_{k\sigma}^{\dagger}d_{k\sigma} \tag{4.12}$$

and the dispersion relation reads $\varepsilon_k = -2t\cos(k) + \varepsilon_0$.

In order to analyze $Q(V)$ we have transformed $H_{\text{nano-el}}$ according to Eq. (4.11). Then, we have calculated the correlation functions given by Eq. (4.9) with

$$\hat{\rho}(Q) = \sum_{i\sigma} \cos(QR_i)d_{i\sigma}^{\dagger}d_{i\sigma} = \sum_{kp\sigma} \mathcal{B}_Q(k,p)d_{k\sigma}^{\dagger}d_{p\sigma} \tag{4.13}$$

and

$$\mathcal{B}_Q(k,p) = \frac{1}{2}\left(\delta_{p,k-Q} - \delta_{p,Q-k} + \delta_{p,k+Q} - \delta_{p,2\pi-(k+Q)}\right). \tag{4.14}$$

Using the equations of motion we have obtained the following form of the correlation function (4.9)

$$\chi(Q,\omega \to 0) = \sum_{kpq\sigma} \frac{\mathcal{B}_Q(k,p)}{\varepsilon_p - \varepsilon_k}\left(\mathcal{B}_Q^*(p,q)\langle d_{k\sigma}^\dagger d_{q\sigma}\rangle - \mathcal{B}_Q^*(q,k)\langle d_{q\sigma}^\dagger d_{p\sigma}\rangle\right) + \chi' \tag{4.15}$$

The second term of the above formula, χ', is proportional to $\langle\langle d^\dagger c|d^\dagger d\rangle\rangle$ and it can be neglected for a weak coupling between the nanosystem and the leads. In order to proceed with the analytical discussion we have also neglected the off–diagonal elements of Γ_{kp}^α. Consequently, $\langle d_{k\sigma}^\dagger d_{q\sigma}\rangle \sim \delta_{kq}$ and the investigated correlation function takes on a simple form

$$\chi(Q,\omega \to 0) = \sum_{kp\sigma} |\mathcal{B}_Q(k,p)|^2 \frac{\langle n_{k\sigma}\rangle - \langle n_{p\sigma}\rangle}{\varepsilon_p - \varepsilon_k}. \tag{4.16}$$

The above equation resembles the Lindhard's form of the correlation function that is well known from the equilibrium physics. However, there exist two important differences. The first one originates from the open boundary conditions, what shows up in the presence of the factor $\mathcal{B}_Q(k,p)$. The second one accounts for the fact that the nanowire is not in equilibrium with a single thermostat. Therefore, the average $\langle n_{k\sigma}\rangle$ is determined by a lesser Green function and cannot be expressed by a single Fermi function. In order to make these differences more visible, we have assumed that $\Gamma_{kp}^R = a_\Gamma \Gamma_0 \delta_{kp}$ and $\Gamma_{kp}^L = \frac{1}{a_\Gamma}\Gamma_0 \delta_{kp}$. Here, we have introduced the parameter $a_\Gamma \in (1,\infty)$ that describes the degree of asymmetry of the coupling to the electrodes. For $\Gamma_0 \to 0$ (which is equivalent to the weak coupling regime) one can explicitly calculated the integral over frequencies in Eq. (4.5). The obtained value of $\langle n_{k\sigma}\rangle$ reads:

$$\langle n_{k\sigma}\rangle = \frac{a_\Gamma}{1+a_\Gamma^2}\left(a_\Gamma f_R(\varepsilon_k) + \frac{1}{a_\Gamma} f_L(\varepsilon_k)\right) \tag{4.17}$$

and the correlation function becomes a sum of two functions, stemming from both the electrodes:

$$\chi(Q,\omega \to 0) = \frac{a_\Gamma^2}{1+a_\Gamma^2}\chi^R(Q) + \frac{1}{1+a_\Gamma^2}\chi^L(Q), \tag{4.18}$$

where

$$\chi^{R(L)}(Q) = \sum_{kp\sigma} |\mathcal{B}_Q(k,p)|^2 \frac{f_{R(L)}(\varepsilon_k) - f_{R(L)}(\varepsilon_p)}{\varepsilon_p - \varepsilon_k}. \tag{4.19}$$

The Fermi distribution functions of the left and right electrodes depend on the bias voltage. Namely, $f_{R(L)}(\varepsilon_k) = f(\varepsilon_k - \mu_{R(L)})$, with $f(x) = (exp(x)+1)^{-1}$

and $\mu_L - \mu_R = eV$. In the case of a symmetric coupling one usually puts $\mu_L = eV/2$ and $\mu_R = -eV/2$. However, for $a_\Gamma \neq 1$ it may be invalid. Therefore, we have introduced a second parameter, a_V, that determines the asymmetry of the potential drop that takes place at contacts between the nanowire and the electrodes. Namely, we assume that $\mu_L = eVa_V$ and $\mu_R = -eV(1 - a_V)$ and consider $a_V \in (0, 1)$.

A comment on the applied approximations ($\Gamma_{kp} \sim \delta_{kp}$ and $\Gamma_0 \rightarrow 0$) is necessary. These crude approximations may affect many properties of the nanowire. In particular, they are inapplicable in the investigations of the current–voltage characteristics. However, as will be demonstrated in the following discussion, the wave-vectors of the oscillations determined from the Fourier transform of Eq. (4.5) and from Eq. (4.18) are the same. It means, that the wave-length of the density oscillations is determined predominantly by the fact that the nanowire is coupled to two particle reservoirs with different chemical potentials.

First, we discuss the Friedel oscillations in the presence of the asymmetric coupling between the nanowire and the leads. This asymmetry occurs when the $a_\Gamma \neq 1$ or $a_V \neq 1/2$. Figure 4.1 shows the spatial distribution of charge carriers as well as correlation function χ for the case when the asymmetry originates only from a_Γ. We have found that a_Γ does influence the correlation function. It means that the wave-length of the density oscillations is independent of this quantity and is exactly the same as obtained for a symmetric coupling. This conclusion can also be drawn from the presented electron distribution. Although, $\langle n_{i\sigma} \rangle$ depends quantitatively on a_Γ, the periodicity of the density fluctuations remains unchanged. For a vanishing bias voltage, χ takes on the maximal value for $Q = \pi$. Since in the half–filled case $k_F = \pi/2$

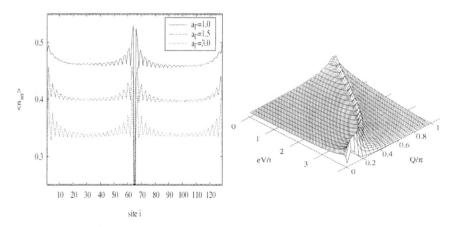

Fig. 4.1. Results obtained for $a_V = 0.5$, $\varepsilon_0 = 0$ and various a_Γ. Left panel shows occupation of sites of a 129-site nanowire for $eV = 2t$. Right panel shows the correlation function $\chi = \chi(Q, V)$, Eq. (4.18), calculated for $a_\Gamma = 3$.

this result remains in agreement with a a standard relation, that holds in the
equilibrium case $Q = 2k_F$. However, for a finite bias voltage the situation is
very different. The correlation function reaches its maximum for a wave-vector
Q that monotonically decreases with V and vanishes when the bias voltage
exceeds the band width of an isolated nanowire. An explicit form of the voltage dependence of Q can be found from Eqs. (4.18) and (4.19) [9]. Namely, for
$a_V = 1/2$ and $\varepsilon_0 = 0$ the maximum of the correlation function χ occurs for
$Q = 2\arccos(eV/4t)$. In the symmetric case, $a_\Gamma = 1$, the bias voltage changes
the wave-length of the Friedel oscillations but the average occupation of sites
is independent of V. On the other hand, for a fixed bias voltage, a_Γ modifies
the occupation number but does not influence the wave-length of the oscillations. These results visibly differ from the equilibrium case, when the wave-
length of the density fluctuation is uniquely determined by the occupation
number.

A different form of the Friedel oscillations occurs for the asymmetric potential drop (see Fig. 4.2). Contrary to the previous case, the wave-length of the
oscillations depends on a_V, provided the bias voltage is finite. One can see,
that the correlation function achieves two local maxima, which occur for different values of Q. However, the results presented in Fig. 4.2 do not allow
one to distinguish between the following possibilities: (i) the Friedel oscillations can be described by a single wave-vector, e.g., by Q that corresponds to
larger value of the investigated correlations functions; (ii) the Friedel oscillations represent a superposition of two oscillations with different values of Q.
Carrying out Fourier transform of $\langle n_{i\sigma} \rangle$ we have found that the second case

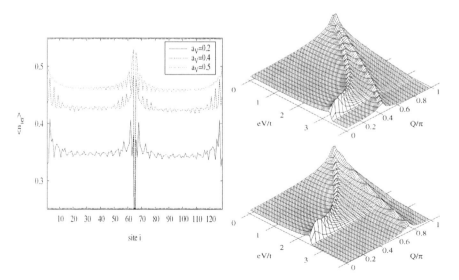

Fig. 4.2. The same as in Fig. 4.1 but for $a_\Gamma = 1$ and various a_V The right upper and
the right lower panels show $\chi = \chi(Q, V)$ obtained for $a_V = 0.4$ and 0.2, respectively.

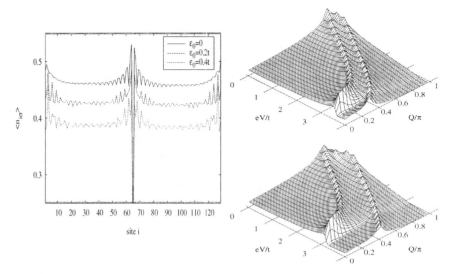

Fig. 4.3. The same as in Fig. 4.1 but for $a_\Gamma = 1$ and various ε_0 The right upper and the right lower panels show $\chi = \chi(Q, V)$ obtained for $\varepsilon_0 = 0.2t$ and $0.4t$, respectively.

is actually realized in the system under investigations. Figure 4.3 shows that a similar situation occurs also for $\varepsilon_0 \neq 0$. The above discussion allows one to draw the following conclusion: The Friedel oscillations can be characterized by a single wave-vector Q provided $\mu_L + \mu_R = \varepsilon_0 = 0$. This result holds independently of the asymmetry of Γ. Otherwise, the spatial dependence of the electron concentration becomes a superposition of oscillations with two different wave-vectors Q.

In the above discussion, we have assumed that there is no potential drop inside the nanowire. In the following we relax this assumption, and consider a linear potential drop that modifies the on-site energy levels:

$$\varepsilon_i = \varepsilon_0 + \frac{eV}{2}\eta\left(1 - \frac{2i}{N+1}\right) \qquad (4.20)$$

For $\eta = 0$ there is no potential drop, whereas for $\eta = 1$ the entire drop takes place inside the nanowire. The assumed linear dependence may be invalid in the presence of impurities. Therefore, we have put $U = 0$ and investigated the density oscillations that occur at the edges of the nanowire. For $\varepsilon_i \neq const$, i.e., for $\eta \neq 0$ H_{nano} cannot be diagonalized by the transformation given by Eq. (4.11). In order to proceed with the discussion, we have solved the arising eigenproblem numerically. Namely, we have introduced new fermionic operators $a_{a\sigma}$

$$d_{i\sigma} = \sum_m U_{im}a_{m\sigma}, \qquad (4.21)$$

where the unitary matrix U leads to a diagonal form of H_{nano}

$$\mathcal{H}_{\mathrm{nano}} = \sum_{n\sigma} \varepsilon_n a_{n\sigma}^{\dagger} a_{n\sigma}. \tag{4.22}$$

Then, we have repeated the previous analysis and applied the same approximation scheme. In particular we have calculated $\chi(Q, \omega \to 0)$. The correlation function in this case takes on the form

$$\chi(Q, \omega \to 0) = \sum_{ij\sigma} \cos(QR_i) \cos(QR_j) \sum_{mn} U_{im} U_{in}^* U_{jn} U_{jm}^* \frac{\langle n_{n\sigma} \rangle - \langle n_{m\sigma} \rangle}{\varepsilon_m - \varepsilon_n}. \tag{4.23}$$

Figure 4.4 shows the numerical results. One see that the character of the Friedel oscillations strongly depends on the potential drop, that takes place inside the nanowire. The maximum of the correlation function occurs for Q that strongly depends on the bias voltage, similarly to the previously discussed cases. However, for $\eta \neq 0$ the value of this maximum decreases when V increases. For sufficiently high voltage, Q-dependence of the correlation functions smears out. In order to understand this result, we have divided the nanowire into small sections. Then, we have carried out the Fourier transform of $\langle n_{i\sigma} \rangle$ independently for each section. Surprisingly, we have found that Q depends on the position of the section of the nanowire. Therefore, for $\eta \neq 0$ the standard formula describing the Friedel oscillations $\cos(Qx + \phi_0)/x^\delta$ holds only locally, i.e., it holds at the lenghscale much smaller than the size of the

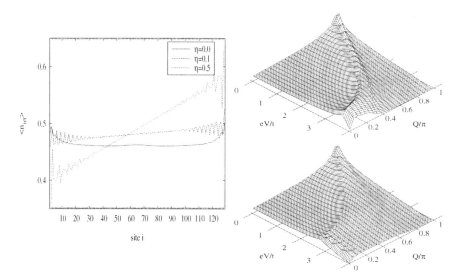

Fig. 4.4. Results obtained for $a_\Gamma = 1$, $a_V = 0.5$ and $\varepsilon_0 = 0$. The left panel shows the occupation of sites of a 129-site nanowire for $eV = 2t$. The right panels show $\chi = \chi(Q, V)$, Eq. (4.23). The upper and lower ones have been obtained for $\eta = 0.1$ and 0.5, respectively.

nanowire. Otherwise, one should consider Q as a slowly varying function of the position.

To conclude, we have analyzed the Friedel oscillations in the nanowire. Using the formalism of the Keldysh Green functions, we have focus on the nonequilibrium case that occurs for finite bias voltage. Generally, the wavelength of the oscillations decreases when the bias voltage increases. The spatial character of the Friedel oscillations strongly depends on the asymmetry of the coupling between nanowire and leads as well as on the potential drop inside the nanowire. As we have shown, experimental investigations of the Friedel oscillations may provide important information concerning the coupling between the nanowire and the leads.

Acknowledgements

This work has been supported by the Polish Ministry of Education and Science under Grant No. 1 P03B 071 30.

References

1. Bedürftig, G., Brendel, B., Frahm, H., and Noack, R. M. Friedel Oscillations in the Open Hubbard Chain, Phys. Rev. B **58**, 10225–10235 (1998).
2. Chen, J., Reed, M. A., Rawlett, A. M., and Tour, J. M. Large On-Off Ratios and Negative Differential Resistance in a Molecular Electronic Device, Science **286**, 1550–1552 (1999).
3. Cohen, A., Richter, K., and Berkovits, R. Spin and Interaction Effects on Charge Distribution and Currents in One-Dimensional Conductors and Rings within the Hartree-Fock Approximation, Phys. Rev. B **57**, 6223–6226 (1998).
4. Donhauser, Z. J., Mantooth, B. A., Kelly, K. F., Bumm, L. A., Monnell, J. D., Stapleton, J. J., Price Jr., D. W., Rawlett, A. M., Allara D. L., Tour, J. M., and Weiss, P. S. Conductance Switching in Single Molecules through Conformational Changes, Science **292**, 2303–2307 (2001).
5. Eggert S. Scanning Tunneling Microscopy of a Luttinger Liquid, Phys. Rev. Lett. **84**, 4413–4416 (2000).
6. Emberly, E. G. and Kirczenow, G. Current-Driven Conformational Changes, Charging, and Negative Differential Resistance in Molecular Wires, Phys. Rev. B **64**, P125318 (2001).
7. Friedel, J. Metallic Alloys, Nuovo Cimento (Suppl.) **7**, 287–311 (1958).
8. Gittins, D. I., Bethell, D., Schiffrin, D. J., and Nichols, R. J. A Nanometre-Scale Electronic Switch Consisting of a Metal Cluster and Redox-Addressable Groups, Nature (London) **408**, 67–69 (2000).
9. Gorczyca, A., Maśka, M. M., and Mierzejewski, M. (2007) The Friedel Oscillations in the Presence of Transport Currents in a Nanowire, Phys. Rev. B **76**, 165419 (2007).
10. Grishin, A., Yurkevich, I. V., and Lerner, I. V. Functional Integral Bosonization for an Impurity in a Luttinger Liquid, Phys. Rev. B **69**, P165108 (2004).

11. Kostyrko, T. and Bułka, B. R. Hubbard Model Approach for the Transport Properties of Short Molecular Chains, Phys. Rev. B **67**, P205331 (2003).
12. Krive, I. V., Rozhavsky, A. S., Mucciolo, E. R., and Oxman, L. E. Electron Transport Through a Mesoscopic Metal-CDW-Metal Junction, Phys. Rev. B **61**, 12835–12841 (2000).
13. Kwapiński, T. Charge Fluctuations in a Perfect and Disturbed Quantum Wire, J. Phys.: Condens. Matter **18**, 7313–7326 (2006).
14. Lee, J., Eggert, S., Kim, H., Kahng, S. -J., Shinohara, H., and Kuk, Y. Real Space Imaging of One-Dimensional Standing Waves: Direct Evidence for a Luttinger Liquid, Phys. Rev. Lett. **93**, P166403 (2004).
15. Mantel, O. C., Bal, C. A. W., Langezaal, C., Dekker, C., and van der Zant, H. S. J. Sliding Charge-Density-Wave Transport in Micron-Sized Wires of $Rb_{0.30}MoO_3$, Phys. Rev. B **60**, 5287–5294 (1999).
16. Mierzejewski, M. and Maśka, M. M. Transport Properties of Nanosystems with Conventional and Unconventional Charge Density Waves, Phys. Rev. B **73**, P205103 (2006).
17. Oxman, L. E., Mucciolo, E. R., and Krive, I. V. Transport in Finite Incommensurate Peierls-Fröhlich Systems, Phys. Rev. B **61**, 4603–4607 (2000).
18. Park, H., Park, J., Lim, A. K. L., Anderson, E. H., Alivisatos, A. P., and McEuen, P. L. Nanomechanical Oscillations in a Single-C(sub 60) Transistor, Nature (London) **407**, 57–60 (2000).
19. Ringland, K. L., Finnefrock, A. C., Li, Y., Brock, J.D., Lemay, S. G., and Thorne, R. E. Sliding Charge-Density Waves as Rough Growth Fronts, Phys. Rev. B **61**, 4405–4408 (2000).
20. Rommer, S. and Eggert, S. Spin- and Charge-Density Oscillations in Spin Chains and Quantum Wires, Phys. Rev. B **62**, 4370–4382 (2000).
21. Schmitteckert, P. and Eckern, U. Phase Coherence in a Random One-Dimensional System of Interacting Fermions: A Density-Matrix Renormalization-Group Study, Phys. Rev. B **53**, 15397–15400 (1996).
22. Weiss, Y., Goldstein, M., and Berkovits, R. Friedel Oscillations in Disordered Quantum Wires: Influence of Electron-Electron Interactions on the Localization Length, Phys. Rev. B **75**, P064209 (2007).
23. van der Zant, H. S. J., Marković, N., and Slot, E. Submicron Charge–Density–Wave Devices, Usp. Fiz. Nauk (Suppl.) **171**, 61–65 (2001).

SPIN ORBIT INTERACTION INDUCED
SPIN-SEPARATION IN PLATINUM NANOSTRUCTURE

KOONG CHEE WENG[1,2], N. CHANDRASEKHAR[1,2],
C. MINIATURA[1,3], AND BERTHOLD-GEORG
ENGLERT[1,4]

[1] *Department of Physics, National University of Singapore,*
Singapore 117542
[2] *IMRE, 3 Research Link, Singapore 117602.*
n-chandra@imre.a-star.edu.sg
[3] *INLN, UNS, CNRS, 1361 route des Lucioles, F-06560 Valbonne,*
France
[4] *Centre for Quantum Technologies, National University*
of Singapore, Singapore 117543

Abstract. Hirsch (1999) proposed a mechanism and geometry for the observation
of the spin-Hall effect. In this work, we present a novel realization of the Hirsch geometry in a platinum (Pt) nanostructure, which is an increasingly important material
for spintronics applications. Measurements were made in a non-local geometry to
avoid spurious effects. The measurements show the large spin Hall conductivity of
Pt. The results are compared with gold (Au) and aluminum (Al). Possible theoretical
explanations of our observations are briefly mentioned.

Keywords: spin-Hall effect, platinum nanostructure

5.1 Introduction

Spintronics, a subfield of condensed matter physics, promises a new generation
of electronic devices, including reconfigurable device architectures [1]. Conventional electronic devices use only the charge state of an electron, whereas
spintronics exploits the electron's spin degree of freedom to carry, manipulate
and store information. Spintronics requires an understanding of spin injection,
manipulation and detection. Typical spintronic devices (e.g. spin valve) use
a ferromagnetic metal to inject spin polarized carriers into another metal [2].
However spin polarized carrier injection into semiconductors is fraught with
problems due to the interface between the electronically and structurally

dissimilar materials [3]. Therefore, recent work has been targeted towards understanding and exploiting spin polarized currents by intrinsic mechanisms in a single material. Once such mechanism is the spin Hall effect (SHE) [4], which refers to the generation of a transverse spin current in materials with spin-orbit interaction due to an applied longitudinal electrical field. For a conductor of finite size, this causes the electrons with different spin states to accumulate at opposite sides.

The development in semiconductor SHE has limited applications due to the small output and low temperature requirement. However, recent work in paramagnetic metal SHE has shown possible room temperature SHE [5, 6]. Interestingly the metallic spin Hall conductivity effect at room temperature is larger than the semiconductor spin Hall conductivity at 4 K [5]. At present spintronics applications rely heavily on the use of ferromagnetic metals as either a spin detector (e.g. hard drive read head) or as a memory storage (e.g. magnetoresistive random access memory), the smaller conductivity mismatch between paramagnetic metals and ferromagnetic metals allows for effective integration of SHE with ferromagnets to improve present technology or to develop novel applications.

However, from practical consideration, such spin currents are impossible to detect directly by conventional electronic means. Currently the measurement of SHE is done via optical means or by using a ferromagnetic metal as spin detector. In 1999 Hirsch [7] proposed a device geometry for the electrical detection of such spin separation. But, this geometry had attracted little or no experimental interest due to the practical difficulties in fabrication and measurement. In this work, we present a novel realization of the Hirsch geometry using conventional device fabrication techniques. Measurements were made in a non-local geometry to remove spurious contributions to the readings. Measurement results for Pt are contrasted with results for Au and Al for a qualitative comparison. Possible theoretical explanations of our observations are briefly mentioned.

5.2 Spin-orbit interaction and its effect on electron spin

Spin-orbit interaction, as a mechanism of spin relaxation or dephasing was first predicted by D'yakonov and Perel' [8]. This mechanism applies to bulk III-V and two dimensional III-V structures. The situation in paramagnetic metals, on the other hand, is completely different. Hirsch [7] was the first to postulate that either skew spin scattering by impurities, or strong spin orbit interaction could generate a spin current in paramagnetic metals. This has been termed the spin Hall effect (SHE), and has not been observed until recently. SHE did not attract much attention until Murakami et al. [9] predicted a dissipationless spin current that could originate from the band structure, and since sparked extensive theoretical and experimental studies [5, 6, 10–15]. The arguments of Murakami et al. are based on analogies with the quantum Hall effect (QHE), and apply in particular to the split off heavy hole band in III-V semiconductors. Rashba [16] has pointed out that such spin currents, caused by spin orbit interaction, are not transport currents which could be employed for transporting spins or spin injection, in thermodynamic equilibrium. Subsequently two landmark experiments have reported the observation of spin accumulation. In n-doped bulk GaAs and InGaAs [10], the observation was attributed to skew scattering of spins by impurities [17], referred to as the extrinsic effect which also included the side-jump mechanism [18]. However, observations on spin accumulation of a two-dimensional hole gas (2DHG) in a p-doped GaAs structure [11] were attributed to intrinsic effects, such as the mechanism proposed by Murakami et al. [9].

More recently, there has been significant experimental development in SHE and inverse SHE (spin current producing charge accumulation) in metallic system. Valenzuela and Tinkham [12] have reported the first electrical detection of the SHE in aluminum strips at 4.2 K. In their experiment, a spin polarized charge current is injected from a ferromagnetic electrode into the Al strips and the pure spin current is "filtered" out using a non-local measurement geometry. Similarly, Kimura et al. [5] generated and detected spin accumulation in Pt using lateral ferromagnetic-nonmagnetic (FM/NM) devices at room temperature (RT) and 77 K.

Spurred by these observations, first-principles calculations by Yao and Fang [13], and Guo et al. [14], have shown that simple metals like tungsten, gold and platinum can have larger intrinsic spin Hall conductivity and are more robust against disorder compared to semiconductors such as GaAs and Si. The calculated intrinsic SHE is insensitive to disorder [13, 14]. Their calculations are based on standard density functional theory using accurate full potential linearized augmented plane wave (FLAPW) and full-potential linear-muffin-tin-orbital (FPLMTO) methods respectively. The relativistic spin-orbit coupling has been treated fully self-consistently and the intrinsic spin Hall conductivity is calculated by using a generalized Kubo formula. The intrinsic effect is interpreted as the integral of the Berry phase curvature, $\mathbf{\Omega}(\mathbf{k})$ over the occupied electronic states, which acts as an effectively spin dependent magnetic field in crystal momentum space to produce a transverse velocity proportional to $\mathbf{E} \times \mathbf{\Omega}(\mathbf{k})$ [19–21] (Refer to [22] for a comprehensive review of Berry phase). Several other related studies in anomalous Hall effect (AHE) have demonstrated the validity of such an approach [23]. Other numerical studies have provided numerical evidence that the intrinsic spin Hall effect can survive weak disorder in the mesoscopic diffusive regime [15, 24]. It must be stressed that, while the precise physical mechanism for the spin-orbit interaction induced spin separation in semiconductor has reached general consensus among the community, the spin separation in paramagnetic metals remains unknown or obscure at this point. Despite these controversial results, the various advancements mentioned above spurred widespread interest in the SHE in semiconductor and metallic structures.

The prospect of using metallic SHE based spin devices is a potentially attractive alternative to the current paradigm. Currently a ferromagnet (FM) is the *de facto* spin source and/or spin detector in the study of metallic (NM) spin transport. However the use of FM as spin source poses a problem, namely the conductivity mismatch at the FM/NM interface which disrupts the transfer of spin polarization across the interface [3]. Furthermore the position of the FM spin detector is required to be within the NM spin diffusion length, as measured from the spin source. It is evident that FM/NM based spin device design would be spatially restricted by the NM spin diffusion length which is of the order of a few tens of nanometers and depends on the operating temperature. In contrast, the requirement of a minimal shielding distance would limit the proximity of the neighboring magnetic electrodes. Additionally, the ferromagnet based spin devices require magnetic control of multiple ferromagnets, thereby posing a difficult problem. In our work, we propose an alternative method of testing the metallic SHE without the use of ferromagnet. As such, our scheme circumvents the FM/NM spin injection problems and also avoids other spurious magnetoresistance effects (e.g. anisotropic magnetoresistance and anomalous Hall effect).

5.3 Experimental procedure

The spin Hall effect postulates a force which is perpendicular to both velocity and spin orientation for carriers moving in a conductor with non-zero spin-orbit interaction. Clearly, in accordance with Onsager's reciprocal relations [25], the same spin dependent force would cause the electrons in a pure spin current to be deflected to one side, thereby generating a transverse charge current. This "Hall-like" behavior results in charge accumulation at the sides, enabling a voltage signal to be measured as a signature of what was initially a pure spin current. This is the essential physics of the device proposed by Hirsch. In order to exploit the reciprocal relation governing the conversion of the spin current to a charge current and the associated potential difference, severe geometrical constraints have to be imposed on the device configuration. This entails the fabrication of quantum point contacts at opposite transverse locations on the sample, laying down an insulating layer, and having the same paramagnetic metal as the third layer, which is electrically connected to the quantum point contacts (see Hirsch's paper for a complete description [7]). The potential difference is manifested in the longitudinal direction of this top third layer. In our work, we design and test a non-magnetic lateral structure based on the principle of SHE, and successfully demonstrate the electrical generation and detection of electron spin in a platinum nanostructure.

We have adopted a two dimensional version of Hirsch's experimental geometry. Our single layer design avoids the fabrication complexity of Hirsch's multilayer structure while maintaining the essence of his proposal. A similar geometry was proposed by Hankiewicz et al. to measure the spin Hall

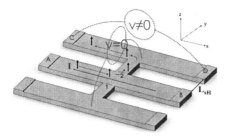
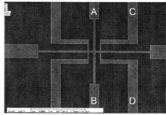

Fig. 5.1. The spin Hall effect and the inverse spin Hall effect. (a) When I_{BA} is applied, the traversing spin-down (spin-up) electrons are deflected to the left (right) which lead to spin current in the y direction. The spin current diffuses along the bridging conductor (transverse channel), and these carriers are again deflected in accordance with the Onsager relations. The resulting effect of the spin imbalance is a spin current induced a charge accumulation between the measurement probes C and D. The signal is termed the spin Hall voltage, V_{sH}. (b) Scanning electron microscopy image of a representative device.

conductivity in mesoscopic-scale system [15]. We create and detect electron spin using the SHE, in order to measure the spin Hall conductivity and the spin diffusion length with a non-local four-terminal setup. The device consists of three parallel conductors with a perpendicular bridging conductor. The devices are prepared with conventional e-beam lithography techniques, DC sputtering and lift-off process. The Pt wire is 200 nm wide with a thickness of 50 nm. The separation,L_{sH}, between each Pt channel varied from 180 to 320 nm (Fig. 5.1a). The current-voltage (I-V) data are measured in a cryostat using a Keithley sourcemeter and nanovoltmeter. The experiment is also repeated for Au and Al in order to compare the SHE in metals with different spin-orbit interaction strength.

The reservoirs A and B are biased to produce a current in the x direction. The distance between A and B is large such that we can assume that the current density is uniform longitudinally. Due to spin-orbit interaction, a transverse spin current is created and diffuses through the bridging conductor in our experimental configuration (see Fig. 5.1a). Then the pure spin current would again be deflected due to the spin-orbit interaction and lead to a charge accumulation on the sides of the bridging conductor. To put it simply, a charge current is passed between electrode A and B, i.e. I_{BA} and leads to a pure spin current, j_s, in the y direction. Subsequently, the spin current leads to a charge accumulation which is measured as a voltage between electrode C and D.

Following the analysis presented by Adagideli et al. [26], we assume that the charge and spin transport is diffusive and that the spin current, $j^z_{y,max}$, reaches its maximum value at the interface at region 2 and 3 shown in Fig. 5.1a. We can write down the expression of the spin current along the transverse or bridging conductor as

$$j_s(y') = j_{y,max}^z \exp\left(\frac{-y'}{\lambda_{sf}}\right) \qquad \text{where} \qquad y' = \begin{cases} y - \frac{w}{2} : y \geq \frac{w}{2} \\ y + \frac{w}{2} : y \leq \frac{w}{2} \end{cases} \qquad (5.1)$$

where λ_{sf} is the spin diffusion length, w is the centre conductor width and y is the lateral distance measured from the centre axis.

In general, the charge current in a conductor is [27]

$$\mathbf{j}_c(\mathbf{r}) = \sigma \mathbf{E}(\mathbf{r}) + \frac{\sigma_{sH}}{\sigma}(\hat{\mathbf{z}} \times \mathbf{j}_s), \qquad (5.2)$$

where the first term arises from the applied electrical field and the second term from the inverse spin Hall effect (ISHE), where we only consider the spin polarization in the z direction. ISHE is the conversion of a spin current into an electric current as a consequence of Onsager's reciprocal relations between spin current and charge current. σ_{sH} is the spin Hall conductivity and σ is the Drude conductivity.

By substituting j_s from Eq. (5.1) into (5.2), and setting $E(y) = 0$ we have

$$E_x(y') = -\frac{\sigma_{sH}}{\sigma^2} j_s(y') = -\frac{\sigma_{sH}}{\sigma^2} j_{y,max}^z \exp\left(\frac{-y'}{\lambda_{sf}}\right). \qquad (5.3)$$

At $y' = L_{sH}$, the spin Hall voltage V_{sH} measured using electrode C and D is:

$$V_{sH} = -dE_x(L_{sH}) = d\frac{\sigma_{sH}}{\sigma^2} j_{y,max}^z \exp\left(\frac{-L_{sH}}{\lambda_{sf}}\right), \qquad (5.4)$$

where d is the width of the transverse conductor.

The spin Hall resistance $R_{sH} = V_{sH}/I$ is: formula (5.3).

$$\ln R_{sH} = \left(\frac{-L_{sH}}{\lambda_{sf}}\right) + \ln\left(d\frac{\sigma_{sH}}{\sigma^2}\frac{j_{y,max}^z}{I}\right). \qquad (5.5)$$

By measuring $\ln R_{sH}$ for various L_{sH}, the value of λ_{sf} can be extracted using Eq. (5.5). However the value of σ_{sH} and $j_{y,max}^z$ could not be individually determined, i.e. only the product of the two quantities can be calculated. However, the value of σ_{sH} can be obtained using optical methods similar to the Kerr rotation method used by Kato et al. [10]. The optical work will be explained in a later publication. Given the value of σ_{sH}, the value of $j_{y,max}^z$ can be calculated.

5.4 Results and discussion

Figure 5.2 shows representative V-I results for our devices for $L_{sH} = 237$ nm. At room temperature the measured voltage does not show any discernible

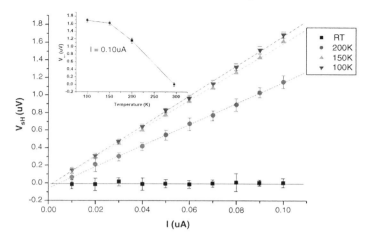

Fig. 5.2. V_{sH} as a function of current I for various temperature. Inset, V_{sH} as a function of temperature for $I = 0.1\,\mu\text{A}$. RT $= 298\,\text{K}$.

dependence on the current. At lower temperature, a linear relationship appears.

Before we discuss these results, some comments are in order. As a consequence of our experimental geometry, the measured voltage cannot arise due to any misalignment or defects in our structures, nor can it arise due to any magnetic field, since no external field was switched on. The influence of stray fields can be ignored, since Hall probe measurements in the vicinity of the measurement region yield fields of the order of the earth's magnetic field. These results clearly indicate a voltage that scales with the current.

Figure 5.3 shows the measured value of R_{sH} as a function of the distance, L_{sH}, between the two channel. Consistent with Eq. (5.4), R_{sH} decreases as a function of L_{sH}. By linear fitting of the graph to Eq. (5.4) the value of λ_{sf} and $\sigma_{sH} j^z_{y,max}$ are obtained. Using the value of σ measured from our sample ($4.06 \times 10^6\ \Omega^{-1}\text{m}^{-1}$ at 100 K), the value of λ_{sf} is found to be 16 nm. Our calculated value of λ_{sf} is of the same order of magnitude as reported elsewhere [28] (14 nm at 4 K). Similarly the value of $\sigma_{sH} j^z_{y,max}$ is found to be $8.9 \times 10^{-9}\ \Omega^{-1}\text{m}^{-3}\text{A}$. We take the literature value of $2.4 \times 10^4\ \Omega^{-1}\text{m}^{-1}$ at RT [5] as a lower limit estimate for σ_{sH} , and we get $j^z_{y,max}$ to be 3.7×10^{-15} Am^{-2}. This sets the lower limit for the spin current j_s for our devices.

The experimental results for Au and Al are shown in Fig. 5.4. The signal for Au shows a linear response similar to Pt, while the signal for Al shows no noticeable signal above instrumental noise. This is not expected as Al, a low atomic number metal, has a weak spin-orbit interaction and Au, a high atomic number metal, should have stronger spin-orbit interaction since the spin-orbit interaction is expected to scale with the fourth power of the atomic number [29]. However, the signal difference, R_{sH}, between Pt ($\sim 10\Omega$) and Au ($\sim 0.02\Omega$) is huge even though the atomic numbers of Au (79) and Pt (78)

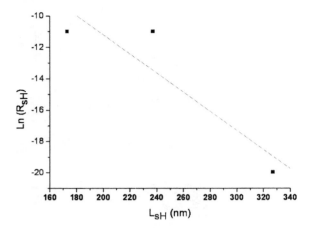

Fig. 5.3. R_{sH} as a function of L_{sH}.

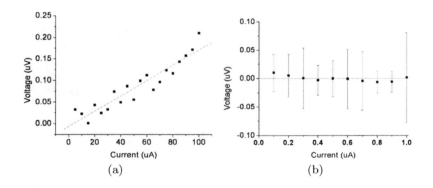

Fig. 5.4. The current-voltage measurement for (a) Au and (b) Al.

are similar. If we consider the SHE to be intrinsic in nature, where the band structure plays an important role, the unconventional result can be explained in the following context. As the band structure near the fermi level for Au is mainly S-type while for Pt is mainly D-type [14], it is predicted that the intrinsic SHE is larger for Pt than Au. However further experimental studies need to be conducted to confirm this hypothesis.

5.5 Conclusion

We have demonstrated the electrical generation and detection of spin via the SHE and inverse SHE in Pt. We also compared the signal of Pt with Au and

Al. The spin diffusion length without the use of ferromagnetic material. The numbers are found to be consistent with other reports in the literature. Spin orbit interaction can be a useful method to induce spin separation in metallic nanostructures of high atomic number metals. Whether this spin separation can be put to any practical use, remains to be seen.

We thank G. Y. Guo for useful discussions. This work was supported by A*Star (Grants No. 012-104-0040) and by National University of Singapore (Grants No. WBS:R-144-000-179-112).

References

1. I. Žutić, J. Fabian and S. Das Sarma, Reviews of Modern Physics **76**, 323 (2004).
2. F.J. Jedema, A.T. Filip and B.J. van Wees, Nature **410**, 345–348 (2001).
3. A.T. Filip, F.J. Jedema, B.J. van Wees and G. Borghs, Physica E: Low - Dimensional Systems And Nanostructures **10**, 478–483 (2001).
4. M. Glazov and A. Kavokin, Journal Of Luminescence **125**, 118–125 (2007).
5. T. Kimura, Y. Otani, T. Sato, S. Takahashi and S. Maekawa, Physical Review Letters **98**, 156601 (2007).
6. T. Seki, Y. Hasegawa, S. Mitani, S. Takahashi, H. Imamura, S.M.J. Nitta, and K. Takanashi, Nature Materials **7**, 125–129 (2008).
7. J.E. Hirsch, Physical Review Letters **83**, 1834–1837 (1999).
8. M.I. D'yakonov and V.I. Perel', Physics Letters A **35**, 459–460 (1971).
9. S. Murakami, N. Nagaosa and S.C. Zhang, Science **301**, 1348–1351 (2003).
10. Y.K. Kato, R.C. Myers, A.C. Gossard, and D.D. Awschalom, Science **306**, 1910–1913 (2004).
11. J. Wunderlich, B. Kaestner, J. Sinova and T. Jungwirth, Physical Review Letters **94**, 047204 (2005).
12. S.O. Valenzuela and M. Tinkham, Nature **442**, 176–179 (2006).
13. Y. Yao and Z. Fang, Physical Review Letters **95**, 156601 (2005).
14. G.Y. Guo, S. Murakami, T.-W. Chen and N. Nagaosa, cond-mat 07050409 (2007).
15. E.M. Hankiewicz, L.W. Molenkamp, T. Jungwirth and J. Sinova, Physical Review B **70**, 241301 (2004).
16. E.I. Rashba, Physical Review B **68**, 241315 (2003).
17. J. Smit, Physica **24**, 39–51 (1958).
18. L. Berger, Physical Review B **2**, 4559–4566 (1970).
19. M. Onoda and N. Nagaosa, Physical Review Letters **90**, 206601 (2003).
20. T. Jungwirth, Q. Niu, and A.H. MacDonald, Physical Review Letters **88**, 207208 (2002).
21. Y. Taguchi, Y. Oohara, H. Yoshizawa, N. Nagaosa and Y. Tokura, Science **291**, 2573–2576 (2001).
22. N.P. Ong and W.-L. Lee, cond-mat/0508236 (2005).
23. Y. Yao, L. Kleinman, A.H. MacDonald, J. Sinova, T. Jungwirth, D.-S. Wang, E. Wang and Q. Niu, Physical Review Letters **92**, 037204 (2004).
24. B.K. Nikolić, L.P. Zarbo and S. Souma, Physical Review B **72**, 075361 (2005).
25. L. Onsager, Physical Review **38**, 2265–2279 (1931).

26. I. Adagideli and G.E.W. Bauer, Physical Review Letters **95**, 256602 (2005).
27. Zhang, S. Physical Review Letters **85**, 393–396 (2000).
28. H. Kurt, R. Loloee, K. Eid, W.P. Pratt Jr. and J. Bass, Applied Physics Letters **81**, 4787–4789 (2002).
29. S. Geier and G. Bergmann, Physical Review Letters **68**, 2520–2523 (1992).

6

THE PROBLEM OF TRUE MACROSCOPIC CHARGE QUANTIZATION IN THE COULOMB BLOCKADE

I.S. BURMISTROV[1] AND A.M.M. PRUISKEN[2]

[1] *L.D. Landau Institute for Theoretical Physics, Kosygina street 2, 117940 Moscow, Russia.* burmi@itp.ac.ru

[2] *Institute for Theoretical Physics, University of Amsterdam, Valckenierstraat 65, 1018XE Amsterdam, The Netherlands.* pruisken@science.uva.nl

Abstract. Based on the Ambegaokar-Eckern-Schön approach to the Coulomb blockade we develop a complete quantum theory of the single electron transistor. We identify a previously unrecognized physical observable in the problem that, unlike the usual average charge on the island, is robustly quantized for any *finite* value of the tunneling conductance as the temperature goes to the absolute zero. This novel quantity is fundamentally related to the non-symmetrized current noise of the system. Our results display all the super universal topological features of the θ-angle concept that previously arose in the theory of the quantum Hall effect.

Keywords: Coulomb blockade, charge quantization

6.1 Introduction

The Ambegaokar-Eckern-Schön (AES) model [1] is the simplest approach to the Coulomb blockade problem [2–5] that has attracted a considerable amount of interest over the years, especially after the experimental observation of the so-called macroscopic charge quantization [6]. The standard experimental set-up is the single electron transistor (SET) [7] which is a mesoscopic metallic island coupled to a gate and connected to two metallic reservoirs by means of tunnelling contacts with a total conductance g (see Fig. 6.1). Even though the physical conditions of the AES model are limited and well known, [5, 8, 9] the theory nevertheless displays richly complex and fundamentally new behavior, much of which has not been understood to date. To explain the observed tunnelling phenomena with varying temperature T and gate voltage V_g one usually considers an isolated island obtained by putting the tunnelling

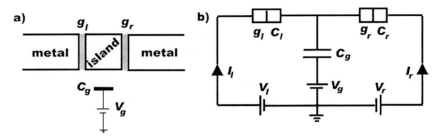

Fig. 6.1. (a) Sketch of the SET device. (b) Equivalent circuit of the SET.

conductance g equal to zero. The AES model then leads to the standard semiclassical or electrostatic picture of the Coulomb blockade which says that at $T = 0$ the average charge (Q) on the island is *robustly* quantized (in units of e) except for very special values of the gate voltage $V_g^{(k)} = e(k + 1/2)/C_g$ with integer k and C_g denoting the gate capacitance. At these very special values a *quantum phase transition* occurs where the average charge Q on the island changes from $Q = k$ to $Q = (k + 1)$.

The experiments on the SET always involve *finite* values of the tunnelling conductance g, however, and this dramatically complicates the semiclassical picture of the Coulomb blockade. Despite ample theoretical work on both the strong [5,10] ($g \ll 1$) and weak [11] ($g \gg 1$) coupling sides of the problem, the matter of "macroscopic charge quantization" still lacks basic physical clarity since the averaged charge Q is known to be *un*-quantized at $T = 0$ for any finite value of the tunnelling conductance g [12], no matter how small. This raises the fundamental question whether a new physical quantity exists in the problem that, unlike the average charge on the island, displays *true macroscopic charge quantization* at low temperatures and for finite values of the tunnelling conductances.

In this paper we present a complete quantum theory of the SET that is motivated by the formal analogies that exist between the AES theory on the one hand, and the theory of the quantum Hall effect [13] on the other. Each of these theories describe an interesting experimental realization of the topological issue of a θ vacuum that originally arose in QCD [14]. In each case one deals with different physical phenomena and therefore different quantities of physical interest. What has remarkably emerged over the years is that the basic scaling behavior is always the same, independent of the specific application of the θ angle that one is interested in [13]. Within the grassmannian $U(m+n)/U(m) \times U(n)$ non-linear σ model, for example, one finds that quantum Hall physics, in fact, a *super universal* topological feature of the theory for all values of m and n. It is therefore of interest to know whether super universality is retained in the AES theory where physical concepts such as the Hall conductance and θ renormalization have not been recognized.

In direct analogy with the theory of the quantum Hall effect we develop, a quantum theory of *observable parameters* g' and q' for the AES model

obtained by studying the *response* of the system to changes in the boundary
conditions. Here, g' is identified as the SET conductance, whereas the q' is a
novel physical quantity that is fundamentally related to the current noise in
the SET. The q' is in all respects same as the Hall conductance in the quantum
Hall effect and, unlike the averaged charge Q on the island, it is *robustly
quantized* in the limit $T \to 0$ independent of the tunneling conductance g.
Our results provide the complete conceptual framework in which the various
disconnected pieces of existing computational knowledge of the AES theory
can in general be understood. Some of the results of this paper have been
reported in a brief manner in Ref. [15].

6.2 AES model

The action involves a single abelian phase $\phi(\tau)$ describing the potential fluctuations on the island $V(\tau) = i\dot{\phi}(\tau)$ with τ denoting the imaginary time [1].
The theory is defined by

$$Z = \int \mathcal{D}[\phi] e^{-S[\phi]}, \quad S[\phi] = S_d + S_t + S_c. \tag{6.1}$$

The action S_d describes the tunneling between the island and the reservoirs

$$S_d[\phi] = \frac{g}{4} \int_0^\beta d\tau_1 d\tau_2 \, \alpha(\tau_{12}) e^{-i[\phi(\tau_1) - \phi(\tau_2)]}. \tag{6.2}$$

Here, $\beta = 1/T$, $\tau_{12} = \tau_1 - \tau_2$ and $g = g_l + g_r$ where $g_{l,r}$ denotes the
dimensionless bare tunneling conductance between the island and left/right
reservoir (see Fig. 6.1). The kernel $\alpha(\tau)$ is usually expressed as $\alpha(\tau) =
(T/\pi) \sum_n |\omega_n| e^{-i\omega_n \tau}$ with $\omega_n = 2\pi T n$. The part S_t describes the coupling
between the island and the gate and S_c is the effect of the Coulomb interaction
between the electrons

$$S_t[\phi] = -2\pi i q W[\phi], \quad S_c[\phi] = \frac{1}{4E_c} \int_0^\beta d\tau \, \dot{\phi}^2. \tag{6.3}$$

Here, $q = C_g V_g / e$ is the external charge and $W[\phi] = 1/(2\pi) \int_0^\beta d\tau \dot{\phi}$ is the
winding number or *topological charge* of the ϕ field. For the system in equilibrium the winding number is strictly an integer [9] which means that Eq. 6.3
is only sensitive to the *fractional* part $k - 1/2 < q \leqslant k + 1/2$ of the external
charge q. The main effect of S_c in Eq. 6.3 is to provide a cut-off for large
frequencies. Eq. 6.2 has classical finite action solutions $\phi_W(\tau)$ with a non-zero
winding number that are completely analogous to Yang-Mills instantons. The
general expression for winding number W is given by [16,17]

$$e^{i\phi_W(\tau)} = e^{-i2\pi T\tau} \sum_{a=1}^{|W|} \frac{e^{i2\pi T\tau} - z_a}{e^{-i2\pi T\tau} - z_a^*}. \tag{6.4}$$

For instantons ($W > 0$) the complex parameters z_a are all inside the unit circle and for anti-instantons ($W < 0$) they are outside. Considering $W = \pm 1$ which is of interest to us, one identifies $\arg z/2\pi T$ as the *position* (in time) of the single instanton whereas $\lambda = (1 - |z|^2)\beta$ is the *scale size* or the duration of the potential pulse $i\dot{\phi}_{\pm 1}(\tau)$. The classical action $S_d[\phi_W] + S_t[\phi_W] = g|W|/2 - 2\pi q W i$, is finite and independent of z_a. Therefore, the set of parameters $\{z_1, \ldots, z_{|W|}\}$ corresponds to the $2|W|$ zero modes. The charging contribution is also finite: $S_c[\phi_W] = (\pi^2 T/E_c)\sum_{a,b=1}^{|W|}(1+z_a z_b^\star)/(1-z_a z_b^\star)$. but explicitly depend on the set $\{z_1, \ldots, z_{|W|}\}$. Thus, the instanton configurations with $|z_a| \to 1$ is suppressed.

6.3 Kubo formulae for the observable parameters

To develop a theory of *observable* parameters of the SET we consider the topological sector with the winding number $W - n$ where n is an arbitrary natural number. Let us employ the following shift $\phi(\tau) \to \phi(\tau) - \tilde{\phi}(\tau)$ with $\tilde{\phi}(\tau) = \omega_n \tau$. We emphasize that the background field $\tilde{\phi}(\tau)$ satisfies the classical equation of motion for the AES action 6.1 which includes the charging term S_c. Write $\exp(-S'[\tilde{\phi}]) = \int \mathcal{D}[\phi]\exp(-S[\tilde{\phi} + \phi])/Z$, then a detailed knowledge of $S'[\tilde{\phi}]$ generally provides complete information on the low energy dynamics of the system [13, 15]. The effective action $S'[\tilde{\phi}]$ is properly defined in terms of a series expansion in powers of ω_n. Retaining only the lowest order terms in the series we can write

$$S'[\tilde{\phi}] = \beta\left[\frac{g'}{4\pi}|\omega_n| - iq'\omega_n + O(\omega_n^2)\right]. \qquad (6.5)$$

The quantities of physical interest are g' and q' (with $k - 1/2 < q' \leqslant k + 1/2$) that are formally given in terms of Kubo-like expressions [15, 18]

$$g' = 4\pi\Im\frac{\partial K^R(\omega)}{\partial \omega}\bigg|_{\omega=0}, \qquad q' = Q + \Re\frac{\partial K^R(\omega)}{\partial \omega}\bigg|_{\omega=0}. \qquad (6.6)$$

Here, the function $K^R(\omega)$ is obtained from the Matsubara correlation function

$$K(i\omega_n) = -\frac{g}{4\beta}\int_0^\beta d\tau_1 d\tau_2 e^{i\omega_n\tau_{12}}\alpha(\tau_{12})D(\tau_{21}), \qquad D(\tau_{12}) = \langle e^{i[\phi(\tau_2)-\phi(\tau_1)]}\rangle \qquad (6.7)$$

by the analytic continuation $i\omega_n \to \omega + i0^+$. Here, the expectation is with respect to the theory of Eq. 6.1 and $Q = q + i\langle\dot{\phi}\rangle/(2E_c)$ is the average charge on the island. The function $K^R(\omega)$ can be written in terms of the retarded propagator $D^R(\omega)$ as [18]

$$K^R(\omega) = g\int\frac{dE dE'}{4\pi^3}E'\Im D^R(E)\frac{n_B(E') - n_B(E)}{E - E' + \omega + i0^+}, \qquad (6.8)$$

where $n_B(E) = [\exp(\beta E) - 1]^{-1}$ denotes the Bose-Einstein distribution function.

The main advantage of the background field formalism of Eqs. 6.5 and 6.6 is that it unequivocally determines the renormalization of the AES model while retaining the close contact with the physics of the SET. To see this we notice first that by expanding the effective action of Eq. 6.5 in powers of ω_n we essentially treat the discrete variable ω_n as a continuous one. This means that the quantities g' and q' in Eq. 6.6 are, by construction, a measure for the response of the system to infinitesimal changes in the boundary conditions.

Before embarking on the further details, it is important to emphasize that the physical observables in Eq. 6.6 are precisely the same quantities that one normally would obtain in ordinary linear response theory [15, 18]. For example, g' is exactly same as the Kubo formula [11, 19, 20] relating a small potential difference V between the reservoirs to the current $\langle I \rangle$ across the island: $\langle I \rangle = e^2 G V/h$ where the SET conductance $G = g_l g_r g'/(g_l + g_r)^2$ and h is Planck's constant. To understand the new quantity q' we notice that the last piece in q' is related to the so-called quantum current noise and a more transparent expression is obtained by writing

$$q' = Q - \frac{(g_l + g_r)^2}{2g_l g_r} i \frac{\partial}{\partial V} \int_{-\infty}^0 dt \langle [I(0), I(t)] \rangle \bigg|_{V=0}. \tag{6.9}$$

6.4 Weak coupling regime, $g' \gg 1$

By evaluating Eq. 6.6 in a series expansion in powers of $1/g$, one obtains the well-known perturbative lowest order results [21–24] for $g'(T)$. The quantity q' is generally unaffected by the quantum fluctuations: $q'(T) = 0$, and to establish the renormalization of q' with temperature it is necessary to include instantons. Following the detailed methodology of Ref. [13], we find [18]

$$\Im D^R(\omega) = \pi \beta \omega \delta(\omega) \left[1 - \frac{2}{g} \ln \frac{g E_c e^\gamma}{2\pi^2 T} \right] + \Im \left(\frac{2\pi i/g}{\omega + i0^+} - \frac{2\pi i/g}{\omega + i g E_c/\pi} \right)$$
$$- \frac{g^2 E_c}{\pi^2 T} e^{-g/2} \Im i e^{-i2\pi q} \left[\pi \beta \omega \delta(\omega) - \frac{1}{\omega + i0^+} + \frac{1}{\omega + i2\pi T} \right], \tag{6.10}$$

where $\gamma = 0.577 \ldots$ denotes the Euler constant. The first line of Eq. 6.10 contains the one-loop perturbative contribution to $\Im D^R(\omega)$ whereas the second line [terms proportional to $\exp(-g/2)$] is due to instantons with $W = \pm 1$. Next, by using Eqs. 6.8 and 6.10, we obtain

$$K^R(\omega) = \frac{i\omega g}{4\pi} \left[1 - \frac{2}{g} \ln \frac{e g E_c}{2\pi^2 T} + \frac{2}{g} \psi \left(1 - \frac{i\omega}{2\pi T} \right) \right] - \frac{g^3 E_c}{2\pi^2} e^{-g/2} e^{i2\pi q}$$
$$\left[\psi(1) - \psi \left(1 - \frac{i\omega}{2\pi T} \right) \right] - \frac{g^3 E_c}{2\pi^2} e^{-g/2} \cos 2\pi q \sum_{n>1} \frac{1}{n} - \sum_{n>0} \frac{g E_c T}{2\pi^2 T n + g E_c}, \tag{6.11}$$

where $\psi(z)$ denotes the Euler di-gamma function and sums in the last line of the equation are restricted by the cut-off $n_{\max} \sim gE_c/T$. Finally, from Eq. 6.6 we obtain the desired results for the temperature dependence of the physical observables in the weak coupling regime:

$$g'(T) = g - 2\ln\frac{gE_ce^{\gamma+1}}{2\pi^2T} - \frac{g^3E_c}{6T}e^{-g/2}\cos 2\pi q, \qquad (6.12)$$

$$q'(T) = q - \frac{g^3E_c}{24\pi T}e^{-g/2}\sin 2\pi q. \qquad (6.13)$$

This results describe the Coulomb blockade oscillations [7] of the physical observables g' and q' with the external charge q in the weak coupling regime, $g' \gg 1$. The amplitude of this oscillations is inversely proportional to the temperature. However, Eqs. 6.12 and 6.13 are valid only at not too low temperatures $T \gg g^3E_ce^{-g/2}$ such that the amplitude of oscillations is still small. We notice that Eq. 6.12 coincides with the result of Ref. [11]. The q' oscillates much stronger than the average charge Q alone in the same temperature range [24–26]:

$$Q(T) = q - \frac{g^2}{\pi}e^{-g/2}\ln\left(\frac{E_c}{2\pi^2e^{\gamma}T}\right)\sin 2\pi q. \qquad (6.14)$$

The Coulomb blockade oscillations 6.12 and 6.13 are analogous to the quantum Hall (instanton or topological) oscillations [27–29] measured experimentally in the quantum Hall regime recently [30].

6.5 Strong coupling regime, $g' \ll 1$

Near the point $g = 0$ and $q = k + 1/2$ we can simplify considerations by projecting the model onto the low-energy states with $Q = k$ and $Q = k+1$ [12]. Then, the AES theory is mostly elegantly described by the spin $1/2$ effective action [18]:

$$S = \beta\, E_cq^2 + \beta\frac{\Delta}{2} + \int_0^\beta d\tau\bar\psi\left(\partial_\tau - \eta + \frac{\Delta}{2}\sigma_z\right)\psi$$

$$+ \frac{g}{4}\int_0^\beta d\tau_1 d\tau_2\alpha(\tau_{12})[\bar\psi(\tau_1)\sigma_-\psi(\tau_1)][\bar\psi(\tau_2)\sigma_+\psi(\tau_2)], \qquad (6.15)$$

where $\psi,\bar\psi$ are the Abrikosov's two-component pseudofermion fields [31,32], $\Delta = E_c(2k + 1 - 2q) > 0$ is the gap at $g = 0$ between the charging levels with $Q = k$ and $Q = k + 1$, σ_j with $j = x, y, z$ stands for the Pauli matrices and $\sigma_\pm = (\sigma_x \pm i\sigma_y)/2$. In order to eliminate the contributions to the physical quantities from non-physical states with number of pseudofermions $N_{pf} \neq 1$ we have introduced the chemical potential η which we should tend to minus infinity: $\eta \to -\infty$, at the end of all calculations. The physical partition function Z as well as correlation functions can be obtained from the partition function Z_{pf} and correlation functions for pseudofermions according to the standard prescription [31, 32]. Comparing Eqs. 6.15 and 6.1, we identify the operators $\bar{\psi}(\tau)\sigma_\pm\psi(\tau)$ with the operators $\exp(\pm i\phi(\tau))$ projected onto the states with $Q = k$ and $Q = k+1$. It is worthwhile mentioning that the effective action 6.15 is similar to the XY case of the Bose-Kondo model for the spin $S = 1/2$. [33, 34] As a final remark here, we notice that states with average charge Q different from k and $k + 1$ yield regular corrections in powers of g to both the g and Δ; in the strong coupling limit $g \ll 1$ they can be safely neglected. [20, 35]

Summing the leading logarithms that corresponds to the one-loop approximation of the renormalization group procedure of Refs. [34], we find

$$\Im D^R(\omega) = \frac{\pi}{\gamma^2}\delta(\omega - \Delta')\tanh\frac{\beta\Delta'}{2}, \qquad \gamma^2 = 1 + \frac{g}{2\pi^2}\ln\frac{\Lambda}{\max\{\Delta', T\}}. \quad (6.16)$$

Here, $\Delta' = \Delta/\gamma^2$ corresponds to the renormalized energy gap between the ground and excited states, and Λ denotes high energy cut-off which we assume of the order of E_c. By using Eq. 6.8, we obtain

$$K^R(\omega) = \frac{g}{\gamma^2}\frac{\omega + \Delta'}{4\pi^2}\left[\psi\left(1 - i\frac{\omega + \Delta'}{2\pi T}\right) - \psi\left(1 - \frac{i\Delta'}{2\pi T}\right) - \frac{2\pi}{\Delta'}Y(\Delta')\right]\tanh\frac{\beta\Delta'}{2},$$
$$(6.17)$$

where $Y(\Delta') = T\sum_{\omega_n>0}\omega_n\Delta'/(\omega_n^2 + \Delta'^2)$. Given Eq. 6.17, it is possible to evaluate integrals in Eq. 6.6; then, we find the following temperature dependence for the physical observables of the SET:

$$g'(T) = \frac{g}{2\gamma^2}\frac{\beta\Delta'}{\sinh\beta\Delta'}, \qquad q'(T) = Q(T) - \frac{\gamma^2 - 1}{2\gamma^2}\tanh\frac{\beta\Delta'}{2}. \quad (6.18)$$

Here, the temperature dependence of the average charge on the island is given as [10]

$$Q(T) = \frac{1}{2}\left(1 - \frac{1}{\gamma^2}\tanh\frac{\beta\Delta'}{2}\right). \quad (6.19)$$

It is worthwhile mentioning that at $T = 0$ we find $Q(T = 0) = \frac{g}{4\pi^2}\ln\frac{\Lambda}{\Delta'}/(1 + \frac{g}{2\pi^2}\ln\frac{\Lambda}{\Delta'})$, which is the familiar result of Matveev [12]: in the presence of tunneling between the island and the reservoirs ($g \neq 0$) the $Q(T)$ *does not quantized* at zero temperature. Given Eq. 6.19, Eq. 6.18 implies that

$$q'(T) = k + \frac{1}{1 + e^{\beta\Delta'}}, \quad (6.20)$$

i.e., for all values of the external charge except the degeneracy points $q = k+1/2$, the novel physical quantity q' is *robustly quantized* at zero temperature independent of the tunneling conductance g. We notice that the result for $g'(T)$ in Eq. 6.18 coincides with the result found in Ref. [10] by other means.

6.6 Summary and conclusions

Following the detailed methodology of Ref. [13], the temperature dependence of the physical observables g' and q' can be expressed in terms of renormalization group $\beta_g(g', q') = dg'/d\ln\beta$ and $\beta_q(g', q') = dq'/d\ln\beta$ functions [18]. In the weak coupling regime, $g' \ll 1$, we obtain from Eqs. 6.12, 6.13

$$\beta_g = -2 - \frac{\pi^2 e^{-\gamma-1}}{3} g'^2 e^{-g'/2} \cos 2\pi q', \quad \beta_q = \frac{\pi e^{-\gamma-1}}{12} g'^2 e^{-g'/2} \sin 2\pi q'. \tag{6.21}$$

The results Eq. 6.21 indicate that instantons are the fundamental objects of the theory that facilitate the *cross-over* between the metallic phase with $g' \gg 1$ at high T and the Coulomb blockade phase with $g' \lesssim 1$ that generally appears at a much lower T only. According to Eq. 6.21, the q' flows always to the integer value at low temperatures except for the special case $q' = k + 1/2$.

In the strong coupling regime, $g' \ll 1$, from Eqs. 6.18 and 6.20 we find near $q' = k + 1/2$

$$\beta_g = -g'^2/\pi^2, \qquad \beta_q = (q' - k - 1/2)\left(1 - g'/\pi^2\right). \tag{6.22}$$

This result indicates that $q' = k + 1/2$ and $g' = 0$ is the *critical* fixed point of the AES theory with g' a marginally irrelevant scaling variable. Equation 6.22, together with the weak coupling results of Eq. 6.21 allow us to construct the unifying scaling diagram for the observable theory, g' and q', which is illustrated in Fig. 6.2.

As we have already pointed out, the physical observables g' and q' are, by construction, a measure for the response of the system to infinitesimal changes in the boundary condition. It immediately leads us to a general criterion for

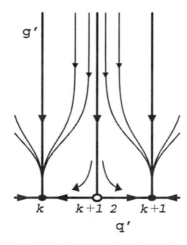

Fig. 6.2. Unified scaling diagram of the Coulomb blockade in terms of the SET conductance g' and the q'. The arrows indicate the scaling toward $T = 0$ (see text).

the strong coupling (*Coulomb blockade phase*) of the SET. More specifically, the general statement which says that the SET scales toward the *Coulomb blockade phase* as $T \to 0$ implies that the physical observables g' and the fractional part of q' all render equal to zero except for corrections that are exponentially small in β. Since the AES theory is invariant under the shift $q \to q + k$ and $q' \to q' + k$ for integer k, we conclude that the AES theory on the strong coupling side generally displays the *Coulomb blockade*: the SET conductance $g' = 0$, whereas the q' is *integer* quantized, unlike the averaged charge Q on the island. This phenomenon which is depicted in Fig. 6.2 by the infrared stable fixed points located at integer values $q' = k$, is fundamentally different from semiclassical picture of the Coulomb blockade since it elucidates the discrete nature of the electronic charge which is independent of tunneling.

To summarize, based on the new concept of θ (or q') renormalization we assign a universal significance to the Coulomb blockade in the SET that previously did not exist beyond the semiclassical picture. We have shown that the AES model is, in fact, an extremely interesting and exactly solvable example of a θ vacuum that displays all the *super universal* topological features that have arisen before in the context of the quantum Hall liquids [13,36] as well as quantum spin liquids [37]. These include not only the existence of *gapless* or critical excitations at $q' = k + 1/2$ (or $\theta = \pi$) but also the *robust* topological quantum numbers that explain the integer quantization of the q' in the SET at zero temperature but finite values of g.

Acknowledgements

The authors are indebted to A. Abanov for bringing the AES model to their attention and for valuable conversations. One of us (I.S.B.) is grateful to O. Astafiev, A. Lebedev and Yu. Makhlin for helpful discussions. The research was funded in part by the Dutch National Science Foundations *NWO* and *FOM*, the EU-Transnational Access program (RITA-CT-2003-506095), CRDF, the Russian Ministry of Education and Science, Council for Grant of the President of Russian Federation, RFBR, Dynasty Foundation and the Program of RAS "Quantum Macrophysics". One of us (ISB) acknowledges the hospitality of the Institute for Theoretical Physics of University of Amsterdam where a part of this work was performed.

References

1. V. Ambegaokar, U. Eckern, and G. Schön, Phys. Rev. Lett. **48**, 1745 (1982)
2. G. Schön and A.D. Zaikin, Phys. Rep. **198**, 237 (1990)
3. H. Grabert, M. Devoret. *Single Charge Tunneling*, ed. by H. Grabert and M.H. Devoret (Plenum, New York, 1992)
4. H. Grabert and H. Horner (eds.), Z. Phys. B **85**, 317 (1991)

5. L.I. Glazman and M. Pustilnik in: *New Directions in Mesoscopic Physics Towards to Nanoscience*, ed. by R. Fazio, G.F. Gantmakher and Y. Imry (Kluwer, Dordrecht, 2003)
6. P. Lafarge, H. Pothier, E.R. Williams, D. Esteve, C. Urbina, and M.H. Devoret, Z. Phys. B **85**, 327 (1991)
7. T.A. Fulton and G.J. Dolan, Phys. Rev. Lett. **59**, 109 (1987)
8. G. Falci, G. Schon, and G. Zimanyi, Phys. Rev. Lett. **74**, 3257 (1995); Physica B **203**, 409 (1994)
9. I.S. Beloborodov, K.B. Efetov, A. Altland, and F.W.J. Hekking, Phys. Rev. B **63**, 115109 (2001); K.B. Efetov, and A. Tschersich, Phys. Rev. B **67**, 174205 (2003)
10. H. Schoeller and G. Schön, Phys. Rev. B **50**, 18436 (1994)
11. A. Altland, L.I. Glazman, A. Kamenev, and J.S. Meyer, Ann. Phys. (N.Y.) **321**, 2566 (2006)
12. K.A. Matveev, Sov. Phys. JETP **72**, 892 (1991)
13. A.M.M. Pruisken and I.S. Burmistrov, Ann. of Phys. **316**, 285 (2005)
14. R. Rajaraman, *Instantons and solitons*, (Amsterdam, North-Holland, 1982); A.M. Polyakov, *Gauge fields and strings*, (Harwood Academic Publishers, Shur, 1987)
15. I.S. Burmistrov, and A.M.M. Pruisken, arXiv: cond-mat/0702400
16. S.E. Korshunov, JETP Lett. **45**, 434 (1987)
17. S.A. Bulgadaev, Phys. Lett. A **125**, 299 (1987)
18. I.S. Burmistrov, and A.M.M. Pruisken, in preparation
19. E. Ben-Jacob, E. Mottola, and G. Schön, Phys. Rev. Lett. **51**, 2064 (1983)
20. C. Wallisser, B. Limbach, P. vom Stein, R. Schäfer, C. Theis, G. Göppert, and H. Grabert, Phys. Rev. B **66**, 125314 (2002)
21. F. Guinea and G. Schön, Europhys. Lett. **1**, 585 (1986)
22. S.A. Bulgadaev, JETP Lett. **45**, 622 (1987)
23. W. Hofstetter and W. Zwerger, Phys. Rev. Lett. **78**, 3737 (1997)
24. I.S. Beloborodov, A.V. Andreev, and A.I. Larkin, Phys. Rev. B **68**, 024204 (2003)
25. S.V. Panyukov and A.D. Zaikin, Phys. Rev. Lett. **67**, 3168 (1991)
26. X. Wang and H. Grabert, Phys. Rev. B **53**, 12621 (1996)
27. D.E. Khmelniskii, Phys. Lett. A **106**, 182 (1984)
28. A.D. Mirlin, D.G. Polyakov, and P. Wölfe, Phys. Rev. Lett. **80**, 2429 (1998); F. Evers, A.D. Mirlin, D.G. Polyakov, and P. Wölfe, Phys. Rev. B **60**, 8951 (1999)
29. A.M.M. Pruisken and I.S. Burmistrov, Phys. Rev. Lett. **95**, 189701 (2005)
30. S.S. Murzin, A.G.M. Jansen, and I. Claus, Phys. Rev. Lett. **92**, 016802 (2004); S.S. Murzin and A.G.M. Jansen, Phys. Rev. Lett. **95**, 189702 (2005)
31. A.A. Abrikosov, Physics **2**, 21 (1965)
32. see, e.g., Yu. A. Izyumov and Yu. N. Skryabin, *Statistical mechanics of magneto-ordered systems*, (Moskva, Nauka, 1987) (in Russian)
33. A.I. Larkin and V.I. Melnikov, Sov. Phys. JETP **34**, 656 (1972)
34. L. Zhu and Q. Si, Phys. Rev. B **66**, 024426 (2002); G. Zaránd and E. Demler, Phys. Rev. B **66**, 024427 (2002)
35. G. Gröppert, H. Grabert, Eur. Phys. J. B **16**, 687 (2000)
36. A.M.M. Pruisken, I.S. Burmistrov and R. Shankar, cond-mat/0602653 (unpublished)
37. A.M.M. Pruisken, R. Shankar and N. Surendran, Phys. Rev. B **72**, 035329 (2005)

Part II

Superconductivity

HIGH-FIELD FLUX DYNAMICS IN DISORDERED TWO-BAND SUPERCONDUCTIVITY

J.M. KNIGHT AND M.N. KUNCHUR
Department of Physics and Astronomy
University of South Carolina,
Columbia, SC 29208, USA.
knight@physics.sc.edu, kunchur@sc.edu

Abstract. The mixed state response of disordered magnesium diboride has a conspicuous absence of the usual flux-density (B) dependent broadening of conductance-temperature and voltage-current curves. This B independence implies a vortex viscosity that rises linearly with B. We show that this this anomalous viscosity function can arise from the quasiparticle sea of the quenched weaker band and the unusually long electric field penetration depth that delocalizes the vortex electric fields.

Keywords: flux dynamics, two-band superconductivity, magnesium diboride

7.1 Introduction

In a type-II superconductor, an applied flux density B has two effects on transport characteristics such as conductance-versus-temperature $G(T)$ or voltage-versus-current $V(I)$ curves: (1) suppression of superconductivity is manifested by a reduction in the effective critical temperature or critical current and (2) the transition becomes increasingly broadened due to flux motion. For example, near T_c, the shift in transition temperature is approximately linear in B, and the $G(T)$ curve becomes increasingly shallow [1,2] with a $1/B$ field dependence that persists even in the case of highly driven non-linear and unstable flux flow [3,4].

While clean MgB$_2$ conforms to the conventional B proportional to R or V response, the presence of impurity scattering in this system transforms the mixed-state response so that $G(T)$ and $V(I)$ curve shapes becomes anomalously B independent as if flux motion has become arrested or increasingly quenched with increasing B [5–8]. Figure 7.1, showing data from Ref. [8], provides one example of this behavior and the cited references provide additional data. Such B independence in disordered MgB$_2$ implies a viscous drag

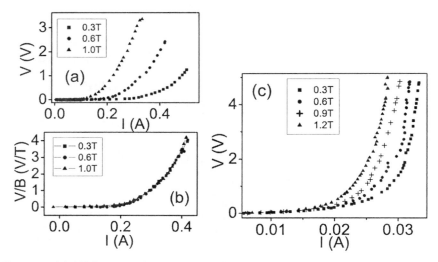

Fig. 7.1. (a) $V(I)$ curves for a clean HPCVD *in-situ* epitaxial MgB$_2$ film showing conventional B dependent ($V \propto B$) broadening due to flux motion. $T = 10$ K. (b) Above V values divided by B and then shifted horizontally and made to collapse. (c) $V(I)$ curves for a disordered carbon-doped HPCVD *in-situ* MgB$_2$ film. $T = 6$ K; the curve shapes are approximately independent of B. (Data from Ref. [8].)

coefficient η that rises linearly with B instead of the B independent η that occurs for conventional flux dynamics. We show here that $\eta \propto B$ is expected when the electric field penetration depth exceeds the coherence length and intervortex spacing, and when a sea of quasiparticles from a nearly normal π band causes vortex dissipation to spread throughout the entire volume instead of being localized near the core of the vortices.

7.2 Electric field penetration depth

The electric field penetration depth ζ has been used to describe the injection of charge carriers from a normal conductor where an electric field is present into a superconductor where the field decays spatially [9, 10]. The same concept serves to describe the penetration of an electric field from the core of a moving vortex into the surrounding superconducting medium [11]. The mechanism governing the process in superconductors with non-zero gap is the relaxation of unequal populations in the electron-like and hole-like excitations (branch imbalance). Tinkham [9] showed that this relaxation is mediated by inelastic scattering in a superconductor with an isotropic gap, but that it can be accomplished by elastic scattering if there is anisotropy. The electric field penetration depth is given by $\zeta = \sqrt{D\tau_Q}$ in terms of the diffusion constant D and the branch imbalance relaxation time τ_Q. The part of τ_Q due to phonons is related to the electron-phonon scattering time τ_{ph}. There is reason to expect that the inelastic scattering by phonons is exceptionally small in MgB_2, and therefore, to the extent that other mechanisms for branch relaxation can be neglected, that the electric field penetration depth is exceptionally large. The evidence comes from the high Debye temperature ($\approx$1,000 K) in MgB_2, indicating that few phonons are present at superconducting temperatures, and from the fact that there is little evidence of temperature dependent resistivity near superconducting temperatures in the samples used in the experiment.

In the work of Hu and Thompson [11] on the structure of moving vortices in single-band superconductors, an arbitrary electric field penetration depth was used, even though they were considering the case of gapless superconductors where ζ is $1/\sqrt{12}$ times the coherence length ξ. Numerical work based on their formulas for $\zeta \gg \xi$ shows penetration of both charge and electric field into the region outside the vortex core, and shows a generally decreased local electric field for a given vortex velocity. The decrease in the local field occurs because spreading the field over a larger region must result in a decrease because the spatially averaged field must still be equal to the macroscopic electric field E determined by the vortex velocity and the external B field, $E = vB$. This is the field measured by the voltage across the sample.

In adapting the single-band results qualitatively to the two-band case of MgB_2 in the range of temperatures and fields where the anomalous behavior is observed, we must take note of the fact that the vortices are sustained by the σ-band electron pairs and that the π-band electrons form a normal conducting background. The π electrons will react to the electric field of the moving vortex both inside and outside of the core, adding to the dissipation of the σ electrons, which occurs primarily near the core.

In a single-band superconductor, τ_Q can be expressed in terms of the electron-phonon scattering time τ_{ph} when elastic scattering and electron-electron scattering do not contribute. The expression is [12]:

$$\tau_Q = \frac{2\hbar T}{\pi |\Delta|^2} \sqrt{1 + \frac{4\||\Delta|^2 \tau_{ph}^2}{\hbar^2}}. \tag{7.1}$$

The temperature dependence of ζ arises from this factor, both from the explicit factor $T = tT_c$ and from the temperature dependence of the gap Δ and of the electron-phonon scattering time τ_{ph}. In the numerical estimates below, we take the standard empirical temperature dependence $\Delta = \Delta_0 \sqrt{1-t}$ for the gap and assume that τ_{ph}^{-1} is proportional to the number of phonons present, i.e., proportional to t^3. With these assumptions, Eq. (7.1) indicates that ζ approaches infinity as $1/\sqrt{1-t}$ at the transition temperature, the same temperature dependence as the coherence length. At $T = 0$, ζ approaches infinity as $1/T$ because of the infinite electron-phonon scattering time in Eq. (7.1).

In calculating the reduced temperature t, we must account for the decrease of T_c with B. This can be obtained with sufficient accuracy by making a linear fit to the experimental data near $B = 0$. The part of ζ determined by elastic scattering in the presence of gap anisotropy and the part due to scattering by π-band electrons are assumed to be temperature independent.

If the value of ζ exceeds the vortex spacing $l_\phi \approx \sqrt{\Phi_0/B}$, the electric field will extend throughout the sample. In that case, the π electrons will contribute fully to the dissipation while the σ band dissipation will still be restricted to the region near the vortex core.

In the next section. these ideas are taken as the basis for a simple model for the viscous coefficient of a moving vortex in dirty samples of MgB$_2$. In the absence of a full treatment of vortex dissipation in a two-band system, we expect that this model can give an indication of the dependence of this coefficient on the temperature and on the magnetic field. A full treatment would be necessary if we are to to understand the system quantitatively and to deal with such questions as the effects of π-band screening on the electric field penetration.

7.3 Viscous drag coefficient

The average electric field E_v near a vortex can be estimated by setting its spatial average equal to the average of the macroscopic field E:

$$E_v \pi \zeta^2 \approx E l_\phi^2 \approx E \frac{\Phi_0}{B}. \tag{7.2}$$

The viscous drag coefficient η can then be obtained by equating the total energy dissipated per unit time per unit vortex length to the drag force ηv per unit length times the velocity:

$$\eta v^2 = \sigma_\sigma \pi \xi^2 E_v^2 + \sigma_\pi \pi \zeta^2 E_v^2. \tag{7.3}$$

Note that the σ band contributes to the dissipation in a cylinder of radius ξ while the π band contributes within a radius ζ. The drag coefficient then follows using the relation $v = E/B$ and Eq. (7.2):

$$\eta = \frac{\Phi_0^2}{\pi \zeta^2} \left(\sigma_\sigma \frac{\xi^2}{\zeta^2} + \sigma_\pi \right). \tag{7.4}$$

The last equation holds as long as $\zeta < l_\phi$. If that is not the case, the electric field spreads throughout the material, and the π band will produce almost as much dissipation as in the normal state. Then ζ in Eq. (7.4) can be replaced by the vortex spacing and the dissipation due to the σ band can be neglected when the coherence length is much less than the vortex spacing. This gives the estimate

$$\eta \approx \frac{\sigma_\pi \Phi_0 B}{\pi}, \tag{7.5}$$

which is in contrast with the B-independent Bardeen-Stephen result

$$\eta \approx \frac{\sigma_n \Phi_0 H_{c2}}{\pi}. \tag{7.6}$$

Thus the flux flow conductivity $\sigma_{\mathrm{ff}} = \eta/B\Phi_0$, which is proportional to B^{-1} in the conventional case, becomes independent of B for the drag coefficient in Eq. (7.5). This change in field dependence could account for the experimentally observed anomalies if certain conditions on the conductivities, discussed below, are satisfied. Next we discuss the values of the parameters that enter the model.

7.4 Parameters

Eq. (7.1) shows that ζ becomes very large near T_c, so that field-independent $I(V)$ and $G(T)$ curves would be expected there. To assess whether this behavior is likely to be realized at temperatures far from the transition temperature, we estimate the value of ζ in MgB_2 in that temperature range. The estimate requires knowledge of the diffusion constant $D = v_F^2 \tau$ and the branch imbalance relaxation time τ_Q. Estimates of these quantities depend on the carrier density n. Measurements of the Hall coefficient [13] $R = 1/ne$ indicate a carrier density of about 150 holes nm^{-3}. However, the zero-temperature magnetic penetration depth $\lambda(0) = \sqrt{mc^2/4\pi ne^2}$, quoted as 152 nm in [14], gives a value of 1.22 nm^{-3}. Both of these seem extreme in comparison to the number of unit cells per nm^3, which is 34. In the estimates, we therefore take a presumed value of 2 carriers per unit cell, or 70 carriers nm^{-3}. The transport relaxation time τ can now be obtained from the measured resistivity, $\rho = 14\,\mu\Omega$ cm, using the Drude formula $\rho = m/ne^2\tau$. The resulting transport relaxation time is 3.6×10^{-15} s^{-1}. Fermi velocities $v_{F\sigma} = 4.4 \times 10^5$ ms^{-1} and $v_{F\pi} = 8.2 \times 10^5$ ms^{-1} are quoted in [15]. Using an average value of 6.4×10^5 m/s, we find a diffusion constant $D = 1.5 \times 10^{-3}$ m^2 s^{-1} and a mean free path of 2.3 nm.

An estimate of the electric field penetration depth using Eq. (7.1) requires an estimate of the electron-phonon scattering time τ_{ep}. A direct estimate of this time from the Bloch-Grüneisen formula [16], using parameters appropriate to MgB_2 and neglecting details of the Fermi surface and phonon spectrum, gives $\tau_{\mathrm{ep}} = 2.2 \times 10^{-10}$ s^{-1}. This results in an electric field penetration depth ζ of the order of 300 nm. The estimate is crude and very sensitive to the Debye temperature. An earlier estimate of ζ given by Tinkham requires extrapolating the high temperature value of τ_{ep}, where the resistivity is proportional to T, back to the Debye temperature. If we again use the Bloch-Grüneisen formula to do the extrapolation, we obtain a value about three orders of magnitude smaller, $\zeta \sim 0.3$ nm. Fortunately, the details of the anomalous regime do not depend sensitively on the value of ζ. All that is required for the anomalous flux dynamics is for ζ to exceed $l_\phi \approx 45/\sqrt{B}$ nm, where B is in Tesla. This condition is satisfied in MgB_2 for zeta values in the higher part of the estimated range, especially for large B.

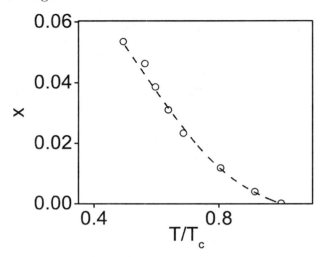

Fig. 7.2. Fit of experimental data for H_{c2} vs T to the theory of Gurevich cited in the text. The abscissa is the reduced temperature T/T_c and the ordinate is the dimensionless critical field $x = H_{c2}D_\sigma/2\Phi_0 T_c$. Parameters are the electron-phonon coupling constants $\lambda_{\sigma\sigma} = 0.81$, $\lambda_{\pi\pi} = 0.285$, $\lambda_{\sigma\pi} = 0.150$, and $\lambda_{\pi\sigma} = 0.115$, and the ratio of diffusion constants $D_\pi/D_\sigma = 30$.

We have made these estimates without differentiating the contributions of the two bands to the diffusion constant or the carrier density. We are not aware of any experimental evidence that could separate the invividual contributions. There is, however, evidence that in the superconducting state the π band is considerably cleaner than the σ band. Measurements of the critical field H_{c2} have been made on the sample that show lack of broadening and lack of field dependence in conductance and voltage curves. They can be fit to the theory of Gurevich [17], which is obtained by adapting the Eilenberger and Usadel equations to the case of a two-band superconductor. It turns out that the shape of the H_{c2} vs. T curve is very sensitive to the ratio of the diffusion constants of the two bands, which is roughly proportional to the ratio of scattering times. A fit to the experimental curve on (Fig. 7.2) can be made with a ratio D_π/D_σ of 30. In obtaining this fit, we found it necessary to increase the interband electron-phonon coupling constants slightly compared to the commonly accepted ones [17–19]. The coupling constants giving the best fit are $\lambda_{\sigma\sigma} = 0.81$, $\lambda_{\pi\pi} = 0.285$, $\lambda_{\sigma\pi} = 0.150$, and $\lambda_{\pi\sigma} = 0.115$. The last two are greater than the ones used by Gurevich by a factor of 1.25.

7.5 Discussion

Experimentally it has been shown [5,8] that disordered MgB_2 shows $V(I)$ and $G(T)$ curves that have strangely B independent shapes, which merely shift as a function of B. In this paper, we have considered a new regime of flux dynamics that can be expected in a two-band superconductor with delocalized vortex electric fields. In a conventional single-band mixed state, vortex dissipation is localized to a region $R \sim \xi$ because usually $\zeta \sim \xi$ and because far from the upper-critical phase boundary, the quasiparticle density diminishes rapidly for $r > \xi$ and hence there is a natural cutoff. In this conventional case, the viscous coefficient is independent of B leading to a flux-flow conductance that is inversely proportional to B. In the delocalized scenario, dissipation extends outside the core region and fills the whole volume because of the availability of the quasiparticles of the quenched (π) band and because $\zeta > l_\phi > \xi$. In this latter delocalized case, the viscous coefficient is proportional to B leading to a flux-flow conductance that is independent of B as observed experimentally.

This model naturally gives rise to a cross over field that must be exceeded for the anomalous behavior to occur. Two conditions must be met for the delocalized scenario: (1) $\zeta > l_\phi$ in order to have delocalized electric fields and (2) the π band's participation in the superconductivity must be quenched (i.e., the "virtual H_{c2}" must be exceeded) so as to provide a sea of quasiparticles that will extend the dissipation into the region between the vortex cores. Interestingly both conditions give a cross over field of $\sim$3,000 G, which agrees with the experimental observation that there is a conventional B dependence of $R(I)$ curves below a few thousand Gauss [8].

A further requirement for the transport characteristics to have a dramatic indifference to B is that the π quasiparticle conductivity be much higher than the σ quasiparticle conductivity. The observed shape of the $H_{c2}(T)$ indeed confirms that. At present it is not known what causes the π band to have such a high conductivity in the superconducting state.

References

1. Michael Tinkham, *Introduction to Superconductivity* (McGraw Hill, New York 1996).
2. A. I. Larkin and Yu. N. Ovchinnikov, in *Nonequilibrium Superconductivity*, D. N. Langenberg and A. I. Larkin, Eds. (Elsevier, Amsterdam 1986), Chapter 11.
3. A. I. Larkin and Yu. N. Ovchinnikov, Zh. Eksp. Teor. Fiz. **68**, 1915 (1975)[Sov Phys. JETP **41**, 960 (1976)].
4. M. N. Kunchur, Phys. Rev. Lett. **89**, 137005 (2002).
5. D. H. Arcos and M. N. Kunchur, Phys. Rev. B **71**, 184516 (2005).
6. J. R. Thompson, K. D. Sorge, C. Cantoni, H. R. Kerchner, D. K. Christen, and M. Paranthaman, Supercond. Sci. Technol. **18**, 970 (2005).
7. A. Kohen et al., Appl. Phys. Lett. **86**, 121503 (2005).
8. M. N. Kunchur, et al., Physica C **437–438**, 171 (2006).
9. M. Tinkham, Phys. Rev. B **6**, 1747 (1972).
10. A. Schmid and G. Schön. J. Low Temp. Phys. **20**, 207 (1975).
11. C. R. Hu and R. S. Thompson, Phys. Rev. B **6**, 110 (1972); and R. S. Thompson and C. R. Hu, Phys. Rev. Lett. **27**, 1352 (1971).
12. Nikolai B. Kopnin, *Theory of Nonequilibrium Superconductivity* (Oxford University Press, Oxford, 2001) p. 220.

13. W. N. Kang, C. U. Jung, Kijoon H. P. Kim, Min-Seok Park, S. Y. Lee, Hyeong-Jin Kim, Eun-Mi Choi, Kyung Hee Kim, Mun-Seog Kim, and Sung-Ik Lee, Appl. Phys. Lett. **79**, 982 (2001).
14. Mun-Seog Kim, John A. Skinta, Thomas R. Lemberger, W. N. Kang, Hyeong-Jin Kim, Eun-Mi Choi, and Sung-Ik Lee, Phys. Rev. B **66**, 064511 (2002).
15. Thomas Dahm, in *Frontiers in Superconducting Materials*, A. V. Narlikar, ed., (Springer Verlag, Berlin, 2005).
16. J. M. Ziman, *Electrons and Phonons* (Oxford University Press, Oxford 1960).
17. A. Gurevich, Phys. Rev. B **67**, 184515 (2003).
18. A. A. Golubov et. al., J. Phys.: Condens. Matter, **14**, 1353 (2002).
19. Hyoung Joon Choi, David Roundy, Hong Sun, Marvin L. Cohen, Steven G. Louis, Phys. Rev. B **69**, 056502 (2004), and references therein.

8

SUPERCONDUCTIVITY IN THE QUANTUM-SIZE REGIME

A.A. SHANENKO[1,2], M.D. CROITORU[3], AND F.M. PEETERS[1]

[1] *TGM, Departement Fysica, Universiteit Antwerpen, Groenenborgerlaan 171, B-2020 Antwerpen, Belgium.*
arkady.shanenko@ua.ac.be
[2] *Bogoliubov Laboratory of Theoretical Physics, Joint Institute for Nuclear Research, 141980 Dubna, Russia*
[3] *EMAT, Departement Fysica, Universiteit Antwerpen, Groenenborgerlaan 171, B-2020 Antwerpen, Belgium*

Abstract. Recent technological advances resulted in high-quality superconducting metallic nanofilms and nanowires. The physical properties of such nanostructures are governed by the size-quantization of the transverse electron spectrum. This has a substantial impact on the basic superconducting characteristics, e.g., the order parameter, the critical temperature and the critical magnetic field. In the present paper we give an overview of our theoretical results on this subject. Based on a numerical self-consistent solution of the Bogoliubov-de Gennes equations, we investigate how the superconducting properties are modified in the quantum-size regime.

Keywords: superconducting nanowires and nanofilms, quantum-size regime, Bogoliubov-de Gennes equations

8.1 Introduction

Recent advances in nanofabrication technology resulted in high-quality metallic nanoscale structures like single-crystalline films with thickness down to a few monolayers [1–3] and wires (both single-crystalline and made of strongly coupled grains) with width of about 10 nm [4–6]. The electron mean free path was estimated as being about or larger than the specimen thickness [2,4] and, so, such nanofilms and nanowires exhibit quantum confinement effects for the transverse electron motion. It means that nonmagnetic impurities can only influence the electron motion parallel to the specimen and, then, the Anderson theorem [7] makes it possible to conclude that the longitudinal scattering of electrons on such nonmagnetic imperfections cannot have a significant effect on the superconducting characteristics.

Most of previous studies concerned high-resistivity films and wires being strongly disordered or granular. In that case the Anderson theorem is violated due to the enhancement of the Coulomb repulsion [8] and because of mesoscopic fluctuations [9] (see discussion in Ref. [10]). In particular, the Coulomb interaction effects in the presence of the diffusive character of the electron motion lead to the suppression of the critical temperature with increase of disorder [11]. On contrast, the present investigation is focused on recently fabricated low-resistivity specimens with minimal disorder like, e.g., Pb single-crystalline nanofilms reported in [1–3]. Notice that these extremely thin nanofilms show no significant indications of defect- or phase-driven suppression of superconductivity. The same is true for the high-quality nanowires (except of the most narrow aluminum specimen with width ≈ 8 nm [4]). The major mechanism governing the superconducting properties in this situation is the transverse quantization of electron motion. This opens up the possibility to investigate *the interplay of quantum confinement and superconductivity.*

Increasing the critical temperature T_c, critical current J_c and critical magnetic field H_c of a superconductor has been a major challenge since the discovery of superconductivity. On the one hand we can look for chemically complex materials exhibiting higher critical parameters. Such a search has been very successful over the last 20 years, and new high-temperature superconducting materials have been developed. On the other hand microstructuring of a superconductor is a different and very promising way which can result in the enhancement of the critical superconducting parameters. In earlier works on micro-structuring of superconductors in the mesoscopic regime, enhancement of J_c was found due to trapping of vortices [12]. Also a significant increase of H_c was realized through such mesoscopic structuring [13], which is a consequence of enhanced surface superconductivity. However, in both cases T_c at zero magnetic field, was unaltered.

Contrary to the mesoscopic regime, in high-quality superconducting nano wires and nanofilms T_c (together with J_c and H_c) exhibits *quantum-size oscillations* with significant resonant enhancements (*superconducting resonances*). The first prediction of such oscillations goes back to the pioneering work of Blatt and Thompson [14], where the thickness-dependent oscillations of the energy gap in a superconducting slab were calculated within the multiband BCS model in the clean limit. The subsequent 40 years after this work, very few experimental papers reported the observation of possible signatures of such behavior in tin and lead films [15], but structural defects were a serious problem at that time and prevented a definite conclusion concerning the existence of such quantum-size oscillations. For decades atomic nuclei were the only system where the interplay of quantum confinement with pairing of fermions could be studied experimentally and where the predictions of Blatt and Thompson were confirmed for for nucleons (see, for instance, [16]). Modern developments in nanofabrication have finally resulted in high-quality nanostructures, and quantum-size oscillations of T_c and H_c (perpendicular to a film) have recently been observed in Pb atomically uniform nanofilms at a new level of experimental precision and sophistication [1, 2]. Our theoretical results [17] indicate that the thickness-dependent increase of T_c recently found in Al and Sn superconducting nanowires [4–6] is also a manifestation of the quantum-size regime.

Here we present an overview of our recent theoretical results on superconducting nanofilms and nanowires in the quantum-size regime. Our study is based on a numerical self-consistent solution of the Bogoliubov-de Gennes (BdG) equations [18, 19] *in the clean limit* and in the presence of quantum confinement of electrons.

8.2 Bogoliubov-de Gennes equations

In the presence of quantum confinement translational invariance in the confined directions is broken and the superconducting order parameter will depend on the position $\Delta = \Delta(\mathbf{r})$. To investigate the equilibrium superconducting properties in this case, one needs to solve the BdG equations [18, 19]. These equations can be written in the form

$$E_i|u_i\rangle = \hat{H}_e|u_i\rangle + \hat{\Delta}|v_i\rangle, \tag{8.1}$$
$$E_i|v_i\rangle = \hat{\Delta}^*|u_i\rangle - \hat{H}_e^*|v_i\rangle, \tag{8.2}$$

where ε_i is the quasiparticle energy, $|u_i\rangle$ and $|v_i\rangle$ are the particle-like and hole-like ket vectors. In the clean limit the single-electron Hamiltonian in Eqs. (8.1) and (8.2) reads

$$\hat{H}_e = \frac{1}{2m_e}\left(\hat{p} - \frac{e}{c}\mathbf{A}\right)^2 + V_{\text{conf}}(\hat{\mathbf{r}}) - E_F, \tag{8.3}$$

with E_F the Fermi level, m_e the electron band mass (set to the free electron mass below) and $V_{\text{conf}}(\hat{\mathbf{r}})$ the confining potential. In turn, $\hat{\Delta}$ stands for the gap operator whose matrix element can be represented as

$$\langle\mathbf{r}|\hat{\Delta}|\mathbf{r}'\rangle = \Delta(\mathbf{r})\delta(\mathbf{r} - \mathbf{r}'), \tag{8.4}$$

with $\delta(x)$ the Dirac δ−function. The order parameter can generally be complex but it is chosen as a real quantity throughout this paper. We remark that the normal mean-field interaction should in general be taken into account in Eq. (8.3) [18]. However, it has a very little effect on the results in most applications and can be ignored. As a mean-field theory, the BdG equations should be solved in a self-consistent manner. The self-consistency relation reads

$$\Delta(\mathbf{r}) = g\sum_{i\in\mathcal{C}}\langle\mathbf{r}|u_i\rangle\langle v_i|\mathbf{r}\rangle(1 - 2f_i), \tag{8.5}$$

where g is the coupling constant, $f_i = 1/(e^{\beta E_i} + 1)$ is the Fermi function [$\beta = 1/(k_B T)$ with T the temperature and k_B the Boltzmann constant]. In Eq. (8.5) $\mathcal{C}$ indicates the set of the quantum numbers corresponding to the single-electron energy ξ_i located in the Debye window $\xi_{i \in \mathcal{C}} \in [-\hbar\omega_D, \hbar\omega_D]$ (ω_D is the Debye frequency), where the single-electron energy ξ_i (measured from the Fermi level) is defined as

$$\xi_i = \langle u_i | \hat{H}_e(0) | u_i \rangle + \langle v_i | \hat{H}_e(0) | v_i \rangle, \qquad (8.6)$$

with $\hat{H}_e(0)$ being the single-electron Hamiltonian at zero magnetic field (see, for example, Ref. [18]); this selection involves *the canonical momentum*. Another important limitation concerns the quasiparticle energies E_i which should normally be positive in Eq. (8.5). However, this limitation is not correct for the regime of gapless superconductivity, where the ground state is modified so that unpaired electrons appear at zero temperature. This regime can be realized, e.g., when a strong magnetic fields (supercurrent) is present [18]. Here one should control shifts of the quasiparticle energies due to the magnetic field (supercurrent). This is in order to exclude nonphysical solutions of the BdG equations that acquire positive energies, while being negative in energy for zero magnetic field (supercurrent).

Below we assume that the confining interaction is zero inside the superconducting sample and infinite outside. Hence, the quantum-confinement boundary condition reads

$$\langle \mathbf{r} | u_i \rangle \Big|_{\mathbf{r} \in S} = \langle \mathbf{r} | v_i \rangle \Big|_{\mathbf{r} \in S} = 0, \qquad (8.7)$$

at the sample surface, i.e. $\mathbf{r} \in S$. In addition, periodic boundary conditions can be applied in the direction parallel to the nanowire/nanofilm.

Expanding $|u_i\rangle$ and $|v_i\rangle$ in terms of the eigenfunctions of $\hat{H}_e(0)$, one can convert Eqs. (8.1) and (8.2) into a matrix equation. Then, the numerical problem can be solved by diagonalizing the relevant matrix and invoking iterations, in order to account for the self-consistency relation given by Eq. (8.5). Our procedure of numerically solving the BdG equations consists of three steps. At the first step, a bulk value of the order parameter is used as the initial guess, and the BdG equations are solved by substituting this value for $\Delta(\mathbf{r})$ in Eqs. (8.1) and (8.2). At the second step, the obtained eigenvectors u_i and v_i are inserted into Eq. (8.5) together with the corresponding quasiparticle energies E_i. At the third step, solving the BdG equations with the order parameter found at the previous step yields new eigenfunctions and quasiparticle energies that are used in the next iteration. The second and third steps should be repeated till the maximum difference $\delta = \max |\Delta^s(\mathbf{r}) - \Delta^{s-1}(\mathbf{r})|$ between the order-parameter values at consecutive iterations is sufficiently small to satisfy the adopted accuracy $\delta < 10^{-5}$. When increasing the thickness of a specimen, the order parameter found for a smaller thickness, can be used as an initial guess in the iteration process. The same is related to changing other relevant parameters such as a magnetic field or a supercurrent. Such a procedure has

Table 8.1. The effective Fermi energy.

	Al	Sn	Pb
E_F (eV)	0.90	2.27	0.93

the advantage that it decreases the required computer time and allows us to track the superconducting state in the metastable region and find possible hysteresis behavior. Being far from the area of the superconducting-to-normal transition, our algorithm is quite stable and rapidly convergent (about 70–100 iterations to reach $\delta < 10^{-5}$) with no significant dependence on the initial guess. Near the critical region the algorithm becomes very sensitive to the initial guess and to numerical viscosity. Here the number of required iterations increases rapidly before convergence is reached.

When expanding $|u_i\rangle$ and $|v_i\rangle$ in eigenfunctions of $\hat{H}_e(0)$, a correct choice of appropriate eigenfunctions is of great importance. The criterion of selecting quasiparticle states in Eq. (8.5) ($\xi_i \in [-\hbar\omega_D, \hbar\omega_D]$) dictates that we should take at least those eigenfunctions of $\hat{H}_e(0)$ whose eigenvalues are in the Debye window. Numerical analysis shows that such an approximation leads, as a rule, to results with an accuracy of a few percent.

Notice that, to obtain the correct period for the quantum-size oscillations of the physical properties in nanosuperconductors within the parabolic band approximation (based on the band mass m_e), one should use the effective Fermi level E_F rather than the true one (see, for details, Refs. [20–22,24]). In general, the effective Fermi level depends on the interplay between the crystal and confinement directions. The used values of E_F are given in Table 8.1. E_F for Al is justified from a good agreement with the experimental data found in Ref. [17]. The data for Pb and Sn are extracted from the experimental results reported in Refs. [22] and [23].

8.3 Quantum-size oscillations and resonances

The physics of quantum-size superconducting resonances can be outlined as follows. The superconducting order parameter is not simply the wave function of an ordinary bound state of two fermions but the wave function of a bound fermion pair in a medium [19,24]. In the homogeneous case the Fourier transform of the Cooper-pair wave function is suppressed for the wavenumbers less than the Fermi one due to the presence of the Fermi sea [25]. Therefore, the Fourier transform of the Cooper-pair wave function appears to be essentially nonzero only in the vicinity of the Fermi wavenumber. Generally, the superconducting order parameter strongly depends on N_D, the number of single-electron states (per spin projection) situated in the Debye window around the Fermi level ($\xi_i \in [-\hbar\omega_D, \hbar\omega_D]$). More precisely, the mean energy density of these states taken per unit volume $n_D = N_D/(2\hbar\omega_D V)$ is the key quantity.

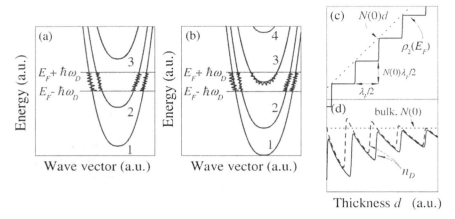

Thickness d (a.u.)

Fig. 8.1. (a) Single-electron subbands versus the wave vector of the motion parallel to the nanofilm/nanowire for the off-resonant case. (b) The same as in panel (a) but for the resonant thickness. (c) The $2D$ energy density of states at the Fermi level $\rho_2(E_F)$ versus the film thickness d. (d) The mean energy density of states in the Debye window n_D versus d: solid curve is for a constant Fermi level, the dashed curve results from taking account of a change in E_F for narrow nanofilms.

In the presence of quantum confinement the band of single-electron states in a clean nanofilm/nanowire is split up in a series of subbands. While the specimen thickness increases (decreases), these subbands move down (up) in energy. Note that the position of the bottom of any subband scales as $1/d^2$, with d the nanofilm/nanowire thickness. Each time when the bottom of a parabolic subband passes through the Fermi surface, the density n_D increases abruptly. In Figs. 8.1a and b, single-electron subbands are schematically plotted versus the wave vector of the quasi-free electron motion parallel to a nanofilm/nanowire. The single-electron states located in the Debye window (making a contribution to the superconducting characteristics) are highlighted with the broken lines. In Fig. 8.1a the bottoms of all subbands $1, 2$ and 3 are situated outside the Debye window, and we are in the off-resonance regime. However, when increasing the thickness of the sample, the bottom of subband 3 moves down so that it enters the Debye window (as shown in Fig. 8.1b), the mean density of states n_D increases and a superconducting resonance develops. This leads to a sequence of peaks in n_D as a function of the sample thickness d (see Fig. 8.1c) and, as a consequence, any superconducting quantity exhibits quantum-size oscillations with remarkable resonant enhancements. Such superconducting resonances are significant in nanoscale samples but smoothed out with an increase in d (see Fig. 8.1d), when n_D slowly approaches its bulk limit $N(0) = mk_F/ = (2\pi^2\hbar^2)$, with k_F the bulk Fermi wavenumber.

Figure 8.1d shows schematically quantum-size oscillations of n_D in the case of nanofilms. To physically understand these oscillations, one can use a reasonable simplified approximation

$$n_D \sim \rho_2(E_F)/d, \quad \rho_2(E) = \frac{m}{2\pi\hbar^2} \sum_{j=0}^{\infty} \theta\left(E - \frac{\hbar^2(j+1)^2\pi^2}{2md^2}\right), \qquad (8.8)$$

where $\theta(x)$ is the Heaviside function, and the factor $1/2$ appears in the expression for the two-dimensional density of states $\rho_2(E)$ due to neglecting the electron spin. As seen from Eq. (8.8) and Fig. 8.1c, $\rho_2(E_F)$ exhibits a sequence of equidistant steps with period $\lambda_F/2 = \pi/k_F$ (λ_F is the Fermi wavelength). The step magnitude is $N(0)\lambda_F/2$. The two-dimensional density of states per unit thickness $\rho_2(E_F)/d$, gives a good insight into the results presented in Fig. 8.1d. However, we stress that n_D is not exactly the same as $\rho_2(E_F)/d$. Strictly speaking, n_D results from the integral of $\rho_2(E)/d$ over the Debye window. This is why n_D does not exhibit jumps at the resonant points. Another result of this integration is that all peaks in n_D appear to be somewhat below the bulk limit $N(0)$ (contrary to the peaks in $\rho_2(E_F)/d$ being exactly $N(0)$). The solid curve in Fig. 8.1d represents n_D calculated when E_F is taken to be independent of the film thickness. The dashed curve in Fig. 8.1d is obtained when the Fermi level is allowed to vary with the film thickness but the electron density is kept constant. For nanowires these quantum-size oscillations in the mean density of states look similar but with a few exceptions. The most important of them are that the resonances are not equidistant in this case, even for a constant Fermi level, and the smoothed jumps in n_D are transformed into sharp peaks appearing due to the Van Hove singularities.

Notice that n_D on the average is smaller than its bulk value. However, as shown below, the superconducting characteristics, e.g., the order parameter, the critical temperature and energy gap, are enhanced at the resonant points as compared to their bulk values. A reason is that the single-electron wave functions are no longer three-dimensional plane waves, and this change plays a significant role on the nanoscale.

8.4 Superconducting nanofilms in the quantum-size regime

To simulate a nanofilm, we investigate a model system with fixed dimensions $L_x = L_y = 0.5 \div 50\,\mu m$, while L_z is varied on the nanoscale, being the film thickness d. For the x and y directions we use periodic boundary conditions ($L_x = L_y \gg \lambda_F$, with $\lambda_F \sim 1\,nm$). The electron motion in the z direction is quantum confined (see Eq. (8.7)). Depending on the thickness, L_x can be changed, in order to reach optimal conditions for the calculations. For instance, to approach the limit $L_x, L_y \to \infty$ of the superconducting order parameter, we take $L_x > 0.5 \div 2\,\mu m$ at a resonant thickness. Off-resonant points require, as a rule, larger values of L_x. The optimal regime depends also on the superconducting quantity in question. In particular, to investigate the critical magnetic field H_c, we found that we should take $L_x > 10\,\mu m$.

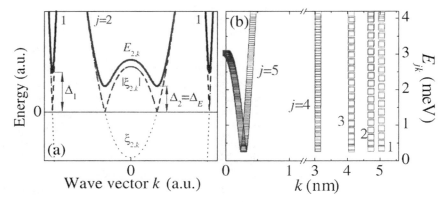

Fig. 8.2. (a) The quasiparticle energy E_{jk} versus k shown schematically together with the single-electron energy ξ_{jk} and its absolute value $|\xi_{jk}|$ for two subbands $j = 1, 2$. (b) Quasiparticle energies at zero temperature for an aluminum nanofilm with $d = 3.05$ nm from numerically solving the BdG equations.

Before the main discussions on our results for nanofilms, it is instructive to make several remarks concerning the quasiparticle energies E_i. Our numerical investigation of the BdG equations shows that E_i can be well approximated by

$$E_{jk} = \sqrt{\xi_{jk}^2 + \Delta_j^2}, \qquad (8.9)$$

with $i = \{j, k\}$, where j is the quantum number (or the set of quantum numbers) controlling the transverse motion of electrons in a nanofilm/nanowire and k is the wave vector of the quasi-free electron motion parallel to the sample. Δ_j stands for the energy-gap associated with subband j. Notice that this approximation is correct only in the absence of a magnetic field or a supercurrent. In the opposite situation an extra term (depending on the magnetic field or the superfluid velocity) should be added to the square root in Eq. (8.9) (see, for instance, Sec. 8.7). In Fig. 8.2a E_{jk} is schematically shown together with the single-electron energy ξ_{jk} and its absolute value $|\xi_{jk}|$ for subbands with $j = 1, 2$. The zero-energy level can here be treated as the Fermi level because ξ_{jk} is measured from E_F, by definition. For the superconducting solution ($\Delta_1, \Delta_2 \neq 0$) we have a gap in the quasiparticle spectrum consisting of the two branches with $j = 1$ and 2. The total gap Δ_E is defined as $\Delta_E = \min \Delta_j$ and in the shown example we have $\Delta_E = \Delta_2$. When $\Delta_j \to 0$, then $E_{jk} \to |\xi_{jk}|$. In Fig. 8.2b the quasiparticle energies calculated from the BdG equations for an aluminum superconducting nanofilm with $d = 3.05$ nm, are plotted versus k. An interesting difference as compared to Fig. 8.2a, is that for this case Δ_j is practically independent of j. This is a particular feature of nanofilms. The situation is different for superconducting nanowires where the formation of new Andreev-type states (see Sec. 8.6) results in different subband gaps. These new Andreev-type states induced by quantum confinement play no significant role

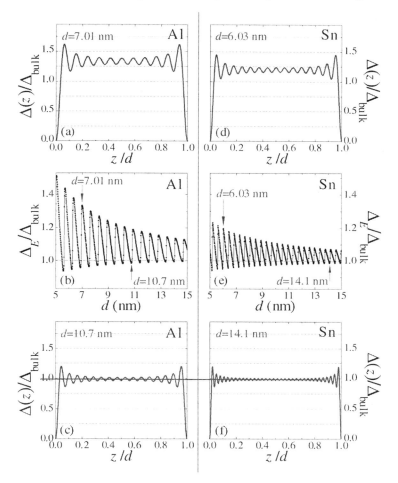

Fig. 8.3. Aluminum nanofilms: (a) the relative energy gap Δ_E/Δ_{bulk} versus d; the relative order parameter $\Delta(z)/\Delta_{bulk}$ as a function of z for (b) the resonant thickness $d = 7.01$ nm and for (c) the off-resonant thickness $d = 10.7$ nm. Tin nanofilms: (d) the energy gap versus d; the order parameter for (e) $d = 6.03$ nm (resonance) and for (f)$d = 14.1$ nm (f).

in nanofilms. It is also of importance to notice that in the presence of a magnetic field (or a supercurrent) Δ_E is no longer equal to min Δ_j. For example, this is the case for the gapless superconductivity when $\Delta_E = 0$ while $\Delta_j \neq 0$.

Figure 8.3 presents more detailed results of a numerical solution of the BdG equations for superconducting metallic nanofilms (panels (a)–(c) for Al [$\hbar\omega_D = 32.31$ meV, $gN(0) = 0.18$], panels (d)–(f) for Sn [$\hbar\omega_D = 16.8$ meV, $gN(0) = 0.25$]. These results are calculated for $T = 0$, zero magnetic field and zero supercurrent (we discuss magnetic effects in Sec. 8.7). As seen from Figs. 8.3a and b, the energy gap Δ_E increases well above its bulk value Δ_{bulk} at the resonant points but stays near Δ_{bulk} in the off-resonant regime. Quantum-size oscillations in Δ_E are more significant for Al as compared to Sn. The point is that these oscillations appear as a result of the interplay between pair condensation and quantum confinement: the larger the condensation energy, the smaller the resonances. Thus quantum-size effects are significant on the nanoscale and washed out when approaching the mesoscopic (bulk) regime. The resonant peaks are equidistant with period of about $\Delta d = \lambda_F/2$, in agreement with the expectations based on Figs. 8.1c and d. The most surprising is that though n_D is mostly below its bulk limit $N(0)$, the energy gap is mostly enhanced as compared to bulk. We remark again that this is a consequence of the change in the transverse single-electron wave functions. They are no longer plane waves in the presence of quantum confinement, and such a difference is of importance on the nanoscale. As seen from Figs. 8.3b and c, the superconducting order parameter is also enhanced at the resonant points and close to Δ_{bulk} in the off-resonant case (panels (c) and (f)). For a film the order parameter does not depend on the longitudinal coordinates x and y, but varies with the transverse coordinate z. However, these spatial variations are significant only near the film edges, where $\Delta(z) = 0$ due to quantum confinement. In the center $\Delta(z)$ exhibit small oscillations with period of about $\lambda_F/2$ (it is exactly $\lambda_F/2$ only for the resonant thicknesses, which can be clearly observed for $d < 5\,\text{nm}$). Hence, $\Delta(z) \approx \Delta_E$ is a reasonable approximation for nanofilms, in both the resonant and off-resonant regimes. The only important feature of the resonant points is that the profile of $\Delta(z)$ is simply shifted up as a whole. At the same time the number of oscillations increases by one (see discussion in Ref. [20]).

It is worth noting that $\Delta(z)$ cannot be interpreted as the Cooper pair density when $d \ll \xi_0$ (on the nanoscale) with ξ_0 the bulk Cooper-pair radius (the Pippard length). This is demonstrated in Fig. 8.4, where the order

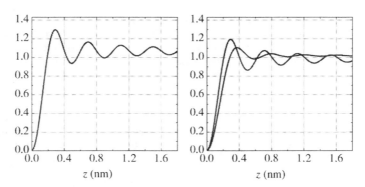

Fig. 8.4. The order parameter $\Delta(z)$ (in units of Δ_{bulk}) together with the electron density (in units of the bulk density n_{bulk}) for (a) the resonant thickness $d = 14.6\,\text{nm}$ and for (b) the off-resonant thickness $d = 14.9\,\text{nm}$, as calculated for tin.

parameter (plotted in units of Δ_{bulk}) is compared with the electron density $n(z)$ (given in units of n_{bulk}, the bulk electron density) at zero tempera-ture for the resonant (a) and off-resonant (b) thicknesses: $d = 14.6\,\text{nm}$ and $d = 14.9\,\text{nm}$, respectively. Contrary to $\Delta(z)$, the spatially averaged value of $n(z)$ does not vary, being n_{bulk}. Another thing is that both quantities $\Delta(z)/\Delta_{\text{bulk}}$ and $n(z)/n_{\text{bulk}}$ exhibit oscillations with period $\lambda_F/2$ (or about $\lambda_F/2$, see previous paragraph). However, the oscillations in $n(z)$ are less pro-nounced and decay quickly when moving towards the center of the sample. The tiny oscillations in $\Delta(z)$ are also a quantum-size effect. However, they are not so important and interesting as the superconducting resonances resulting in enhancement of the spatially averaged order parameter (the energy gap and the critical temperature).

To observe quantum-size superconducting oscillations, experimental groups grow thin single-crystal films with atomic-scale uniformity in thickness. Nanofilms made of most of the elemental superconductors grow in a granular or amorphous fashion on insulating substrates. However, Pb is very good in this sense because under certain growth conditions it forms crystalline and extraordinarily clean films [1–3]. Photo-electron spectroscopy demonstrates that in Pb films grown on silicon (111) the period of quantum-size oscil-lations in the density of single-electron states at the Fermi level is about 2 ML [1, 21, 22]. So, one can expect oscillations in the superconducting prop-erties between Pb(111) films with even and odd numbers of monolayers. This expectation was recently confirmed in the experiments [1, 3], and experimen-tal results were well reproduced by our group using the BdG equations [20]. For a clear observation of the quantum-size superconducting oscillations, experimentalists use very narrow films with thickness of about $N \sim 10$ mono-layers ($N = d/a$, where the lattice spacing $a = 0.286\,\text{nm}$ [22] for Pb(111) films). In this case a substrate and possible protecting coverage can change the electron-phonon coupling (with respect to bulk) due to interface effects [26]. For instance, in ultrathin films ($N < 12 - 16\,\text{ML}$) of Ag on Fe(100) and V(100) substrates the electron-phonon coupling was found to be significantly larger than in bulk and decreases down to the bulk value as the film thickness increases [26,27]. Deviations of the coupling constant from its bulk limit follow approximately an overall $1/N$-dependence [26] and exhibit damped oscilla-tions with film thickness [26,27]. This $1/N$-dependence can be understood (see [28]) as due to the fact that the relative number of film atoms at the interface is proportional to $1/N$. In the case of Pb(111) ultrathin flat terraces grown on sil-icon (111), the electron-phonon coupling gradually increases towards the bulk value as the film gets thicker [1]. Oscillations of the coupling constant with film thickness are clearly present here, as well. Keeping in mind these results, we can assume that for Pb(111) ultrathin films grown on a silicon substrate the dependence of the electron-phonon coupling constant on the film thickness can generally be approximated (for more details, see our work [17]) as

$$g = g_0 - \frac{g_1(2k_F a N)}{N}, \qquad (8.10)$$

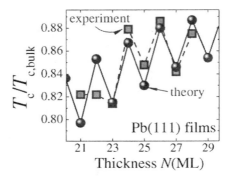

Fig. 8.5. The even-odd oscillations of the relative critical temperature $T_c/T_{c,\text{bulk}}$ in Pb nanofilms grown on silicon (111): empty circles represent the experimental data from Ref. [1], solid circles are numerical results from the BdG equations.

where $g_0 N(0) = 0.39$ (this provides the correct bulk limit for Pb) and $g_1(x)$ is a function oscillating with period 2π. Note that for $d > 5 \div 10\,\text{nm}$ the coupling approaches its bulk value. Equation (8.10) makes it possible to reproduce the even-odd oscillations observed in Pb on silicon (111) (see Fig. 8.5 and Ref. [20]). However, recently another experimental group reported about persistent quantum-size superconducting oscillations in Pb films grown on silicon (111) down to the lowest thickness five monolayers without any sign of suppression. This suggests that the decrease in the electron-phonon coupling in ultrathin Pb(111) nanofilms reported in Ref. [1] may be of a different origin than the one due to a silicon substrate (for instance, the influence of small disorder). The situation is even more controversial because a suppression of the critical temperature in Pb(111) nanofilms with thickness 5–18 monolayers was recently reported in Ref. [2]. This problem deserves a further study.

Another interesting systems are films made of strongly coupling grains. In this case some of the physical properties can be nearly independent of the grain size and, to some extent, such nanofilms can be treated as highly crystalline. In particular, there is no indication that the grain size has a strong influence on the energy gap measured in very recent experiments with Al superconducting nanofilms [29]. The energy gap was shown to enhance with decreasing film thickness. The film thickness in the experiment was not uniform on the atomic scale, and only its averaged value was known. Hence, to explain these experimental data, theoretical results found for a slab being uniform in thickness, should be somewhat averaged over its thickness variations. Such averaging procedure requires much additional information. Strictly speaking, one should solve the BdG equations for a complex geometry modelling a film with relevant thickness fluctuations. Fortunately, the energy gap is larger than its bulk value at the resonant points but stays about Δ_{bulk} for the off-resonant thicknesses. Thus, it is reasonable to assume that the averaged

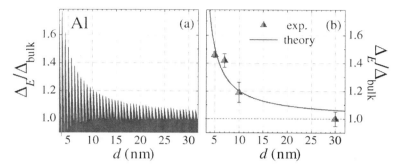

Fig. 8.6. The thickness-dependent energy gap in Al nanofilms: (a) theoretical results from the BdG equations, and (b) experimental results from Ref. [29] (triangles) together with the solid curve being the general trend of the superconducting resonances.

theoretical Δ_E should exhibit an enhancement (when decreasing film thickness) in correlation with the resonant amplitudes. Figure 8.6 shows that the experimental data from Ref. [29] are indeed in good agreement with the solid curve being the general trend of the resonant enhancements calculated for aluminum nanofilms with uniform thickness ar zero temperature. We remark that when plotting the experimental data in Fig. 8.6b, we used the measured value of the energy gap for a thick 30 nm nanofilm [29] as the bulk limit.

8.5 Superconducting nanowires in the quantum-size regime

For nanowires we considered a square-cross-section sample and a cylinder one. The quantum-confinement boundary conditions given by Eq. (8.7) are applied for the transverse electron motion in the x and y directions, and periodic boundary conditions are used in the longitudinal z direction (typical periodicity of $L_z = 0.5 \div 50\,\mu\text{m}$). Figure 8.7a shows quantum-size oscillations of Δ_E for an Al square-cross-section nanowire ($D = L_x = L_y$ denotes the wire thickness) as calculated from the BdG equations ($T = 0$). The gap exhibits pronounced resonant enhancements: it increases well above the bulk value at the resonant thicknesses and, then, decreases up to the point when a new resonance comes into play. As compared to Δ_{bulk}, resonant enhancements are again more significant than drops below Δ_{bulk} between the superconducting resonances. In general, we conclude that the quantum-size oscillations in superconducting metallic nanowires are much more pronounced in comparison to those in nanofilms. This is in agreement with the expectation that quantum-confinement effects are more significant for nanowires. The quantization of the transverse electron motion results in a nonuniform spatial distribution of the superconducting order parameter. Contrary to nanofilms, this distribution

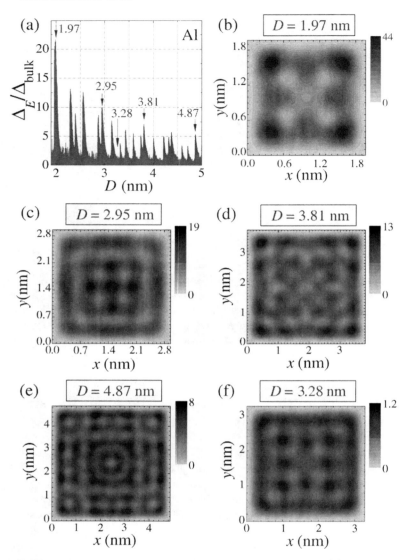

Fig. 8.7. The square-cross-section superconducting Al nanowire: (a) the energy gap Δ_E in units of the bulk gap $\Delta_{\rm bulk}$ versus nanowire thickness D, and the contour plots of the transverse spatial distribution of the order parameter $\Delta(\mathbf{r})$ (in units of $\Delta_{\rm bulk}$) for the resonant thicknesses $D = 1.97$ (b), 2.95 (c), 3.81 (d) and 4.87 nm (e) and for the off-resonant thickness $D = 3.28$ nm (f).

shows spectacular patterns reflecting a complex interplay of pairing with quantum confinement. In Fig. 8.7b–f the contour plots of the transverse distribution of $\Delta(\mathbf{r})/\Delta_{\rm bulk}$ in an Al square-cross-section nanowire are presented for the resonant thicknesses $D = 1.97$, 2.95, 3.81 and 4.87 nm [(b)–(e), respectively] and for the off-resonant point $D = 3.28$ nm (f). In the areas, where the order parameter is enhanced, it significantly exceeds its bulk value for the resonant thicknesses. Even for the off-resonant regime (see Fig. 8.7d), there are domains where $\Delta(\mathbf{r})$ exceeds its bulk value. A distinctive feature of superconducting nanowires is a strongly nonuniform spatial distribution of the pair condensate. The spatially averaged value of the order parameter $\bar{\Delta}$ is enhanced at the resonant points. However, such enhancement now occurs due to the presence of significant local enhancements in $\Delta(\mathbf{r})$, and the approximation $\Delta(\mathbf{r}) \approx \Delta_E$ is not valid (compare with nanofilms). In samples with larger D the pair condensate exhibits a more uniform pattern of spatial distribution. When $D \to \infty$ quantum-size oscillations decay, and we obtain the bulk limit.

It is instructive to compare the results calculated for Al with those for Sn which are presented in Fig. 8.8. Quantum-size oscillations and resonant increases are less significant for Sn as compared to Al by a factor of about 2 (compare Figs. 8.7a and 8.8b). This is agreement with the corresponding results for nanofilms (see Figs. 8.3a and c). The density of the resonances per unit thickness is larger for Sn, as well. However, the superconducting resonances are no longer equidistant owing to the more complicated single-electron spectrum. We also notice that the width of the peaks become smaller when passing to Sn.

In Fig. 8.9 our results from a numerical solution of the BdG equations for T_c of Al and Sn nanocylinders are compared with recent experimental data for high-quality Al and Sn nanowires (D is the diameter here). A remarkable feature of such superconducting nanostructures is that they exhibit a thickness-dependent increase of T_c rather than a suppression typical for sufficiently disordered samples. By analogy with our discussion about aluminum nanofilms, we can expect that inevitable thickness fluctuations in real nanowires will lead to a smoothing of the quantum-size oscillations so that only an overall increase of the experimental critical temperature with decreasing thickness survives. For Al we have at our disposal polycrystalline nanorods made of strongly coupled grains [4, 30], like nanofilms reported in Ref. [29]. There is practically no tunnel barrier between the grains in this case and the electron mean free path is about the nanowire thickness. Such superconducting nanostructures were fabricated by evaporating of pure aluminum on a silicon substrate and using conventional e-beam lithography. Then, Ar^+ ion sputtering was used to progressively reduce the sample cross section [4]. In Figs. 8.9a and b we compare our numerical results with the experimental data on T_c in such Al nanowires [30]. The critical temperature for the thickest wire which had an effective diameter of about $D = 116$ nm is here taken as $T_{c,\text{bulk}}$. Note that in Fig. 8.9 we consider samples with $D > 15$ nm. This allows one to neglect possible modification of the electron-phonon coupling due to the presence of the substrate (see our discussion on Pb nanofilms). For Sn we can use data for single-crystalline samples produced by electrodepositing into porous membranes [5]. Another interesting technique is based on catalitic decomposition of hot acetylene over solid tin dioxide, which yields individual

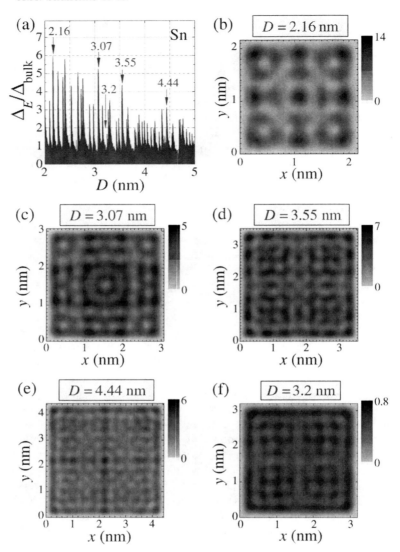

Fig. 8.8. The same as in Fig. 8.1 but now for Sn (the contour plots are given for the resonant points $D = 2.16$ (b), 3.07 (c), 3.55 (d), 4.44 nm (e) and for the off-resonant thickness width $D = 3.2$ nm (f).

micrometer-long carbon nanotubes filled with highly pure, single-crystalline, superconducting tin nanowires [6]. In Figs. 8.9c and d we plot our theoretical results for T_c in tin nanocylinders together with experimental results for both fabrication methods of tin nanowires. The critical temperature for the $D = 70$ nm is treated as $T_{c,\text{bulk}}$ here. As follows from Fig. 8.9, there is good

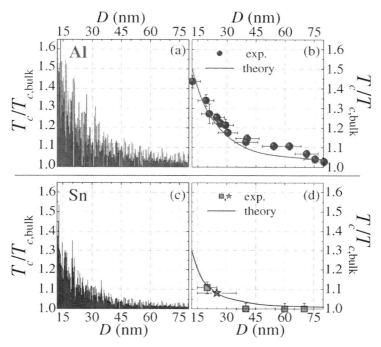

Fig. 8.9. The thickness-dependent enhancement of T_c in metallic nanocylinders: (a) theoretical results from the BdG equations for Al, and (b) experimental results from Ref. [30] (solid circles) versus the general trend of the resonant enhancements given in the previous panel (solid curve); (c) and (d), the same but for Sn (squares and the star are experimental results from Ref. [5] and Ref. [6], respectively).

agreement between the experimental data and the resonant enhancements calculated from the BdG equations for cylindrical nanowires with uniform cross section. This strongly suggests that the thickness-dependent increase of T_c found in highly crystalline Al and Sn nanowires is a signature of the quantum-size regime.

8.6 New Andreev-type states induced by quantum confinement

Quasiparticles "feel" the spatial variation of the superconducting order parameter as a kind of potential barrier. This physical mechanism is the basis for Andreev quantization, and can be referred to as the Andreev mechanism. Andreev states were investigated previously: (i) for an isolated normal region of the intermediate state of a type-I superconductor [31] (or for a similar case of SNS contacts [33]), and (ii) in the core of a single vortex for the mixed state of a type-II superconductor [32]. Both situations are characterized by the bulk superconducting exterior: $\Delta(\mathbf{r}) \to \Delta_{\text{bulk}}$ when $|\mathbf{r}| \to \infty$. The exterior of a nanoscale specimen is different due to quantum confinement. As follows from our numerical results for clean superconducting nanowires, in this case the Andreev mechanism manifests itself through the formation of new Andreev-type states that appear due to spatial inhomogeneity of the superconducting condensate. These new Andreev-type states induced by quantum confinement are mainly located beyond the regions where the order parameter is enhanced. We remark that such states cannot be localized in the domains where the order parameter is significantly suppressed because the characteristic length for spatial variations of the order parameter in the case of interest is about the sample thickness $d \ll \xi$, with ξ the Ginzburg-Landau coherence length.

To discuss this in more detail, let us consider a cylindrical aluminum nanowire. For a cylindrical geometry we have $i = \{j, m, k\}$, where j is associated with the transverse cylindrical coordinate, m is the azimuthal quantum number, and k is the wave vector of the quasi-free electron motion along the nanowire. For the particle-like and the hole-like wave functions we get

$$\langle \mathbf{r}|u_{jmk}\rangle = u_{jmk}(\rho) \frac{e^{im\varphi}}{\sqrt{2\pi}} \frac{e^{ikz}}{\sqrt{L}}, \tag{8.11}$$

$$\langle \mathbf{r}|v_{jmk}\rangle = v_{jmk}(\rho) \frac{e^{im\varphi}}{\sqrt{2\pi}} \frac{e^{ikz}}{\sqrt{L}}, \tag{8.12}$$

with ρ, φ, z the cylindrical coordinates, and $\Delta(\mathbf{r}) = \Delta(\rho)$. The probability density to locate the electron-type and hole-type quasiparticles along the transverse coordinate ρ are given by $|u_{jmk}(\rho)|^2$ and $|v_{jmk}(\rho)|^2$, respectively. Our numerical analysis shows that $u_{jmk}(\rho)$ and $v_{jmk}(\rho)$ are nearly proportional to each other. An example of the transverse distribution of the pair condensate is given in Fig. 8.10a. For the sake of simplicity, here we take the extremely narrow wire with $D = 1.74\,\text{nm}$ (the resonant thickness). In Fig. 8.10a we plot $\Delta(\rho)$ versus ρ/R ($R = D/2$) as obtained from a numerical self-consistent solution of Eqs. (8.1) and (8.2) at $T = 0$. As seen, the $\Delta(\rho)$ is strongly enhanced and nonuniform ($\Delta_{\text{bulk}} = 0.25\,\text{meV}$ for the chosen parameters). In this case only the quasiparticle states from the subbands with $j = 0, m = 0$, $j = 0, |m| = 1$ and $j = 0, |m| = 2$ make a contribution to the order parameter. The dominant contribution comes from the states with $j = 0, m = -2$ and $j = 0, m = 2$ that control the local enhancement of the order parameter at $\rho = 0.58\,R$ ($d = 2R$). Indeed, as follows from Fig. 8.10b, the profile of $|u_{jkm}(\rho)|^2$ for $j = 0, |m| = 2$ is similar to the profile of $\Delta(\rho)$ in the region where $\Delta(\rho)$ is enhanced. Contributions from the quasiparticle states with $j = 0, |m| = 1$ and $j = 0, m = 0$ are minor due to a smaller density of states. These quasiparticles are mainly located outside the area of the great local enhancement of the order parameter and responsible for the values of $\Delta(\rho)$ at $\rho/R < 0.58$, thus forming an exterior for the local peak in $\Delta(\rho)$. We refer to them as the *new Andreev-type states* induced by quantum confinement.

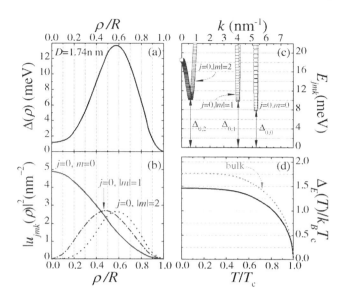

Fig. 8.10. Cylindrical aluminum nanowire at the resonant diameter $D = 1.74$ nm: (a) the order parameter versus ρ/R ($R = D/2$) at $T = 0$; (b) the corresponding quasiparticle transverse distributions given by $|u_{jmk}(\rho)|^2$ for the five relevant branches (the lowest energy state is given for each quasiparticle branch) together with (c) the quasiparticle energies; (d) the temperature-dependent energy gap $\Delta_E(T)$ in units of $k_B T_c$ versus the relative temperature T/T_c.

In Fig. 8.10c the corresponding quasiparticle energies are plotted as function of the wave vector k. As seen, the energy gaps $\Delta_{0,2}$ and $\Delta_{0,-2}$ are larger than $\Delta_{0,0}$, $\Delta_{0,1}$ and $\Delta_{0,-1}$, and this difference is due to the Andreev mechanism. In particular, the quasiparticles with $j = 0, |m| = 1$ are mainly located in the nearest vicinity of the local enhancement of the order parameter, and, as a result, $\Delta_{0,1}$ and $\Delta_{0,-1}$ are below but nearly the same as $\Delta_{0,2}$. As to the states with $j = 0, m = 0$, they are more successive in avoiding the area of the enhancement, and, so, $\Delta_{0,0}$ is smaller than $\Delta_{0,2}$ by 20%. It turns out that an important consequence of the Andreev mechanism is a modification of the ratio of the temperature-dependent energy gap $\Delta_E(T)$ to the critical temperature. In Fig. 8.10c $\Delta_E(T)/k_B T_c$ is plotted versus T/T_c for $D = 1.74$ nm. One can see that $\Delta_E(T)/k_B T_c$ is smaller as compared to its bulk value. The point is that T_c is mainly determined by the states with $j = 0, |m| = 2$ whereas the total energy gap is governed by the Andreev-type states with $j = 0, m = 0$. The ratio $\Delta_E/k_B T_c$ (at $T = 0$) is an oscillating function of d, approaching 1.76 from below with increasing thickness (see our recent work [34] and Fig. 8.11).

For the case of $D \approx 6 - 10$ nm, corresponding to the narrowest superconducting nanowires fabricated in recent experiments [4–6], this ratio still differs with about 10% from the bulk value. Thus, it can be probed, e.g., through

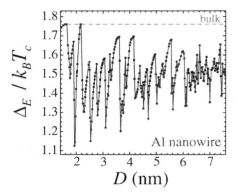

Fig. 8.11. The ratio $\Delta_E/k_B T_c$ versus the diameter D of a cylindrical aluminum nanowire at zero temperature.

tunnelling experiments. Concluding this section, we noticed that the new Andreev-type states play no serious role in superconducting nanofilms (see Fig. 8.2b). As seen from the results discussed in Sec. 8.4, the order parameter in films is nearly uniform as compared with nanowires. This conclusion is in agreement with the experimental results for Pb highly-crystalline nanofilms where the ratio $\Delta_E(T)/k_B T_c$ was found to be very close to the bulk one [3].

8.7 Superconducting-to-normal transition induced by a magnetic field

One more interesting aspect of superconductivity in the quantum-size regime concerns the magnetic-field dependent superconducting-to-normal (SN) phase transition. Let us consider a superconducting nanowire in a parallel magnetic field and limit ourselves to the Meissner state (vortices can hardly appear in nanowires). According to the Ginzburg-Landau (GL) theory [35, 36]: (i) the critical magnetic field should increase as $1/D$ in mesoscopic wires, and (ii) the SN phase transition driven by a parallel magnetic field is of second order for narrow mesoscopic wires while of first order in bulk type-I superconductors [18]. Recent calculations within the BdG equations for wires with $D = 20 \div 200$ nm [37] confirmed the GL conclusion about the second order transition, which is in agreement with recent experimental data for Sn [5, 6] and Zn [30] with $D > 20$ nm.

The situation changes dramatically on the nanoscale. Based on a numerical solution of the Bogoliubov-de Gennes equations for a clean cylindrical metallic nanowire in a parallel magnetic field, we can show that: (1) the transition in question, occurs as a cascade of first-order phase transitions for diameters $D < 10 \div 15$ nm, (2) the critical field is strongly enhanced, and (3) exhibits large quantum-size oscillations. The Pauli paramagnetism is found to be significant for smaller diameters $D < 5\,\mathrm{nm}$. Figure 8.12 shows an example of quantum-size oscillations of the critical parallel magnetic field $H_{c,\|}$ (without taking account of the Pauli paramagnetism) calculated from the BdG equations for aluminum nanocylinder at $T = 0$. Note that $H_{c,\|}$ is set as the magnetic field above which the spatially averaged order parameter $\bar{\Delta}$ drops below $0.01\Delta_{\mathrm{bulk}}$, with Δ_{bulk} the bulk gap. At a resonant point the main contribution to the superconducting quantities comes from the subband (or subbands) whose bottom passes through the Fermi surface. In our case the subbands with the same $|m|$ are degenerate at $H_\| = 0$ and, hence, any size-dependent resonant enhancement of the order parameter (e.g., the energy gap and critical temperature) is specified by the set $(j, |m|)$ at $H_\| = 0$. Due to quantum-size oscillations in the pair-condensation energy, we get corresponding oscillations in the critical magnetic field whose resonances can also be labelled by $(j, |m|)$. As seen from Fig. 8.12, $H_{c,\|}$ exhibits huge enhancements as compared to the bulk critical magnetic field $H_{c,\mathrm{bulk}} = 0.01\,\mathrm{T}$ (to simplify our discussion, here we show our results for extremely narrow wires). Resonances in $H_{c,\|}$ are very dependent on D and $|m|$. The states with large $|m|$ are strongly influenced by $H_\|$ and, so, the resonances in $H_{c,\|}$ governed by large $|m|$ are, as a rule, much less pronounced. In contrast, the resonances controlled by $m = 0$ are very stable (see Eq. (8.13)). For instance, a superconducting solution to Eqs. (8.1) and (8.2) exists at $D = 1.94\,\mathrm{nm}$ (the resonance associated with $(j, |m|) = (1, 0)$) even for an abnormally large magnetic field of about $1{,}000\,\mathrm{T}$. Similar behavior is found for the resonance at $D = 3.21\,\mathrm{nm}$ with $(j, m) = (2, 0)$. Note that in Fig. 8.12 two neighboring resonances with $(j, m) = (2, 0)$ ($D = 3.21\,\mathrm{nm}$) and $(j, |m|) = (1, 5)$ ($D = 3.28\,\mathrm{nm}$) merge and result in one profound increase in $H_{c,\|}$.

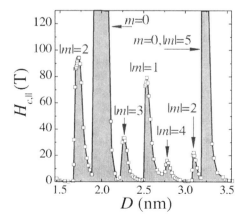

Fig. 8.12. Quantum-size oscillations of $H_{c,\|}$ as calculated from the BdG equations for an aluminum nanocylinder at $T = 0$.

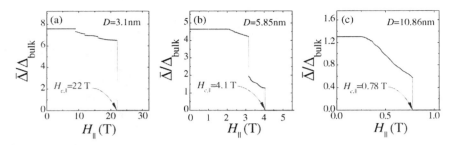

Fig. 8.13. Cylindrical aluminum nanowire for the resonant diameters $D = 3.1$ (a), 5.85 (b) and 10.86 nm (c): the spatially averaged order parameter $\bar{\Delta}$ given in units of Δ_{bulk} (taken for $H_{||} = 0$) versus the parallel critical magnetic field $H_{||}$ at zero temperature.

In Fig. 8.13 the spatially averaged order parameter $\bar{\Delta}$ is plotted versus the parallel magnetic field $H_{||}$ for the resonant diameters $D = 3.1$, 5.85 and 10.86 nm (the Pauli paramagnetism is not included). In the presence of a magnetic field the quasiparticle energies in a cylindrical confining geometry are well approximated by (see also Eqs. (8.11) and (8.12))

$$E_{jmk} = \sqrt{\xi_{jmk}^2 + \Delta_{jm}^2} - m\mu_B H_{||}, \tag{8.13}$$

where μ_B is the Bohr magneton, ξ_{jmk} is the single-electron energy for zero magnetic field (see Eq. (8.6). Due to the term $-m\mu_B H_{||}$ (compare with Eq. (8.9)) the subband energy gaps are no longer Δ_{jm}. This is why the total energy gap Δ_E can be zero even when $\Delta_{jm} \neq 0$. So, it is more convenient to work with the spatially averaged order parameter $\bar{\Delta} = \int \Delta(\rho)\rho d\rho$. As seen from Fig. 8.13, $\bar{\Delta}$ exhibits a jump down to zero at $H_{||} = H_{c,||}$. It is also of interest that there are additional jumps for $H_{||} < H_{c,||}$ (some of them are very small). This *cascade of first order transitions* appears due to the transverse quantization of the electron motion: a jump in $\bar{\Delta}$ arises each time when a new quasiparticle branch touches zero, and the system ground state is modified due to unpairing of electrons from the corresponding subband. It is of importance to note that clear signatures of the cascade of first-order transitions are present even in the off-resonant regime, as seen from Fig. 8.14a where $\bar{\Delta}$ is plotted versus $H_{||}$ for the off-resonant diameter $D = 8$ nm. Figure 8.14b shows the dependence of the superconducting order parameter $\Delta(\rho)$ on the transverse coordinate ρ for various magnetic fields at the same diameter. As seen, diamagnetic current "eat" the order parameter with increasing $H_{||}$, and this process is more intensive near the sample edge, where the diamagnetic current is more significant.

Notice that $H_{c,||}$ has been measured in Sn [5] and Zn [38] wires with diameters down to 20 nm. These wires are still in the mesoscopic regime. It is expected that data on $H_{c,||}$ for $D < 20$ nm will be available soon. For such

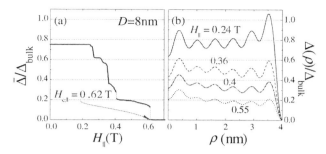

Fig. 8.14. Cylindrical aluminum nanowire at the off-resonant point $D = 8\,\text{nm}$: (a) $\bar{\Delta}/\Delta_{\text{bulk}}$ versus $H_{||}$ and (b) $\Delta(\rho)/\Delta_{\text{bulk}}$ given for $H_{||} = 0.24, 0.36, 0.4$ and $0.55\,\text{T}$.

narrow nanowires we predict the SN transition driven by $H_{||}$ as a cascade of first order transitions and an oscillating (but smoothed) enhancement of $H_{c,||}$.

Concluding this section, we remark that a similar cascade should exist for the SN transition driven by a magnetic field perpendicular to the nanowire. In addition, one can expect the same physics behind a suppression of super-conductivity by a magnetic field in high-quality metallic nanofilms. For perpendicular magnetic field, the situation can be more complicated due to the presence of vortices in this case.

8.8 Conclusion

In conclusion, quantum confinement is the major mechanism governing super-conductivity in high-quality nanoscale samples that are in the clean limit. Based on a numerical solution of the Bogoliubov-de Gennes equations, we investigated how the basic superconducting properties can be modified due to quantum confinement in nanofilms and nanowires. In the case of interest the electron band splits up into a series of subbands, which results in quantum-size oscillations of the superconducting characteristics with profound resonant enhancements. At a resonant point, the superconducting order parameter (the energy gap, the critical temperature and magnetic field) can be increased by more than an order of magnitude as compared to bulk. In nanowires the spatial distribution of the pair condensate exhibits distinct nonuniform patterns leading to the formation of new Andreev-type states. This significantly modifies the ratio of the energy gap to the critical temperature. In nanofilms the spatial distribution of the order parameter is much more uniform, and the new Andreev-type states induced by quantum confinement play no sig-nificant role. The superconducting-to-normal transition driven by a magnetic field becomes a cascade of first-order transitions contrary to the second order transition found in the mesoscopic regime. A similar cascade can be expected in the presence of current.

Acknowledgements

This work was supported by the Flemish Science Foundation (FWO-Vl), the Interuniversity Attraction Poles Programme – Belgian State – Belgian Science Policy (IAP), the ESF-AQDJJ network and BOF-TOP (University of Antwerpen).

References

1. Y. Guo, Y. -F. Zhang, X. -Y. Bao, T. -Z. Han, Z. Tang, L. -X. Zhang, W. -G. Zhu, E. G. Wang, Q. Niu, Z. Q. Qiu, J. -F. Jia, Z. -X. Zhao, and Q. K. Xue, Science **306**, 1915 (2004).
2. M. M. Özer, J. R. Thompson, and H. H. Weitering, Nature Physics **2**, 173 (2006); *ibid.*, Phys. Rev. B **74**, 235427 (2006); M. M. Özer, Y. Jia, Z. Zhang, J. R. Thompson, and H. H. Weitering, Science **316**, 1594 (2007).
3. D. Eom, S. Qin, M. -Y. Chou, and C. K. Shih, Phys. Rev. Lett. **96**, 027005 (2006).
4. M. Savolainen, V. Touboltsev, P. Koppinen, K. -P. Riikonen, and K. Arutyunov, Appl. Phys. A, **79**, 1769 (2004); M. Zgirski, K. -P. Riikonen, V. Touboltsev, and K. Arutyunov, Nano Lett. **5**, 1029 (2005).
5. M. L. Tian, J. G. Wang, J. S. Kurtz, Y. Liu, M. H. W. Chan, T. S. Mayer, and T. E. Mallouk, Phys. Rev. B **71**, 104521 (2005).
6. L. Jankovič, D. Gournis, P. N. Trikalitis, I. Arfaoui, T. Cren, P. Rudolf, M. -H. Sage, T. T. M. Palstra, B. Kooi, J. De Hosson, M. A. Karakassides, K. Dimos, A. Moukarika, and T. Bakas, Nano Lett. **6**, 1131 (2006).
7. P. W. Anderson, J. Phys. Chem. Solids **11**, 26 (1959).
8. B. L. Altshuler, A. G. Aronov, and P. A. Lee, Phys. Rev. Lett. **44**, 1288 (1980).
9. B. L. Altshuler, JETP Lett. **41**, 648 (1985); P. A. Lee and A. D. Stone, Phys. Rev. Lett. **55**, 1622 (1985).
10. M. A. Skvortsov and M. V. Feigel'man, Phys. Rev. Lett. **95**, 057002 (2005).
11. Y. Oreg and A. M. Finkelstein, Phys. Rev. Lett. **83**, 191 (1999).
12. K. Harada, O. Kamimura, H. Kasai, T. Matsuda, A. Tonomura, V. V. Moshchalkov, Science **274**, 1167 (1996); V. V. Moshchalkov, M. Baert, V. V. Metlushko, E. Rosseel, M. J. Van Bael, K. Temst, Y. Bruynseraede, and R. Jonckheere, Phys. Rev. B **57**, 3615 (1998).
13. V. V. Moshchalkov, L. Gielen, and C. Strunk et al., Nature (London) 373, 319 (1995); P. S. Deo, V. A. Schweigert, F. M. Peeters, Phys. Rev. Lett. **79**, 4653 (1997); V. A. Schweigert and F. M. Peeters, Phys. Rev. B **57**, 13817 (1998); A. K. Geim, S. V. Dubonos, I. V. Grogorieva, K. S. Novoselov, F. M. Peeters, and V. A. Schweigert, Nature (London) **407**, 55 (2000).
14. J. M. Blatt and C. J. Thompson, Phys. Rev. Lett. **10**, 332 (1963).
15. Yu. F. Komnik, E. I. Bukhshtab, and K. K. Man'kovskii, Sov. Phys. JETP **30**, 807 (1970); B. G. Orr, H. M. Jaeger, and A. M. Goldman, Phys. Rev. Lett. **53**, 2046 (1984); O. Pfennigstorf, A. Pet-kova, H. L. Guenter, and M. Henzler, Phys. Rev. B **65**, 045412 (2002).
16. S. Hilaire, J. -F. Berger, M. Girod, W. Satula, and P. Schuck, Phys. Lett. B **531**, 61 (2002).

17. A. A. Shanenko and M. D. Croitoru, Phys. Rev. B **73**, 012510 (2006); A. A. Shanenko, M. D. Croitoru, M. Zgirski, F. M. Peeters, and K. Arutyunov, Phys. Rev. B **74**, 052502 (2006).
18. P. G. de Gennes, *Superconductivity of Metals and Alloys* (W. A. Bejamin, New York, 1966).
19. N. N. Bogoliubov, Sov. Phys. Usp. **67**, 549 (1959).
20. A. A. Shanenko, M. D. Croitoru, and F. M. Peeters, Europhys. Lett. **76**, 498 (2006); *ibid*, Phys. Rev. B **75**, 014519 (2007).
21. Y. -F. Zhang, J. -F. Jia, T. -Z. Han, Z. Tang, Q. T. Shen, Y. Guo, Z. Q. Qiu, and Q. -K. Xue, Phys. Rev. Lett. **95**, 096802 (2005).
22. C. M. Wei and M. Y. Chou, Phys. Rev. B **66**, 233408 (2002).
23. B. G. Orr, H. M. Jaeger, and A. M. Goldman, Phys. Rev. Lett. **53**, 2046 (1984).
24. L. P. Gor'kov, Sov. Phys. JETP **7**, 505 (1958).
25. L. N. Cooper, Phys. Rev. **104**, 1189 (1956).
26. D. A. Luh, T. Miller, J. J. Paggel, and T. -C. Chiang, Phys. Rev. Lett. **88**, 256802 (2002).
27. T. Valla, M. Kralj, A. Šiber, M. Milun, P. Pervan, P. D. Johnson, and D. P. Woodruff, J. Phys. C **12**, L477 (2000).
28. T. -C. Chiang, Science **306**, 1900 (2004).
29. N. A. Court, A. J. Ferguson, and R. G. Clark, Supercond. Sci. Technol. **21**, 015013 (2008).
30. M. Zgirski and K. Yu. Arutyunov, Phys. Rev. B **75**, 172509 (2007).
31. A. F. Andreev, Sov. Phys. JETP **22**, 455 (1966).
32. C. Caroli, P. G. de Gennes, and J. Matricon, Phys. Lett. **9**, 307 (1964).
33. P. G. de Gennes, and D. Saint-James, Phys. Lett. **4**, 151 (1963).
34. A. A. Shanenko, M. D. Croitoru, R. G. Mints, and F. M. Peeters, Phys. Rev. Lett. **99**, 067007 (2007).
35. V. P. Silin, Zh. Eksp. Teor. Fiz. **21**, 1330 (1951) (in Russian).
36. O. S. Lutes, Phys. Rev. **105**, 1451 (1957).
37. J. E. Han and V. H. Crespi, Phys. Rev. B **69**, 214526 (2004).
38. J. S. Kurtz, R. R. Johnson, M. Tian, N. Kumar, Z. Ma, S. Xu, and M. H. W. Chan, Phys. Rev. Lett. **98**, 247001 (2007).

9

KONDO EFFECT COUPLED
TO SUPERCONDUCTIVITY
IN ULTRASMALL GRAINS

H. NAGAO[1] AND S.P. KRUCHININ[2]
[1] *Division of Mathematical and Physical Science, Graduate School of Natural Science and Technology, Kanazawa University, Kakuma, Kanazawa 920-1192, Japan.* nagao@wriron1.s.kanazawa-u.ac.jp
[2] *Bogolyubov Institute for Theoretical Physics, The Ukrainian National Academy of Science, Kiev 03143, Ukraine.* skruchin@i.com.ua

Abstract. We investigate the Kondo effect and superconductivity in ultrasmall grains by using a model, which consists of sd and BCS Hamiltonians with the introduction of a pseudofermion. We discuss physical properties of the condensation energy and behavior of the gap function and the spin singlet order parameter corresponding to the Kondo effect in relation to the critical level spacing and co-existence. We find that strong local magnetic moments from the impurities makes the transition temperature for superconductivity reduced. However, weak couplings λ of the superconductivity do not destroy the spin singlet order parameter at all. Finally we derive the exact equation for the Kondo regime in nanosystem and discuss the condensation energy from the viewpoint of the energy level.

Keywords: Kondo effect, superconductivity, ultrasmall grains

9.1 Introduction

The Kondo effect has been much attracted great interest in the properties in semiconductor quantum dots. The Kondo effect can be understood as a magnetic exchange interaction between a localized impurity spin and free conduction electrons [1]. To minimize the exchange energy, the conduction electrons tend to screen the spin of the magnetic impurity and the ensemble forms a spin singlet. In a quantum dot, some exotic properties of the Kondo effect have been observed [2–4]. Recently, Sasaki et al. has found a large Kondo effect in a quantum dots with an even number of electrons [5]. The spacing of discrete levels in such quantum dots is comparable with the strength of electron-electron Coulomb interaction. The Kondo effect in multilevel

J. Bonča, S. Kruchinin (eds.), *Electron Transport in Nanosystems.*
© Springer Science + Business Media B.V. 2008

quantum dots has been investigated theoretically by several groups [6–8]. They have shown that the contribution from multilevels enhances the Kondo effect in normal metals. There are some investigations on the Kondo effect in quantum dots coupled ferromagnetism [9], noncollinear magnetism [10], superconductivity [11] and so on [12, 13].

Properties of ultrasmall superconducting grains have been also theoretically investigated by many groups [15–24]. Black et al. have revealed the presence of a parity dependent spectroscopic gap in tunnelling spectra of nanosize Al grains [15, 16]. In such ultrasmall superconducting grains, the bulk gap has been discussed in relation to physical properties such as the parity gap [21], condensation energy [22], electron correlation [23] with the dependence of level spacing of samples [24]. In previous works [25], we have also discussed physical properties such as condensation energy, parity gap, and electron correlation of two-gap superconductivity in relation to the size dependence and effective pair scattering process. The possibility of new two-gap superconductivity has been also discussed by many groups [26–37].

In a standard s-wave superconductor, the electrons form pairs with antialigned spins and are in a singlet state as well. When the superconductivity and Kondo effect present simultaneously, the Kondo effect and superconductivity are usually expected to be competing physical phenomena. The local magnetic moments from the impurities tend to align the spins of the electron pairs in the superconductor which often results in a strongly reduced transition temperature. Buitelaar et al. have experimentally investigated the Kondo effect in a carbon nanotube quantum dot coupled to superconducting Au/Al leads [11]. The have found that the superconductivity of the leads does not destroy the Kondo correlations on the quantum dot when the Kondo temperature. A more subtle interplay has been proposed for exotic and less well-understood materials such as heavy-fermion superconductors in which both effects might actually coexist [38].

In this paper, we investigate the Kondo effect and superconductivity in ultrasmall grains by using a model, which consists of sd and reduced BCS Hamiltonians with the introduction of a pseudofermion. A mean field approximation for the model is introduced, and we calculate physical properties of the critical level spacing and the condensation energy. These physical properties are discussed in relation to the coexistence of both the superconductivity and the Kondo regime. Finally we derive the exact equation for the Kondo regime in nanosystem and discuss the condensation energy from the viewpoint of the correlation energy.

9.2 Kondo regime coupled to superconductivity

In nanosize superconducting grains, the quantum level spacing approaches the superconducting gap. It is necessary to treat discretized energy levels of the small system. For ultrasmall superconducting grains, we can consider the pairing-force Hamiltonian to describe electronic structure of the system [39] and can know the critical level spacing where the superconducting gap function vanishes at a quantum level spacing [24]. In this section, we present a model for a system in Kondo regime coupled to superconductivity and discuss physical properties such as critical level spacing and condensation energy by using a mean field approximation in relation to gap function, spin singlet order as the Kondo effect, coexisitence and so on.

9.2.1 MODEL

We consider a model coupled to superconductivity for quantum dots to investigate the Kondo effect in normal metals, which can be expressed by the effective low-energy Hamiltonian obtained by the Schrieffer-Wolff transformation [40]:

$$H = H_0 + H_1 + H_2, \qquad (9.1)$$

where

$$H_0 = \sum_{k,\sigma} \varepsilon_k a^\dagger_{k\sigma} a_{k\sigma} + \sum_\sigma E_\sigma d^\dagger_\sigma d_\sigma, \qquad (9.2)$$

$$H_1 = J \sum_{k,k'} \left[S_+ a^\dagger_{k'\downarrow} a_{k\uparrow} + S_- a^\dagger_{k'\uparrow} a_{k\downarrow} + S_z \left(a^\dagger_{k'\uparrow} a_{k\uparrow} - a^\dagger_{k'\downarrow} a_{k\downarrow} \right) \right], \qquad (9.3)$$

$$H_2 = -g \sum_{k,k'} a^\dagger_{k\uparrow} a^\dagger_{k\downarrow} a_{k'\downarrow} a_{k'\uparrow}. \qquad (9.4)$$

$a^\dagger_{k\sigma}$ ($a_{k\sigma}$) and $d^\dagger_\sigma$ (d_σ) are the creation (annihilation) operator corresponding to conduction electrons and the effective magnetic particle as an impurity, respectively. In this study we assume the magnetic particle is fermion $S = 1/2$ for the simplicity. E means an extraction energy given by $E_{\uparrow,\downarrow} = -E_0 \pm E_z$ included the Zeeman effect. The second term in Eq. (9.1) means the interaction between conduction electrons and the spin in a quantum dot. S is the spin operator as $S_+ = d^\dagger_\uparrow d_\downarrow$, $S_- = d^\dagger_\downarrow d_\uparrow$, and $S_z = (d^\dagger_\uparrow d_\uparrow - d^\dagger_\downarrow d_\downarrow)/2$. The third term corresponds to the interaction between conduction electrons from the pairing force Hamiltonian.

Here, we introduce a pseudofermion for the magnetic particle operator [41] as

$$d^\dagger_\uparrow = f_\downarrow, d_\uparrow = f^\dagger_\downarrow,$$

$$d^\dagger_\downarrow = -f_\uparrow, d_\downarrow = -f^\dagger_\uparrow. \qquad (9.5)$$

In this transformation, we have the condition

$$f^\dagger_\uparrow f_\uparrow + f^\dagger_\downarrow f_\downarrow = 1. \qquad (9.6)$$

We can know $|\sigma\rangle = f^\dagger_\sigma |0\rangle$. The spin operator S can be rewritten as $S_+ = f^\dagger_\uparrow f_\downarrow$, $S_- = f^\dagger_\downarrow f_\uparrow$, and $S_z = (f^\dagger_\uparrow f_\uparrow - f^\dagger_\downarrow f_\downarrow)/2$. The Hamiltonian can be rewritten as

$$H_0 = \sum_{k,\sigma} \tilde{\varepsilon}_k c_{k\sigma}^\dagger c_{k\sigma} + \sum_{\sigma} E f_\sigma^\dagger f_\sigma, \qquad (9.7)$$

$$H_1 = J \sum_{k,k',\sigma,\sigma'} f_\sigma^\dagger f_{\sigma'} c_{k'\sigma'}^\dagger c_{k\sigma}, \qquad (9.8)$$

$$H_2 = -g \sum_{k,k'} c_{k\uparrow}^\dagger c_{k\downarrow}^\dagger c_{k'\downarrow} c_{k'\uparrow}, \qquad (9.9)$$

where $c_{k\sigma} = \sum_i U_{ik} a_{i\sigma}$ with $\tilde{\varepsilon}_k = \sum_{i,j} U_{ki}^\dagger [\varepsilon_i \delta_{ij} - J/2] U_{jk}$. For the simplicity, we only focus $E_z = 0$ without an external magnetic field: $E = E_0$.

9.2.2 MEAN FIELD APPROXIMATION

In this section, we introduce a mean field approximation for the present Hamiltonian of Eq. (9.1). Eto et al. have presented the mean field approximation for the Kondo effect in quantum dots [42].

In the mean field approximation, we can introduce the spin singlet order parameter

$$\Xi = \frac{1}{\sqrt{2}} \sum_{k,\sigma} \langle f_\sigma^\dagger c_{k\sigma} \rangle. \qquad (9.10)$$

This order parameter describes the spin couplings between the dot states and conduction electrons. The superconducting gap function can be expressed as

$$\Delta = \sum_k \langle c_{k\downarrow} c_{k\uparrow} \rangle. \qquad (9.11)$$

Using these order parameters in Eqs. (9.8) and (9.9), we obtain the mean-field Hamiltonian

$$H_{\mathrm{MF}} = \sum_{k,\sigma} \tilde{\varepsilon}_k c_{k\sigma}^\dagger c_{k\sigma} + \sum_\sigma \tilde{E} f_\sigma^\dagger f_\sigma + \sqrt{2} J \sum_{k,\sigma} \left[\Xi f_\sigma c_{k\sigma}^\dagger + \Xi^* c_{k\sigma} f_\sigma^\dagger \right]$$

$$- g \sum_k \left[\Delta^* c_{k\downarrow} c_{k\uparrow} + \Delta c_{k\uparrow}^\dagger c_{k\downarrow}^\dagger \right]. \qquad (9.12)$$

The constraint of Eq. (9.6) is taken into account by the second term with a Lagrange multiplier λ. In this study, we assume a constant density of state with the energy region of the Deby energy, and the coupling constants can be expressed as $J = d\tilde{J}$ and $g = d\lambda$.

9.3 Discussion

By minimizing the expectation value of H_{MF} in Eq. (9.12), the order parameters are determined self-consistently. First, we show the Kondo effect without the pairing force part ($g = 0$) in the framework of the mean field approximation. Next, the Kondo effect in the presence of the superconductivity is discussed in relation to the critical level spacing and condensation energy. Finally, we derive the exact equation for the Kondo effect in ultrasmall grains coupled to normal metals and discuss properties such as the condensation energy in relation to Richardson's exact equation for the superconductivity.

9.3.1 CRITICAL LEVEL SPACING IN KONDO EFFECT

In ultrasmall grains such as quantum dots etc., the quantum level spacing approaches order parameters. For ultrasmall superconducting grains, the critical level spacing d_c^{BCS} can be expressed as $d_c^{\mathrm{BCS}} = 4\omega_D e^{\gamma}\exp(-1/\lambda)$ for even number of electrons, where ω_D means the Deby energy. This result suggests that the gap function of a nanosize system with the level spacing d vanishes, when the coupling parameter λ_c is less than the value $(\ln 4\omega_D/d + \gamma)^{-1}$. The bulk gap function Δ_c with λ_c can be expressed as $\Delta_c = \omega_D\mathrm{sh}^{-1}(1/\lambda_c)$.

Figure 9.1a shows the gap function of a nanosize system in the framework of the standard BCS theory. We can find the region where the gap function vanishes, when the coupling becomes less than λ_c. This means the level spacing is larger than gap function in this region.

Here we drive the critical level spacing for only the Kondo regime ($\lambda = 0$). The equation determining the singlet order parameter can be expressed as

$$\Xi = \sum_k \frac{\Xi\left(\xi_k - x\right)}{\left(\xi_k - x\right)^2 + \Xi^2}, \tag{9.13}$$

where $\xi_k = \tilde{\varepsilon}_k - \mu$, $x = [\tilde{\varepsilon}_k + \tilde{E} \pm \sqrt{(\tilde{\varepsilon}_k - \tilde{E})^2 + 4\Xi^2}]/2$ and μ is the chemical potential. For the case of the critical level spacing, the solution has the spin

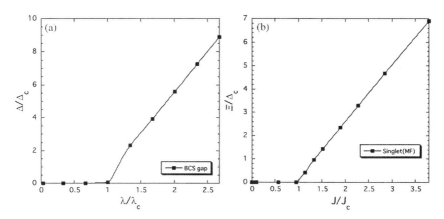

Fig. 9.1. Gap function and spin singlet order: (a) The gap function. The gap function vanishes in the region of smaller λ value than λ_c. (b) Spin singlet order parameter. In the case of $\tilde{J} < \tilde{J}_c$, the singlet order vanishes. The system consists of eight energy levels and eight electrons with the level spacing $d = 1.0$ and $\omega_D = 1.0$.

singlet order vanishes. From Eq. (9.13), we can find the critical level spacing d_c^{Kondo} for the Kondo regime.

$$d_c^{\text{Kondo}} = 4\omega_D e^\gamma \exp\left[-\frac{1}{2\sqrt{2}\tilde{J}}\right]. \tag{9.14}$$

When the coupling parameter $\tilde{J}$ is smaller than $\tilde{J}_c = [2\sqrt{2}(\ln(4\omega_D/d) + \gamma)]^{-1}$, the spin singlet order parameter vanishes.

Figure 9.1b shows the spin singlet order parameter of Eq.(9.10) in the case $g = 0$. In the region of $\tilde{J} < \tilde{J}_c$, the order parameter vanishes. This result suggests the critical level spacing in the Kondo effect.

9.3.2 KONDO EFFECT COUPLED TO SUPERCONDUCTIVITY

In this study, we consider a simple system which consists of eight energy levels and eight electrons and investigate the critical level spacing and the condensation energy of the coupled system between the superconductivity and the Kondo regime in the framwork of the mean field approximation of Eq. (9.12).

Figure 9.2a shows the spin singlet order parameter and the gap function for several cases. We can find the critical level spacings for the gap function and for the spin singlet order parameter. When $\lambda < \lambda_c$ and $\tilde{J} > \tilde{J}_c$, we can find only the spin singlet order parameters. In the region of λ/λ_c from 1.4 to 1.7 with $\tilde{J}/\tilde{J}_c = 0.189$, we can find the coexistence of both the gap function and the spin singlet order parameter. In larger λ/λ_c than 1.7, only the gap function still exists, and the spin singlet order parameter vanishes. In $\tilde{J}/\tilde{J}_c = 0.284$, we can find the coexistence in the region $\lambda/\lambda_c = 1.7$–2.3. These results suggest that strong local magnetic moments from the impurities makes the transition temperature for superconductivity reduced. However, weak couplings λ

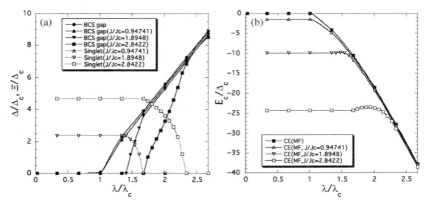

Fig. 9.2. Physical properties in coupled system: (a) Gap function and spin singlet order parameter. (b) Condensation energy. $\tilde{J}/\tilde{J}_c = 0, 0.94741, 1.8948$, and 2.8422. Other parameters are the same.

of the superconductivity do not destroy the spin singlet order parameter at all. These results are a good agreement with the experimental results [11]. We can find that there is the coexistence region for both the superconductivity and the Kondo regime.

Figure 9.2b shows the condensation energy for several λ and $\tilde{J}$ values. We can find the condensation energy of the coupled system between the superconductivity and the Kondo regime becomes lower than that of the pure superconductivity. In the coexistent region, the highest value of the condensation energy appears in all cases.

9.3.3 EXACT SOLUTION FOR KONDO REGIME

The standard BCS theory gives a good description of the phenomenon of superconductivity in large sample. However, when the size of a superconductor becomes small, the BCS theory fails. To investigate physical properties such as the condensation energy, parity gap, etc., it is necessary to take more accurate treatment. For the superconductivity in ultrasmall grains, the exact solution to the reduced BCS Hamiltonian presented by Richardson [39] has been applied to investigate such physical properties [18].

By using the wave function describing all pair electron excitations, we can derive the exact solution for the pairing force (reduced) Hamiltonian

$$2 - \sum_{k=1}^{N} \frac{\lambda}{\tilde{\varepsilon}_k - E_i} + \sum_{l=1, l \neq i}^{n} \frac{2\lambda}{E_l - E_i} = 0, \tag{9.15}$$

where N and n are the number of orbital and the number of the occupied orbital, respectively. E_i corresponds to the exact orbital. Figure 9.3 shows the condensation energy and the pairing energy level for the nanosize superconductivity. Note that physical properties obtained by the mean field approximation give a good description for the high density of state ($d \to \infty$). We can find the different behavior of the condensation energy from that obtained by the mean field approximation as shown in Fig. 9.3a. Figure 9.3b shows qualitative behaivior of the pairing energy level in the ground state. In λ of about 1.6, above two energy levels in Fig. 9.3b are completely paired. The pairing behavior has been already reported by many groups [39, 43].

Let us derive the exact equation for the Kondo regime in ultrasmall grains. We can consider the Hamiltonian $H = H_0 + H_1$ in Eq. (9.1). We introduce a creation operator describing all excited states of the spin singlet coupling between a conduction electron and a pseudofermion.

$$B_j^\dagger = \sum_{k,\sigma} \frac{c_{k\sigma}^\dagger f_\sigma}{\tilde{\varepsilon}_k - E_j}, \tag{9.16}$$

where E_j means the exact eigenenergies in the Kondo regime. The exact eigenstate $|\Psi_n\rangle$ for the Kondo regime can be written as $|\Psi_n\rangle = \Pi_{\nu=1}^{n} B_\nu^\dagger |0\rangle$.

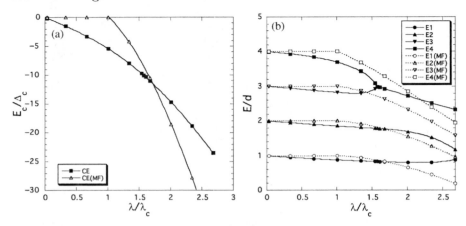

Fig. 9.3. Exact solution for the superconductivity: (a) Condensation energy of the exact solution with that obtained by the mean field approximation (b) Pairing energy level with energy level obtained by the mean field approximation. Eight energy levels, eight electrons, $d = 1.0$, $\omega_D = 4.0$.

Other electrons, which are not related to the spin singlet order, contribute $E_{\text{single}} = \sum_{k=1}^{n} \tilde{\varepsilon}_k$ to the eigenenergy. The ground state energy E_{GS} can be written as $E_{GS} = \sum_{k=1}^{n} [E_k + \tilde{\varepsilon}_k]$.

By operating the Hamiltonian to the exact eigenstate, we obtain the condition as

$$1 + \sum_{k=1}^{N} \frac{\tilde{J}}{\tilde{\varepsilon}_k - E_j} = 0. \tag{9.17}$$

This equation gives the exact solution for the Kondo regime. Note that the creation operator of Eq. (9.16) might be true boson by comparing with the case of the reduced BCS model.

Figure 9.4 shows the condensation energy of the exact solution in the Kondo regime with that obtained by the mean field approximation. We can find the different behavior of the condensation energy from that obtained by the mean field approximation. However, the behavior is similar to that in the superconductivity in nanosize system.

9.4 Concluding remarks

We have investigated properties of the Kondo regime coupled to the superconductivity in ultrasmall grains by using a mean field approximation. In the framework of the mean field approximation, we have found the critical level spacing for the Kondo regime. The result suggests that the Kondo effect vanishes, when the level spacing becomes larger than the critical level spacing.

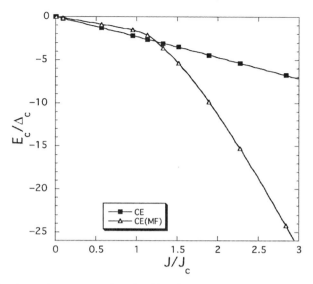

Fig. 9.4. Condensation energy for the Kondo regime: all parameters used in the system are the same: Eight energy levels, eight electrons, and $d = 1.0$, $\omega_D = 4.0$.

We have calculated physical properties of the critical level spacing and the condensation energy of the coupled system by using the mean field approximation. From the results, we have found that strong local magnetic moments from the impurities makes the transition temperature for superconductivity reduced. However, weak couplings λ of the superconductivity do not destroy the spin singlet order parameter at all. These results are a good agreement with the experimental results [11]. We have found that there is the coexistence region for both the superconductivity and the Kondo regime.

Finally we have derived the exact equation for the Kondo regime in nanosystem and have discussed the condensation energy from the viewpoint of the energy level. It might be not easy to find the exact equation for the Kondo regime coupled to superconductivity. The exact properties such as the condensation energy etc. in the Kondo regime by using the exact equation will be presented elsewhere.

In summary, we have investigated the Kondo effect and superconductivity in ultrasmall grains by using a model, which consists of sd and reduced BCS Hamiltonians with the introduction of a pseudofermion. A mean field approximation for the model have been introduced, and we have calculated physical properties of the critical level spacing and the condensation energy. These physical properties have been discussed in relation to the coexistence of both the superconductivity and the Kondo regime. Finally we have derived the exact equation for the Kondo regime in nanosystem and discuss the condensation energy from the viewpoint of the energy level.

Acknowledgment

H.N is grateful for a financial support of the Ministry of Education, Science and Culture of Japan (Research No. 19029014). S.K is grateful for a financial support of the MUES(project M/145-2007). The authors thank Prof. K. Nishikawa, K. Yamaguchi for their continued encouragement helpful discussion.

References

1. L. Kouwenhoven and L. Glazman, Phys. World **14**, No.1, 33 (2001).
2. D. Goldhaber-Gordon, H. Shtrikman, D. Mahalu, D. Abusch-Magder, U. Meirav, and M. A. Kastner, Nature (London) **391**, 156 (1998).
3. W. G. van der Wiel, S. De Franceschi, T. Fujisawa, J. M. Elzerman, S. Tarucha, and L. P. Kouwenhoven, Science **289**, 2105 (2000).
4. T. Inoshita, Science **281**, 526 (1998).
5. S. Sasaki, S. De Franceschi, J. M. Elzerman, W. G. van der Wiel, M. Eto, S. Tarucha, and L. P. Kouwenhoven, Nature (London) **405**, 764 (2000).
6. T. Inoshita, A. Shimizu, Y. Kuramoto, and H. Sakaki, Phys. Rev. B **48**, R14725 (1993).
7. W. Izumida, O. Sakai, and Y. Shimizu, J. Phys. Soc. Jpn. **67**, 2444 (1998).
8. A. Levy Yeyati, F. Flores, and A. Martin-Rodero, Phys. Rev. Lett. **83**, 600 (1999).
9. A. N. Pasupathy, R. C. Bialczak, J. Martinek, J. E. Grose, L. A. K. Donev, P. L. McEuen, D. C. Ralph, Science **306**, 86 (2004).
10. D. Matsubayashi and M. Eto, Phys. Rev. B **75**, 165319 (2007).
11. M. R. Buitelaar, T. Nussbaumer, and C. Schönenberger, Phys. Rev. Lett. **89**, 256801 (2002).
12. A. Ueda and M. Eto, Phys. Rev. B **73**, 235353 (2006).
13. M. Eto and Y. V. Nazarov, Phys. Rev. Lett. **85**, 1306 (2000).
14. P. W. Anderson, J. Phys. Chem. Solids **11**, 28 (1959).
15. C. T. Black, D. C. Ralph, and M. Tinkham, Phys. Rev. Lett. **76**, 688 (1996).
16. D. C. Ralph, C. T. Black, and M. Tinkham, Phys. Rev. Lett. **74**, 3241 (1995).
17. S. Reich, G. Leitus, R. Popovitz-Biro, and M. Schechter, Phys. Rev. Lett. **91**, 147001 (2003).
18. F. Braun and J. von Delft, Phys. Rev. Lett. **81**, 4712 (1998).
19. F. Braun and J. von Delft, Adv. Sol. State Phys. **39**, 341 (1999).
20. B. Jankó, A. Smith, and V. Ambegaokar, Phys. Rev. B **50**, 1152 (1994).
21. K. A. Matveev and A. I. Larkin, Phys. Rev. Lett. **78**, 3749 (1997).
22. V. N. Gladilin, V. M. Fomin, and J. T. Devreese, Solid Sate Comm. **121**, 519 (2002).
23. J. von Delft, A. D. Zaikin, D. S. Golubev, and W. Tichy, Phys. Rev. Lett. **77**, 3189 (1996).
24. R. A. Smith and V. Ambegaokar, Phys. Rev. Lett. **77**, 4962 (1996).
25. H. Nagao, H. Kawabe, and S. P. Kruchinin, *Electron Correlation in New Materials and Nanosystems NATO Science Series II. Mathematics, Physics and Chemistry*, Vol. 241 (Springer/Berlin/Heidelberg, New York, 2007), pp. 117–127.

26. V. A. Moskalenko, Fiz. Met. Metalloved **8**, 503 (1959).
27. H. Suhl, B. T. Matthias, and R. Walker, Phys. Rev. Lett. **3**, 552 (1959).
28. J. Kondo, Prog. Theor. Phys. **29**, 1 (1963).
29. K. Yamaji and Y. Shimoi, Physica C **222**, 349 (1994).
30. R. Combescot and X. Leyronas, Phys. Rev. Lett. **75**, 3732 (1995).
31. H. Nagao, M. Nishino, M. Mitani, Y. Yoshioka, and K. Yamaguchi, Int. J. Quantum Chem. **65**, 947 (1997).
32. P. Konsin, N. Kristoffel, and T. Örd, Phys. Lett. A **129**, 339 (1988).
33. P. Konsin and B. Sorkin, Phys, Rev. B **58**, 5795 (1998).
34. H. Nagao, M. Nishino, Y. Shigeta, Y. Yoshioka, and K. Yamaguchi, J. Chem. Phys. **113**, 11237 (2000).
35. J. Kondo, J. Phys. Soc. Jpn. **70**, 808 (2001).
36. J. Kondo, J. Phys. Soc. Jpn. **71**, 1353 (2002).
37. S. P. Kruchinin and H. Nagao, Phys. Particle Nuclei **36** *Suppl.*, S127 (2005).
38. D. L. Cox and M. B. Maple, Phys. Today **48**, No.2, 32 (1995).
39. R. W. Richardson, Phys. Rev. **141**, 949 (1966).
40. A. C. Hewson, *The Kondo Problem to Heavy Fermion*, (Cambridge University Press, Cambridge; 1993).
41. A. Yoshimori and A. Sakurai, Suppl. Prog. Theor. Phys. **46**, 162 (1970).
42. M. Eto and Y. V. Nazarov, Phys. Rev. B **64**, 085322 (2001).
43. H. Kawabe, H. Nagao, and S. P. Kruchinin, *Electron Correlation in New Materials and Nanosystems NATO Science Series II. Mathematics, Physics and Chemistry*, Vol. 241 (Springer, Berlin/Heidelberg/New York, 2007) pp. 129–139.

10

EMERGING MEASUREMENT TECHNIQUES
FOR STUDIES OF MESOSCOPIC SUPERCONDUCTORS

A. RYDH[1], S. TAGLIATI[1], R.A. NILSSON[1], R. XIE[2],
J.E. PEARSON[2], U. WELP[2], W.-K. KWOK[2], AND
R. DIVAN[3]

[1] *Department of Physics, Stockholm University, AlbaNova University Center, SE-106 91 Stockholm, Sweden.* arydh@physto.se
[2] *Materials Science Division, Argonne National Laboratory, 9700 South Cass Avenue, Argonne, IL 60439, USA*
[3] *Center for Nanoscale Materials, Argonne National Laboratory, 9700 South Cass Avenue, Argonne, IL 60439, USA*

Abstract. Experimental research on mesoscopic systems puts high demands on the measurement infrastructure, including measurement system with associated sample preparation, experimental design, measurement electronics, and data collection. Successful experiments require both the ability to manufacture small samples and to successfully and accurately study their novel properties. Here, we discuss some aspects and recent advancements of general measurement techniques that should benefit several characterization methods such as thermodynamic, magnetic, and transport studies of mesoscopic superconductors.

Keywords: superconductivity, calorimetry, mesoscopic, Si_3N_4 membrane, measurement techniques, lock-in, FPGA, DAQ, ADC

10.1 Introduction

Mesoscopic superconductors, here loosely defined as samples with properties deviating from their regular, bulk behavior due to a size restriction, have been studied extensively both theoretically [1–3] and by means of computer simulations [4–6]. Experimental results, [7–9] however, are comparatively scarce in some areas, such as thermodynamic studies. This is no longer due to a lack of materials. With recent advancements in nanotechnology and the availability of tools such as Focused Ion Beam (FIB), electrochemistry, and Molecular Beam Epitaxy (MBE), several methods now exist to obtain mesoscopic samples with

J. Bonča, S. Kruchinin (eds.), *Electron Transport in Nanosystems.*
© Springer Science + Business Media B.V. 2008

novel shapes [10] and properties. The characterization of such samples often requires accurate measurements at low temperatures, in high magnetic fields, and with a good control of position, heat flow, or power dissipation. Measurement instrumentation and methods, thus, need to be pushed to their limits and adopted to the scales of small samples. Here, we review some fairly simple measurement techniques that can be applied to a variety of setups to improve critical factors such as the signal-to-noise ratio and absolute accuracy.

10.2 Traditional techniques review

In most measurement systems there are components where an electrical voltage or current is applied to a sensor and the sensors response should be monitored. The sensor could be a thermometer, a gauss meter, a strain gauge, or the sample itself. Good methods for such measurements have been known for almost two centuries, but the availability of modern electronics and the application to mesoscopic physics poses new opportunities and limitations.

10.2.1 FOUR-POINT PROBES

The four-point probe setup, shown in Fig. 10.1a, aims to measure an unknown resistance or impedance R_x. It avoids the problem of lead and contact resistances, but assumes a known excitation. To simultaneously determine the excitation, a known reference impedance R_R is usually placed in series with R_x, as shown in Fig. 10.1b. This enables accurate determination of impedance, but also current-voltage $(I - V)$ characterization, differential conductance (dI/dV) measurements, and pulse studies. For accurate dc and dI/dV measurements, thermal electromotive forces (EMF) are easily corrected for by switching the exitation polarity, typically twice for each sampling, to also cancel out drifts in the EMF. In general, the excitation stability can often be

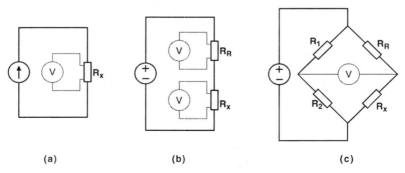

(a)   (b)   (c)

Fig. 10.1. (a) Four-probe setup. (b) Four-probe setup with excitation reference. (c) Wheatstone bridge.

worse than the measurement resolution. Direct measurement of the excitation is therefore often useful. It should be noted, that the use of a current preamplifier instead of the series resistor of Fig. 10.1b is a preferred method to measure small currents, since a large resistor in combination with cable capacitance may cause phase distortions and deteriorated frequency response.

10.2.2 BRIDGES

Bridge circuits are also well suited [11] for measurements of dI/dV and d^2I/dV^2. The Wheatstone bridge, shown in Fig. 10.1c, has its main advantage in setups where two or more impedances are close to balance, and the slight deviation from balance is of interest. If the impedance R_2 is adjustable, the components R_1 and R_2 form a potentiometer that can be adjusted until the voltmeter or galvanometer reads zero. The deviations are then amplified by a comparatively high factor, yielding a lower noise. In a mesoscopic context, a problem may be that the excitation is limited by heating constraints. In such a case, Johnson noise rather than measurement electronics could limit the resolution of a direct measurement over R_x. The use of a bridge is then merely practical.

10.2.3 LOCK-IN AMPLIFIERS

The lock-in amplifier is a widely used instrument that can be seen as a detector, synchronous with an internal or external reference, that applies the orthogonality of different sine waves to determine the amplitude and phase of an ac signal of certain frequency. Due to its very narrow-band detection, the lock-in is ideal for detecting low-level signals in noisy environments. The sampled signal and reference are multiplied in a mixer and the product is averaged by a low-pass filter and/or integrated over a specified time. By using both reference and its quadrature, the signal phase is also obtained. The phase sensitive detection (PSD) benefits greatly from the use of a digital reference. Modern, digital lock-ins are, therefore, built on analog-to-digital converters (ADC) with subsequent digital processing, using a digital signal processor (DSP) or field-programmable gate array (FPGA) [12,13]. In addition to general benefits, such as absence of reference drift and truly orthogonal in-phase and quadrature, the digital lock-in enables, processing power permitting, *simultaneous* detection of signals at higher harmonics (*2f, 3f*), and/or at any other frequency using the *same*, sampled signal from the ADC. Such use is yet uncommon, but should have great potential in systems with frequency-dependent amplitude response. ADC cards for lock-in applications typically have about 14 bit effective resolution at a 500 kS/s sampling rate. With a 10 V range, this corresponds to a noise floor of about $1\mu V/Hz^{1/2}$. A preamplifier is, thus, usually needed to reach the thermal noise floor. The choice of preamplifier should depend on source resistance and required bandwidth. In the limit of very low source resistance, a transformer preamplifier may be used to achieve sub-$nV/\mathrm{Hz}^{1/2}$. The basic lock-in amplifier in combination with a four-point probe or bridge setup is the standard measurement solution for many experimental tasks. In the following we will give some examples on techniques that use fairly small modifications to this concept to further improve resolution and absolute accuracy.

10.3 Differential Micro Calorimetry

Calorimetry is a powerful tool for the general understanding and characterization of materials, for mapping out phase diagrams, and to study phase transitions. Naturally, several methods have been devised to enable the measurement of heat capacity also with small, sometimes sub-microgram, samples [14]. A main benefit of using small samples is the short relaxation times, which lead to faster measurements and the availability of new measurement methods. With smaller samples, the effects of spurious heat flow increases, but simultaneously thermal equilibrium times become shorter. Micro calorimeters, therefore, solve the problem of heat flows by varying the sample temperature at a rate slightly above the systems natural relaxation rate. Here we illustrate the technique by describing a membrane-based, differential micro calorimeter operating in steady-state ac mode.

10.3.1 TECHNIQUE OVERVIEW

In the steady-state ac method [15], the sample temperature is modulated with a small amplitude of the order of 0.1 K, while the average temperature is kept constant or scanned slowly over the temperature range of interest. The sample is placed on a platform, which constitutes a weak, thermal link between sample and a thermal bath of known base temperature. A local heater and thermometer are in close contact with the sample. At any moment, the heat added to the sample by the heater is balanced by the sum of heat flow out through the sample support structure and the change in internal energy. If the heater power is allowed to vary in a periodic manner, $P = P_0 \left(1 - \cos \omega' t\right)$, the temperature oscillation at steady-state can be described by the equation

$$\frac{dT_{ac}(t)}{dt} + \frac{T_{ac}(t)}{\tau} = -\frac{P_0}{2k} \cos \omega' t \qquad (10.1)$$

where $T_{ac}(t)$ is the time dependent part of the sample temperature, k is the thermal conductance between sample and bath, and $\tau = k/C$ is a characteristic relaxation time, where C is the heat capacity. Writing $T_{ac}(t) = T_s \sin \left(\omega' t + \phi_s\right)$, Eq. (10.1) can be solved for C, giving

$$C = \frac{P_0}{\omega' T_s} \sqrt{1 - \left(\frac{T_s}{T_0}\right)^2} \qquad (10.2)$$

Here $T_0 = P_0/k$ is the oscillation amplitude at low frequencies, $\omega'\tau \ll 1$. The working frequency is chosen to be close to $\omega'\tau \approx 10$, corresponding to $T_s/T_0 \approx 1/10$. Using lower frequencies decreases the sensitivity while using higher frequencies requires more heater power for a given temperature modulation amplitude Ts without any gain in resolution.

10.3.2 DEVICE CONSTRUCTION

A differential calorimetry system was built around a Si frame with two free-standing, back-etched Si_3N_4 membranes (1×1 mm, 150 nm thick) acting as sample and reference cells. Thin film meanders and Cu vs. Au-2.1%Co thermocouple crosses were deposited onto the membranes to act as heaters and thermometers, measuring the temperature differences between sample, reference, and Si frame (base). The heaters were driven by ac currents of frequency $\omega = \omega'/2$ while the temperature oscillations were measured at the second harmonic, using lock-in amplifiers with transformer preamplifiers.

10.3.3 MEASUREMENT EXAMPLE

The typical frequency dependence of the device is shown in Fig. 10.2. The scan was made at 60 K with a small (650 ng) single crystal of $YBa_2Cu_3O_{7-\delta}$ on the sample side and leaving the reference side empty. Both T_s and T_r (temperature oscillation amplitude of reference) approach the value $T_0 = P_0/k$ at low frequencies, describing the temperature gradient and heat flow between heaters and base frame rather than heat capacity. When the frequency is

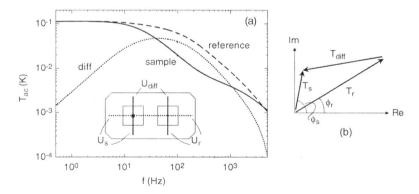

Fig. 10.2. (a): Frequency dependence of temperature oscillation amplitudes. Amplitudes were corrected for transformer distortion effects at low and high frequencies for clarity. The inset illustrates the measurement layout for proper determination of T_s, T_r, and $T_{diff} = T_s - T_r$ using transformer preamplifiers. (b): The amplitude of the temperature difference T_{diff} depends not only on the oscillation amplitudes T_s and T_r, but also on the phase shifts ϕ_s and ϕ_r.

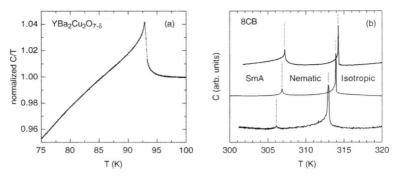

Fig. 10.3. (a): Superconducting transition of a small $YBa_2Cu_3O_{7-\delta}$ crystal. (b): Heat capacity signature of the liquid crystal phase transitions in octocyanobiphenyl (8CB) with decreasing film thickness, going from top to bottom.

increased, the sample oscillation amplitude starts to decrease, reflecting the influence of the sample heat capacity. The two amplitudes approach each other again at very high frequencies. This is because the thermal connection between sample and membrane support is lost.

To separately measure the voltage from the differential temperature oscillation T_{diff} in addition to T_s and T_r, as shown in the inset of Fig. 10.2, has several benefits. One advantage is that the differential signal is very small when a reference is used, which may enable higher amplification and thereby lower noise. The main advantage is, however, that the phase difference $\phi_s - \phi_r$ can be recreated from the three amplitudes, as illustrated in Fig. 10.2b. With a known phase difference, the square root in Eq. (10.2) can be taken into account even without measuring T_0, which simplifies the measurements and increases the absolute resolution.

Figure 10.3a shows the superconducting transition of the $YBa_2Cu_3O_{7-\delta}$ crystal. The slight upturn in C/T at T_c is clearly resolved, indicating the good resolution of the measurements, despite the use of a sample that is 10^3 to 10^5 times smaller than traditionally used samples for specific heat measurements. In panel (b) of Fig. 10.3, the calorimetric signatures of the smectic-A to nematic and nematic to isotropic transitions are shown for a thin layer of octylcyanobiphenyl (8CB) liquid crystal as a function of layer thickness. The measurements were made by placing a microscopic droplet of 8CB onto the membrane and letting it slowly spread with time. When the liquid crystal becomes confined, it is known [16, 17] that the relative peak height changes and the transition temperature decreases, as is also seen in Fig. 10.3. From the temperature shifts and a comparison with the literature, we conclude that the calorimeter is able to resolve both transitions in films down to a thickness of about 100 nm, and that the nematic to isotropic transition should be resolvable in films of just a few nm thickness.

10.4 Synchronization of FPGA-based lock-in amplifiers

If a measurement requires several lock-in amplifiers, the synchronization in-between them may become an issue. This is especially true if the frequency is not constant, if several outputs are to be generated simultaneously, or at short integration times, corresponding to large data transfer rates. To address this issue, a phase-synchronized measurement system was constructed around a user-configurable FPGA with integrated ADCs.

10.4.1 TECHNIQUE OVERVIEW

A schematic of the measurement system is shown in Fig. 10.4. A 1 M gates FPGA module with eight integrated, 200 kS/s ADCs and 1 MS/s DACs (digital-to-analog converters) was configured using National Instruments Lab-VIEW FPGA-module. Sync inputs and output were implemented using 40 MHz DIO ports. Timing-critical FPGA loops were set to run as timed loops at 40 MHz, while data processing loops were running at a multiple of the sampling rate, to enable serial processing of the simultaneously-sampled data. To stay within the constraints of the FPGA size, the PLL feedback routine and low-pass filter of the integrator were simplified as compared with a standard lock-in, but could be extended for a 3 M gates FPGA module. By using a central phase generator, a set of completely simultaneous lock-in amplifiers is obtained. The generated phase is increased in steps that reflect the momentary frequency. This makes it possible to modify the frequency at any time

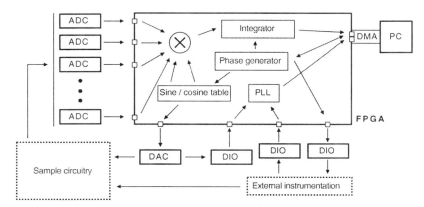

Fig. 10.4. Simplified schematic of measurement system. Digitized signals from eight simultaneous-sampling ADCs with 200 kS/s sampling rate are multiplied with the generated sine/cosine reference and passed to an integrator that acts as a synchronized low-pass filter. Resulting in-phase and quadrature information is streamed to a PC using direct memory access (DMA). Internal and external triggers are sampled at 40 MS/s using digital input/output (DIO) and passed to the phase-locked loop (PLL).

and several times in a single clock cycle without any loss of phase correlation between inputs and/or outputs. If two regular lock-in amplifiers were used (one of them generating the sync signal), a change in frequency would cause loss of correlation between them for at least one cycle period $1/f$.

10.4.2 MEASUREMENT EXAMPLE

To illustrate the system capability, a simple test measurement was setup as shown in the circuit sketch of Fig. 10.5. Two sampling channels were used to measure the amplitudes of an applied voltage U_a and the voltage across a 100 Ω resistor U_x as a function of time. The measurements were taken and analyzed with an effective time constant of 1 period (43 ms), but were then reduced to a 10 s time constant per point for clarity (to lower the amount of white noise). The result is presented in Fig. 10.5. It is seen that both measurements display slow variations in the ac amplitude, possibly due to a small thermal drift in either generator or measurement setup. If a regular lock-in amplifier were used to measure U_x, the drift in U_a would have to be measured with a separate instrument. Otherwise, the amplitude drift would be unknown and add as noise in U_x. However, if U_x is normalized to U_a, shown as the voltage ratio, it is clear that the noise level can be decreased significantly. With a factor of 3.2 for the conversion to a 1 s time constant, the resolution is still better than 1 in 10^5.

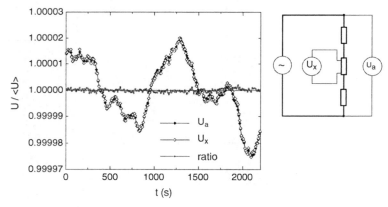

Fig. 10.5. Study of ac amplitude resolution and amplitude correlation using time-synchronized, FPGA-based lock-in amplifiers. Measured amplitudes U_x and U_a were normalized to their respective averages $<U_a> \approx 4.64\ V_{rms}$ and $<U_x> \approx 0.70\ V_{rms}$, taken as the average over the entire 2,200 s measurement period. The test setup was created by applying a nominal 5 V_{rms} (23 Hz) voltage from a Stanford SR830 lock-in amplifier to a 100 Ω resistor (the sample) in series with some current-limiting resistors. The effective time constant in the graph is 10 s per point.

10.5 Summary

In summary, we have presented measurement techniques that can be applied to the study of small samples. The general technique of using an FPGA-based set of well-synchronized lock-in amplifiers for high-resolution measurements should be especially suitable for transport, magnetic, and thermodynamic characterization, which are all important tools for experimental research on mesoscopic superconductivity. A differential micro calorimeter was described, capable of probing sub-microgram samples and thin films. Efforts are on the way to replace the thermocouples of the micro calorimeter with a Wheatstone bridge of thin-film thermometers operated in conjunction with the FPGA-based measurement system. This should enable micro-caloric studies at low temperatures with both high absolute accuracy and good resolution.

Acknowledgments

Support from the Swedish Research Council and the US Department of Energy, Office of Science, Office of Basic Energy Sciences, under contract No. DE-AC02-06CH11357 is gratefully acknowledged.

References

1. J. R. Schrieffer and M. Tinkham, Superconductivity, Rev. Mod. Phys. **71(2)**, S313–S317 (1999).
2. H. J. Fink and A. G. Presson, Magnetic irreversible solution of the Ginzburg-Landau equations, Phys. Rev. **151(1)**, 219–228 (1966).
3. K. Tanaka, I. Robel, and B. Janko, Electronic structure of multiquantum giant vortex states in mesoscopic superconducting disks, PNAS **99(8)**, 5233–5236 (2002).
4. P. S. Deo, V. A. Schweigert, and F. M. Peeters, Magnetization of mesoscopic superconducting disks, Phys. Rev. Lett. **79**, 4653–4656 (1997).
5. V. G. Kogan, J. R. Clem, J. M. Deang, and M. D. Gunzburger, Nucleation of superconductivity in finite anisotropic superconductors and the evolution of surface superconductivity toward the bulk mixed state, Phys. Rev. **B 65**, 094514 (2002).
6. V. R., Misko, V. M. Fomin, J. T. Devreese, and V. V. Moshchalkov, Stable vortex-antivortex molecules in mesoscopic superconducting triangles, Phys. Rev. Lett. **90**, 147003 (2003).
7. V. V. Moshchalkov, L. Gielen, C. Strunk, R. Jonckheere, X. Qiu, C. Van Haesendonck, and Y. Bruynseraede, Effect of sample topology on the critical fields of mesoscopic superconductors, Nature **373**, 319–322 (1995).
8. A. K. Geim, I. V. Grigorieva, S. V. Dubonos, J. G. S. Lok, J. C. Maan, A. E. Filippov, and F. M. Peeters, Phase transitions in individual sub-micrometre superconductors, Nature **390**, 259–262 (1997).

9. Y. Guo, Y. -F. Zhang, X. -Y. Bao, T. -Z. Han, Z. Tang, L. -X. Zhang, W. -G. Zhu, E. G. Wang, Q. Niu, Z. Q. Qiu, J. -F. Jia, Z. -X. Zhao, and Q. -K. Xue, Superconductivity modulated by quantum size effects, Science **306**, 1915–1917 (2004).

10. Z. Zhao, C. Y. Han, W. -K. Kwok, H. -H. Wang, U. Welp, J. Wang, and G. W. Crabtree, *Tuning the architecture of mesostructures by electrodeposition,* J. Am. Chem. Soc. **126**, 2316–2317 (2004).

11. M. V. Moody, J. L. Paterson, and R. L. Ciali, High-resolution dc-voltage-biased ac conductance bridge for tunnel junction measurements, Rev. Sci. Instrum. **50**, 903–908 (1979).

12. P. -A. Probst and A. Jaquier, Multiple-channel digital lock-in amplifier with PPM resolution, Rev. Sci. Instrum. **65(3)**, 747–750 (1994).

13. A. Restelli, R. Abbiati, and A. Geraci, Digital field programmable gate array-based lock-in amplifier for high-performance photon counting applications, Rev. Sci. Instrum. **76**, 093112 (2005).

14. A. Rydh, Calorimetry of Sub-Microgram Grains, in Encyclopedia of Materials: Science and Technology, 2006 Online Update, edited by K. H. J. Buschow, M. C. Flemings, R. W. Cahn, P. Veyssière, E. J. Kramer, and S. Mahajan (Elsevier, Oxford, 2006). Available online at http://www.sciencedirect.com/science/referenceworks/0080431526.

15. P. F. Sullivan and G. Seidel, Steady-state, ac-temperature calorimetry, Phys. Rev. **173**, 679–685 (1968).

16. L. Wu, B. Zhou, C. W. Garland, T. Bellini, and D. W. Schaefer, Heat-capacity study of nematic-isotropic and nematic-smectic-A transitions for octylcyanobiphenyl in silica aerogels, Phys. Rev. **E 51**, 2157–2165 (1995).

17. Z. Kutnjak, S. Kralj, G. Lahajnar, and S. Zumer, Calorimetric study of octylcyanobiphenyl liquid crystal confined to a controlled-pore glass, Phys. Rev. **E 68**, 021705 (2003).

11

INTERPLAY OF MAGNETISM
AND SUPERCONDUCTIVITY IN CeCoIn$_5$

R. MOVSHOVICH, Y. TOKIWA, F. RONNING,
A. BIANCHI, C. CAPAN, B.L. YOUNG,
R.R. URBANO, N.J. CURRO, T. PARK,
J.D. THOMPSON, E. BAUER, AND J.L. SARRAO
Los Alamos National Laboratory, MS K764, Los Alamos, NM 87545
USA. roman@lanl.gov

Abstract. CeCoIn$_5$ is a heavy fermion superconductor which appears to be strad-
dling the boundary between the superconducting and magnetic ground states. At
the superconducting critical field H_{c2} this material displays NFL behavior in transport and thermodynamic properties, pointing at a Quantum Critical Point (QCP) at
H_{c2}, and hinting at the presence of magnetic fluctuations, probably due to an AFM
order superseded by the superconductivity. In the High-Field-Low-Temperature
(HFLT) corner of the superconducting phase of CeCoIn$_5$, within 20% off H_{c2}, an
additional phase appears within the superconducting phase, and the normal-to-
superconducting transition itself becomes first order. This behavior is consistent with
a strong Pauli limited superconductivity, and the low temperature high field phase
being an inhomogeneous superconducting FFLO phase. Recent NMR experiments,
however, point to a long range magnetic order within HFLT state. Experiments on
CeRhIn$_5$ under pressure show magnetic field induced AFM order within the superconducting phase, with some similarities to the phase diagram of CeCoIn$_5$. Could
the HFLT phase transition be due to magnetic order? Importantly, the HFLT phase
does not extend into the normal state above H_{c2}. We need a picture of a magnetism "attracted" to superconductivity to explain the data on the HFLT phase in
CeCoIn$_5$.

Keywords: heavy fermion, superconductivity, FFLO phase

11.1 Introduction

CeCoIn$_5$ is a heavy fermion with a record high (for heavy fermion compounds)
superconducting transition temperature $T_c = 2.3$ K [1]. CeCoIn$_5$ forms in a
tetragonal crystal structure which can be thought of layers of CeIn$_3$ (derived

J. Bonča, S. Kruchinin (eds.), *Electron Transport in Nanosystems.*
© Springer Science + Business Media B.V. 2008

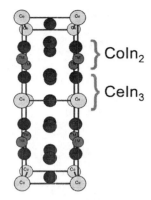

Fig. 11.1. Crystal structure of CeCoIn$_5$, showing the layers of CeIn$_3$ and CoIn$_2$.

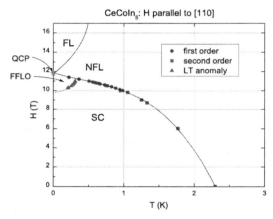

Fig. 11.2. Phase diagram of CeCoIn$_5$ for $H \parallel [110]$. SC transition changes from second to first order at ≈ 1 K and 10 T. A possibly FFLO state occupies the high field - low temperature corner of the superconducting phase, and a Quantum Critical Point (QCP) lies close to the superconducting critical field H_{c2}.

from the parent cubic compound CeIn$_3$) separated by the layers of CoIn$_2$ along the c-axis, as shown in Fig. 11.1.

This perspective on the crystal structure of CeCoIn$_5$ as being quasi-two-dimensional is supported by both band structure calculations [2] and deHaas-van Alphen measurements [3–5], which uncovered large undulating cylindrical parts of the Fermi surface with the axis along the [001] direction. The crystalline anisotropy leads to anisotropy in physical properties, including magnetization (about a factor of two, with an easy axis along [001]). The superconducting critical field H_{c2} is itself anisotropic, with $H_{c2} = 4.95$ T for $H \parallel [001]$ and 11.6 T for $H \parallel [100]$ [6].

CeCoIn$_5$ exhibits a number of fascinating properties in the vicinity of H_{c2}. Figure 11.2 shows the phase diagram of CeCoIn$_5$ with field $H \parallel [110]$,

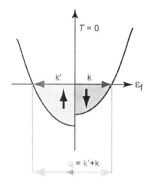

Fig. 11.3. Schematic of the formation of the FFLO state and Pauli limiting of the BCS state.

and introduces the main subjects of this paper. These include a Quantum Critical Point (QCP), coinciding with the superconducting critical field H_{c2}, and a purported Fulde-Ferrell-Larkin-Ovchinnikov (FFLO) state, named after its discoverers (theoretically) over 40 years ago [7, 8], in the high field-low temperature corner of the superconducting phase (the HFLT state). It is the interplay of these two phenomena and underlying interactions that is the main question that we are attempting to address below.

The subject that received the most attention since its discovery [9] is the novel HFLT phase which exists below 300 mK and between 10 T and $H_{c2} = 11.6$ T, with the field applied within the basal plane. It was suggested that this phase might be a realization of the spatially inhomogeneous FFLO superconducting state.

The FFLO phase is a result of a superconducting condensate minimizing the Zeeman energy of electron spins in magnetic field. Figure 11.3 gives a schematic representation of the effect of the magnetic field on both superconducting and normal states via Zeeman splitting of the spin up and spin down electron bands. In a spin-singlet BCS superconductor electrons form Cooper pairs with opposite electron spins. As a result, the two electrons' Zeeman energies cancel each other. In the normal state electron spins would preferentially point along the magnetic field, lowering total energy of the system and leading to Pauli susceptibility. The resulting competition between the superconducting condensate energy and the Pauli energy provides a mechanism for suppression of superconductivity at the Pauli limiting field H_P [10], in addition to the so-called orbital limiting (with characteristic orbital limiting field H_{c2}^0) due to the opposite forces by the magnetic field on the two electrons of the Cooper pair.

There are several types of systems that are traditionally thought to be good candidates to display the Pauli limiting and the FFLO physics. They are the systems that have weak or zero orbital limiting, and include superconducting films with magnetic field in the film's plane, low dimensional organic superconductors, and the heavy fermion materials. CeCoIn$_5$ is quasi-2D and a heavy fermion compound, and therefore is a good system in which to look for an FFLO state.

Phase diagram analysis allows us to estimate H_{c2}^0 (directly from the slope of the critical field at $T_c(H = 0)$) to range between 35 and 50 Tesla [6, 11], to be compared with the experimental superconducting critical field of $H_{c2} =$ 11.6 T. Large suppression of H_{c2} indicates that CeCoIn$_5$ is in a strong Pauli limit, i.e. Pauli limiting largely determines the observed H_{c2}.

The FFLO state takes advantage of the Zeeman energy of the superconducting electrons by pairing electrons at the Fermi energy with non-equal momenta k and k', leading to a Cooper pair with a finite total momentum $q = k + k'$ (see Fig. 11.3). The resulting superconducting state has an order parameter that is modulated in real space, e.g. $\Delta = \Delta_0 \cos(qr)$ for the Larkin-Ovchinnikov state.

It was predicted theoretically that in a strongly Pauli limited superconductor the order of the superconducting transition will change from second to first [12]. This was indeed observed in CeCoIn$_5$ for magnetic field both within the $a - b$ tetragonal plane and perpendicular to it with a variety of experimental tools [6,13,14]. Figure 11.4 shows specific heat for $H \parallel [001]$, when $H_{c2} \approx 5$ T. For fields below 4.5 T the specific heat anomaly has a classic mean field shape, with a sharp step at T_c, followed by a gradual decrease at lower temperature. Magnetic field suppresses T_c and the size of the superconducting anomaly. The 4.8 T data, however, breaks that monotonic evolution: the anomaly narrows down, becomes symmetrical, and the peak hight increases compared to the data for 4.5 T. This behavior demonstrates that the order of the superconducting transition changes from second to first somewhere

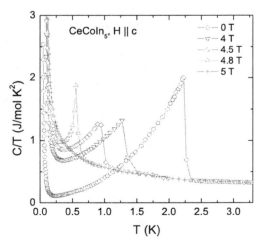

Fig. 11.4. Specific heat of CeCoIn$_5$, showing transition from second order character of the SC anomaly below 4.5 T to first order at 4.8 T.

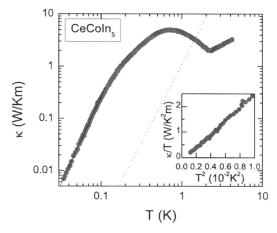

Fig. 11.5. Thermal conductivity of CeCoIn₅. The peak below T_c and the T^3 behavior at low temperature indicate high purity of the sample and a long quasiparticle mean free path in the SC state.

between 4.5 T and 4.8 T. Detailed analysis of additional data pinpointed the change to occur at $T_0 \approx 0.3\ T_c$ [13] for this field orientation. The first order superconducting transition in CeCoIn₅ was taken to be due to strong Pauli limiting effect, consistent with the phase diagram analysis above and in accord with theoretical expectations.

The next natural step was to look for the FFLO state in CeCoIn₅, since it should be driven by the same physics (Pauli limiting) that lead to the first order SC transition. The FFLO state itself is expected to be rather fragile in a sense that it should be easily destroyed by the minute amount of impurities. The last requirement for the formation of the FFLO state is therefore that the sample must be very clean, i.e. electron mean free path should be many times that of the superconducting coherence length. The first indications of the extreme purity of CeCoIn₅ came from the measurements of the thermal conductivity κ in CeCoIn₅ [15]. κ/T displays a very sharp kink at T_c, and rises dramatically by an order of magnitude with decreasing temperature, reaching a maximum at $T/T_c = 0.2$, Fig. 11.5. This, combined with the fact that the number of heat carrying normal quasiparticles is drastically decreased between T_c and $0.2\ T_c$, lead to an estimate of the quasiparticle mean free path of several microns deep within the superconducting state, orders of magnitude greater than the superconducting coherence length (on the order of 100 Å).

Finally, to take advantage of the quasi-2D nature of CeCoIn₅ in the search for the FFLO state, specific heat measurements were performed with the field within the basal $a-b$ plane, where $H_{c2} = 11.6$ T. Some of the specific heat data that lead to the phase diagram in Fig. 11.2 are displayed in Fig. 11.6. In addition to a very sharp first order superconducting transition at T_c, an second anomaly within the superconducting state suggests a transition into an FFLO state.

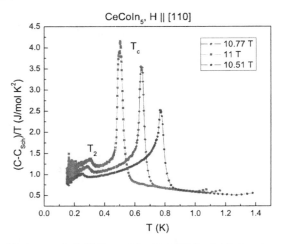

Fig. 11.6. Specific heat of CeCoIn5 with $H \perp [001]$. Additional anomaly (marked as T_2) below a sharp first order transition into superconducting state at T_c indicates transition in to a potentially FFLO state.

11.2 Magnetism in CeCoIn5

In addition to the measurements described above, there has been a number of other reports in support of the FFLO scenario in CeCoIn5, including a number of thermodynamic, transport, and microscopic studies. For a recent review, see Ref. [16]. However, recent NMR investigations [14] showed that there is a long range antiferromagnetic order within the HFLT phase. Figure 11.7 shows NMR spectra taken at 11.1 T for a range of temperatures from 890 to 50 mK. In(1) site lies in the Ce plane, whereas In(2) site lies outside of the Ce plane. The resonance associated with the In(1) site shifts to lower frequency, but remains sharp. The signal from the In(2) resonance disappears at the HFLT transition at about 300 mK, and reappears at much lower temperature below 100 mK as a broad double-peak structure over a range of 2.6 MHz, characteristic of a powder-pattern-like spectrum due to an incommensurate magnetic order. Therefore, the two In sites within a single unit cell experience different magnetic field distribution, which lead the authors to propose that an AFM long range order is present in the HFLT state with the ordering vector $q = (\pi/a - \delta; \pi/a, \pi/b)$ [14].

Is it possible that the HFLT phase is of magnetic origin? Previous studies [17, 18] have shown that there is indeed a nearby magnetic ground state, which manifests itself via non-Fermi-liquid (NFL) behavior in the normal state of CeCoIn5. Figure 11.8 shows specific heat data of CeCoIn5 for field $H \parallel [001]$ at the superconducting critical field $H_{c2} = 5$ T, and above it up to 9 T. Logarithmic divergence of $\gamma = C/T$ vs. T at H_{c2} is a non-Fermi-liquid behavior characteristic of the proximity to a Quantum Critical Point (QCP). Similar behavior is observed above H_{c2} for $H \parallel [100]$ [19].

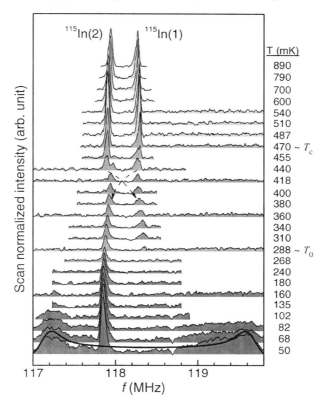

Fig. 11.7. NMR spectra of In(1) and In(2) transitions in CeCoIn₅ at 11.1 T. The In(1) transition shifts down in frequency around $T_c = 470$ mK, whereas the In(2) shifts up in frequency. The broad double-peak structure at 50 mK is the In(2) spectrum, and the solid line is the fit to a model with the AFM ordering vector $q = (\pi/a - \delta; \pi/a, \pi/b)$, as discussed in Ref. [14].

It was suggested [17] that the QCP is due to an AFM ground state that is superseded by superconductivity $(T_c > T_N)$, and when the SC state is formed, all of the Fermi Surface is gapped, leaving no states to participate in the AFM order. Such scenario leads to a phase diagram shown in Fig. 11.9, where magnetic field suppresses superconductivity faster than magnetism, and T_c and T_N are driven to zero by roughly the same magnetic field ($H_{QCP} \approx H_{c2} = 5$ T). Several attempts were made to separate H_{c2} and H_{QCP} and uncover the AFM state. Doping Sn for In suppressed both T_c and T_N roughly equally, with the strongest $\log T$ divergence of γ remaining at H_{c2} [20]. Recent Cd-doping studies, however, finally revealed the underlying AFM state [21]. As Fig. 11.10 shows, Cd doping of about 0.5% suppresses the superconducting state, and stabilizes the AFM ground state. The dashed line represents a possible extrapolation of the AFM phase boundary that would intercept the

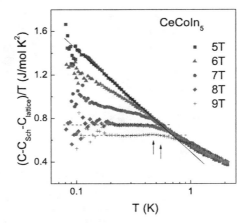

Fig. 11.8. Specific heat of CeCoIn$_5$ with $H \parallel [001]$. C/T diverges logarithmically at H_{c2}. Arrows indicate the onset of the Fermi liquid behavior (constant C/T) at higher fields.

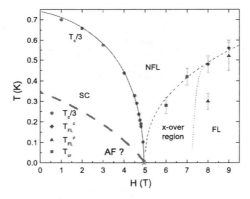

Fig. 11.9. Phase diagram of Co with $H \parallel [001]$, showing a large non-Fermi-liquid (NFL) region. Putative boundary of the avoided AFM state, which leads to the NFL behavior, is indicated by the dashed line.

temperature axis at a positive value. $x = 0$ situation is in fact identical to the $H = 0$ one displayed in Fig. 11.9. Thus, the AFM phase revealed by the Cd doping studies is the "avoided AFM phase" suggested to exist in CeCoIn$_5$ on the basis of earlier specific heat studies [17].

11.3 CeRhIn$_5$ under pressure: Close relative to CeCoIn$_5$

CeRhIn$_5$ is an ambient pressure antiferromagnet. Hydrostatic pressure suppresses AFM state and stabilizes superconductivity [22, 23]. Above the pressure $P_1 = 1.77$ GPa, $T_c > T_{AFM}$, and once superconducting state is stabilized,

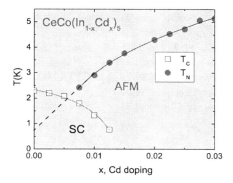

Fig. 11.10. Phase diagram of CeCo(In$_{1-x}$Cd$_x$)$_5$. Cd doping suppresses SC and stabilizes AFM state. Dashed line is a possible extrapolation of the AFM phase boundary to $x = 0$.

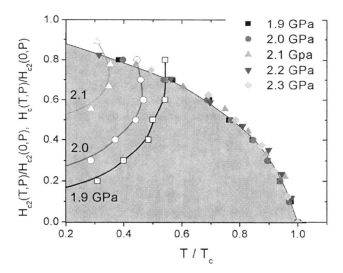

Fig. 11.11. Reduced phase diagram of CeRhIn₅ for several pressures $P > P_1$. When AFM ground state is close. Dashed line is a possible extrapolation of the AFM phase boundary to $x = 0$.

all the Fermi Surface is gapped, and AFM state does not develop below T_c. However, it was shown that application of magnetic field for pressure just above P_1 restores long range magnetic order [24]. Figure 11.11 shows reduced $H - T$ phase diagram of CeRhIn₅ for several pressures above P_1. Magnetic field induces an AFM state within the superconducting state, and the closer the pressure is to P_1, the smaller magnetic field is required to induce the long range magnetic order. This phenomena was explained within the model [25] where AFM correlations set in within the vortex cores, and become long range order as the vortices come close together with increasing magnetic field. The shape of the field-induced AFM phases in CeRhIn$_5$ is reminiscent of that of the HFLT phase in CeCoIn$_5$, however, there is a very important difference. In case of CeRhIn$_5$ AFM state penetrates into the normal state above H_{c2}, whereas in CeCoIn$_5$ the boundary between the HFLT phase and the mixed superconducting vortex state does not cross the SC - normal phase boundary $H_{c2}(T)$. The HFLT phase is integrally connected to superconductivity, as if the AFM order is "attracted" to superconductivity, a unique situation.

11.4 HFLT phase and magnetism in CeCoIn$_5$ under pressure

As mentioned above, recent NMR studies of CeCoIn$_5$ found evidence of a long range AFM order within the HFLT phase. The question arises then whether the avoided AFM state, revealed by the Cd doping studies, is at the core of the HFLT phase? Does quantum criticality at H_{c2} lead to the formation of the HFLT state? Measurements under pressure give us important clues about the interplay of superconductivity and magnetism in CeCoIn$_5$.

Specific heat measurements under pressure with $H \parallel [110]$ revealed that pressure increases the extent of the superconducting phase, with both T_c and H_{c2} growing [11]. In addition, the measurements showed that the phase space occupied by the HFLT phase itself increases as well. To test the connection between the QCP and the HFLT phase we investigated the dependence of the field H_{QCP} of the QCP on pressure [26] via the resistivity measurements.

Figure 11.12 shows the results of the analysis of the resistivity data for several pressures up to 1.3 GPa with field $H \parallel [001]$. For each pressure resistivity was fit to the expression $\rho = \rho_0 + AT^2$. The resulting values of the coefficient A of the quadratic term are shown in Fig. 11.12a. A is proportional

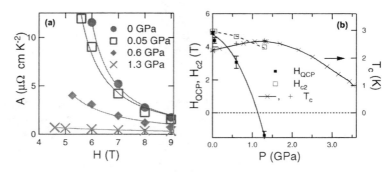

Fig. 11.12. Analysis of the resistivity data of CeCoIn$_5$ under pressure. (a) Coefficient A of the $\rho = \rho_0 + AT^2$ fit to the resistivity data at several pressures. Solid curves are the fits as described in the text, diverging at H_{QCP}. (b) H_{QCP} and H_{c2} as a function of pressure, with H_{QCP} suppressed to zero by 1.3 GPa.

to the square of quasiparticle mass, or, alternatively, the density of states at the Fermi level. It is expected to diverge at a QCP, and we fit it to the expression $A = \alpha(H - H_{QCP})^{\beta}$. The resulting fits are displayed in the Fig. 11.12a. Figure 11.12b shows H_{QCP} obtained from the fit, together with H_{c2} and T_c, as a function of pressure. H_{QCP} is suppressed to zero by about 1.3 GPa, clearly separating from H_{c2}. This is a strong indication that the tendency toward magnetism is suppressed with pressure, while the HFLT state was shown to be enhanced by it. These results support the conjecture that the HFLT state is of non-magnetic, possibly FFLO, origin.

11.5 Conclusions

In this paper we presented results of several measurements that paint a picture of a complex state in CeCoIn$_5$ with subtle balance between tendencies toward magnetic order and superconductivity. As a result, the Quantum Critical Point due to the underlying antiferromagnetic state, coincides with the superconducting critical field in CeCoIn$_5$. In addition, a novel superconducting state develops in the high field and low temperature (HFLT) corner of the state of the superconducting phase with the long range antiferromagnetic order. Regardless of whether the HFLT state is of purely magnetic origin, or magnetism accompanies a fundamentally FFLO state, we are presented here with magnetism that is stabilized in the superconducting state only, and does not extend into the normal state. This is a highly unique situation, exact opposite to the canonical picture of the competition between superconductivity and magnetism, and is worthy of detailed experimental and theoretical studies. Additional investigations of the HFLT state in CeCoIn$_5$ are required before a firm case can be made for its nature.

References

1. C. Petrovic, P.G. Pagliuso, M.F. Hundley, R. Movshovich, J.L. Sarrao, J.D. Thompson, Z. Fisk, J. Phys. Condens. Matter. **13**, L337 (2001)
2. P.M. Oppeneer, S. Elgazzar, A.B. Shick, I. Opahle, J. Rusz, R. Hayn, J. Magnet. Magnetic Mater. **310**(2 SUPPL. PART 2), 1684 (2007)
3. D. Hall, E.C. Palm, T.P. Murphy, S.W. Tozer, Z. Fisk, U. Alver, R.G. Goodrich, J.L. Sarrao, P.G. Pagliuso, T. Ebihara, Phys. Rev. B **64**, 212508 (2001)
4. D. Hall, E.C. Palm, T.P. Murphy, S.W. Tozer, C. Petrovic, E. Miller-Ricci, L. Peabody, C.Q.H. Li, U. Alver, R.G. Goodrich, J.L. Sarrao, P.G. Pagliuso, J.M. Wills, Z. Fisk, Phys. Rev. B (Condens. Matter Mater. Phys.) **64**(6), 064506 (2001)
5. H. Shishido, R. Settai, S. Hashimoto, Y. Inada, Y. Onuki, J. Magnet. Magnet. Mater. **272/276**, 225 (2004)
6. T. Tayama, A. Harita, T. Sakakibara, Y. Haga, H. Shishido, R. Settai, Y. Onuki, Phys. Rev. B **65**, 180504 (2002)

7. P. Fulde, R.A. Ferrell, Phys. Rev. **135**, A550 (1964)
8. A.I. Larkin, Y.N. Ovchinnikov, J. Exptl. Theoret. Phys. (USSR) **47**, 1136 (1964). [Sov. Phys. JETP **20**, 762, (1965).]
9. A. Bianchi, R. Movshovich, C. Capan, P.G. Pagliuso, J.L. Sarrao, Phys. Rev. Lett. **91**, 187004 (2003)
10. A.M. Clogston, Phys. Rev. Lett. **2**, 9 (1962)
11. C.F. Miclea, M. Nicklas, D. Parker, K. Maki, J.L. Sarrao, J.D. Thompson, G. Sparn, F. Steglich, Phys. Rev. Lett. **96**(11), 117001 (2006)
12. K. Maki, T. Tsuneto, Prog. Theoretic. Phys. **31**, 945 (1964)
13. A. Bianchi, R. Movshovich, N. Oeschler, P. Gegenwart, F. Steglich, J.D. Thompson, P.G. Pagliuso, J.L. Sarrao, Phys. Rev. Lett. **89**, 137002 (2002)
14. B.L. Young, R.R. Urbano, N.J. Curro, J.D. Thompson, J.L. Sarrao, A.B. Vorontsov, M.J. Graf, Phys. Rev. Lett. **98**(3), 036402 (2007)
15. R. Movshovich, M. Jaime, J.D. Thompson, C. Petrovic, Z. Fisk, P.G. Pagliuso, J.L. Sarrao, Phys. Rev. Lett. **86**, 5152 (2001)
16. Y. Matsuda, H. Shimahara, J. Phys. Soc. Jap. **76**, 051005 (2007)
17. A. Bianchi, R. Movshovich, I. Vekhter, P.G. Pagliuso, J.L. Sarrao, Phys. Rev. Lett. **91**, 257001 (2003)
18. J. Paglione, M.A. Tanatar, D.G. Hawthorn, E. Boaknin, R.W. Hill, F. Ronning, M. Sutherland, L. Taillefer, C. Petrovic, P.C. Canfield, Phys. Rev. Lett. **91**, 246405 (2003)
19. F. Ronning, C. Capan, A. Bianchi, R. Movshovich, A. Lacerda, M.F. Hundley, J.D. Thompson, P.G. Pagliuso, J.L. Sarrao, Phys. Rev. B **71**(10), 104528 (2005)
20. E. Bauer, C. Capan, F. Ronning, R. Movshovich, J.D. Thompson, J.L. Sarrao, Phys. Rev. Lett. **94**, 047001 (2005)
21. L.D. Pham, T. Park, S. Maquilon, J.D. Thompson, Z. Fisk, Phys. Rev. Lett. **97**, 056404 (2006)
22. H. Hegger, C. Petrovic, E.G. Moshopoulou, M.F. Hundley, J.L. Sarrao, Z. Fisk, J.D. Thompson, Phys. Rev. Lett. **84**, 4986 (2000)
23. M. Nicklas, V.A. Sidorov, H.A. Borges, P.G. Pagliuso, J.L. Sarrao, J.D. Thompson, Phys. Rev. B **70**(2), 020505 (2004)
24. T. Park, F. Ronning, H.Q. Yuan, M.B. Salamon, R.M. Adn J.L. Sarrao, J.D. Thompson, Nature **440**, 65 (2006)
25. E. Demler, S.Y. Zhang, Phys. Rev. Lett. **87**, 067202 (2001)
26. F. Ronning, C. Capan, E.D. Bauer, J.D. Thompson, J.L. Sarrao, R. Movshovich, Phys. Rev. B **73**(6), 064519 (2006)

BIPOLARONIC PROXIMITY AND OTHER UNCONVENTIONAL EFFECTS IN CUPRATE SUPERCONDUCTORS

A.S. ALEXANDROV

Department of Physics, Loughborough University,
Loughborough, United Kingdom. $a.s.alexandrov@lboro.ac.uk$

Abstract. There is compelling evidence for a strong electron-phonon interaction (EPI) in cuprate superconductors from the isotope effects on the supercarrier mass, high resolution angle resolved photoemission spectroscopies (ARPES), a number of optical and neutron-scattering measurements in accordance with our prediction of high-temperature superconductivity in polaronic liquids. A number of observations point to the possibility that high-T_c cuprate superconductors may not be conventional Bardeen-Cooper-Schrieffer (BCS) superconductors, but rather derive from the Bose-Einstein condensation (BEC) of real-space pairs, which are mobile small bipolarons. Here I review the bipolaron theory of unconventional proximity effects, the symmetry and checkerboard modulations of the order parameter and quantum magneto-oscillations discovered recently in cuprates.

Keywords: bipolarons, cuprates, proximity, symmetry, magnetooscillations

12.1 Polarons in high-temperature superconductors

Many unconventional properties of cuprate superconductors may be attributed to the Bose-Einstein condensation (BEC) of real-space pairs, which are mobile small bipolarons [1–7]. A possible fundamental origin of such strong departure of the cuprates from conventional BCS behaviour is the unscreened Fröhlich EPI providing the polaron level shift E_p of the order of 1 eV [3], which is routinely neglected in the Hubbard U and $t - J$ models. This huge interaction with c-axis polarized optical phonons is virtually unscreened at any doping of cuprates. In order to build an adequate theory of high-temperature superconductivity, the long-range Coulomb repulsion and the *unscreened* EPI should be treated on an equal footing with the short-range Hubbard U. When these interactions are strong compared with the kinetic energy of carriers, this

J. Bonča, S. Kruchinin (eds.), *Electron Transport in Nanosystems.*

"Coulomb-Fröhlich" model predicts the ground state in the form of mobile inter-site bipolarons [3, 8].

Nowadays compelling evidence for a strong EPI has arrived from isotope effects [9], more recent high resolution angle resolved photoemission spectroscopies (ARPES) [10], and a number of earlier optical [11–14], neutron-scattering [15] and recent inelastic scattering measurements [16] in cuprates. Whereas calculations based on the local spin-density approximation (LSDA) often predict negligible EPI, the inclusion of Hubbard U in the $LSDA + U$ calculations greatly enhances its strength [17].

A parameter-free estimate of the Fermi energy using the magnetic-field penetration depth found a very low value, $\epsilon_F \lesssim 100$ meV [18] clearly supporting the real-space (i.e. individual) pairing in cuprate superconductors. There is strong experimental evidence for a gap in the normal-state electron density of states of cuprates [2], which is known as the pseudogap. Experimentally measured pseudogaps of many cuprates are $\gtrsim 50$ meV [19]. If following Ref. [20] one accepts that the pseudogap is about half of the pair binding energy, Δ, then the condition for real-space pairing, $\epsilon_F \lesssim \pi\Delta$, is well satisfied in most cuprates (typically the small bipolaron radius is $r_b \approx 0.2-0.4$ nm).

Also magnetotransport and thermal magnetotransport data strongly support preformed bosons in cuprates. In particular, many high-magnetic-field studies revealed a non-BCS upward curvature of the upper critical field $H_{c2}(T)$ (see [21] for a review of experimental data), in accordance with the theoretical prediction for the Bose-Einstein condensation of charged bosons in the magnetic field [22]. The Lorenz number, $L = e^2\kappa_e/T\sigma$ differs significantly from the Sommerfeld value $L_e = \pi^2/3$ of the standard Fermi-liquid theory, if carriers are double-charged bosons [23]. Here κ_e, and σ are electron thermal and electrical conductivities, respectively. Reference [23] predicted a rather low Lorenz number for bipolarons, $L \approx 0.15L_e$, due to the double elementary charge of bipolarons, and also due to their nearly classical distribution function above T_c. Direct measurements of the Lorenz number using the thermal Hall effect [24] produced the value of L just above T_c about the same as predicted by the bipolaron model, and its strong temperature dependence. This breakdown of the Wiedemann-Franz law is apparently caused by excited single polarons coexisting with bipolarons in the thermal equilibrium [25, 26]. Also unusual normal state diamagnetism uncovered by torque magnetometery has been convincingly explained as the normal state (Landau) diamagnetism of charged bosons [27].

However, despite clear evidence for the existence of polarons in cuprates, no consensus currently exists concerning the microscopic mechanism of high-temperature superconductivity. While a number of early (1990s) and more recent studies prove that the Mott-Hubbard insulator promotes doping-induced polaron formation [28], some other works suggest that EPI does not only not help, but hinder the pairing instability. The controversy should be resolved experimentally. Here I argue that the giant (GPE) and nil (NPE) proximity effects provide another piece of evidence for bipolaronic BEC in cuprates [29]. The same bipolaronic scenario also explains the symmetry and checkerboard modulations of the order parameter, and recently observed magnetooscillations in the vortex state of cuprates.

12.2 Unconventional proximity effects

Several groups [30] reported that in the Josephson cuprate SNS junctions supercurrent can run through normal N-barriers with the thickness $2L$ greatly exceeding the coherence length, when the barrier is made from a slightly doped non-superconducting cuprate (the so-called N' barrier), Fig. 12.1.

Using the advanced molecular beam epitaxy, Bozovic et al. [30] proved that GPE is intrinsic, rather than caused by any extrinsic inhomogeneity of the barrier. Resonant scattering of soft-x-ray radiation did not find any signs of intrinsic inhomogeneity (such as charge stripes, charge-density waves, etc.) either [31]. Hence GPE defies the conventional explanation, which predicts that the critical current should exponentially decay with the characteristic length of about the coherence length, $\xi \lesssim 1$ nm in cuprates. Annealing the junctions at low temperatures in vacuum rendered the barrier insulating. Remarkably when the $SN'S$ junction was converted into a superconductor-insulator-superconductor (SIS) device no supercurrent was observed, even in devices with the thinnest (one unit cell thick) barriers [32] (nil proximity effect, NPE).

Both GPE and NPE can be broadly understood as the bipolaronic Bose-condensate tunnelling into a cuprate *semiconductor* [29].

To illustrate the point one can apply the Gross-Pitaevskii (GP)-type equation for the superconducting order parameter $\psi(\mathbf{r})$, generalized by us [33] for

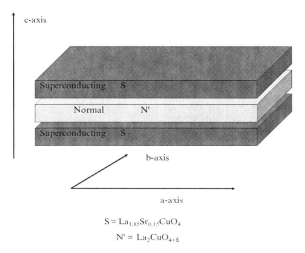

$$S = La_{1.85}Sr_{0.15}CuO_4$$
$$N' = La_2CuO_{4+\delta}$$

Fig. 12.1. SNS cuprate nanostructure.

a charged Bose liquid (CBL), since, as discussed above, many observations including a small coherence length point to a possibility that cuprate superconductors may not be conventional BCS superconductors, but rather derive from BEC of real-space pairs, such as mobile small bipolarons [3],

$$\left[E(-i\hbar\nabla + 2e\mathbf{A}) - \mu + \int d\mathbf{r}' V(\mathbf{r} - \mathbf{r}')|\psi(\mathbf{r}')|^2 \right] \psi(\mathbf{r}) = 0. \tag{12.1}$$

Here $E(\mathbf{K})$ is the center-of-mass pair dispersion and the Peierls substitution, $\mathbf{K} \Rightarrow -i\hbar\nabla + 2e\mathbf{A}$ is applied with the vector potential $\mathbf{A}(\mathbf{r})$.

The integro-differential equation (12.1) is quite different from the Ginzburg-Landau [34] and Gross-Pitaevskii [35] equations, describing the order parameter in the BCS and neutral superfluids, respectively. Here μ is the chemical potential and $V(\mathbf{r})$ accounts for the long-range Coulomb and a short-range composed boson-boson repulsions [4]. While the electric field potential can be found from the corresponding Poisson-like equation [33], a solution of two coupled nonlinear differential equations for the order parameter $\psi(\mathbf{r})$ and for the potential $V(\mathbf{r})$ in the nanostructure, Fig. 12.2, is a nontrivial mathematical problem. For more transparency we restrict our analysis in this section by a short-range potential, $V(\mathbf{r}) = v|\psi(\mathbf{r})|^2$, where a constant v accounts for the short-range repulsion. Then in the absence of the magnetic field Eq. (12.1) is reduced to the familiar GP equation [35] using the continuum (effective mass) approximation, $E(\mathbf{K}) = K^2/2m_c$. In the tunnelling geometry of $SN'S$ junctions, Figs. 12.2, 12.3, it takes the form,

$$\frac{1}{2m_c}\frac{d^2\psi(Z)}{dZ^2} = [v|\psi(Z)|^2 - \mu]\psi(Z), \tag{12.2}$$

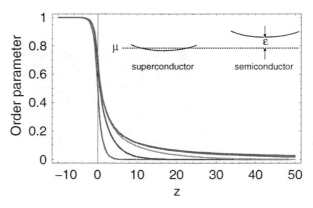

Fig. 12.2. BEC order parameter at the SN boundary for $\tilde{\mu} = 1.0, 0.1, 0.01$ and $\leqslant 0.001$ (upper curve). The chemical potential is found above the boson band-edge due to the boson-boson repulsion in cuprate superconductors and below the edge in cuprate semiconductors with low doping.

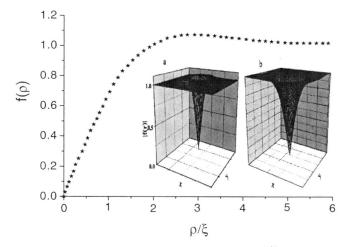

Fig. 12.3. The order parameter profile $f(\rho) = \psi(\mathbf{r})/n_s^{1/2}$ of a single vortex in CBL [33] (symbols). Inset: CBL vortex (a) [33, 49] compared with the Abrikosov vortex (b) [48] (here $\rho = [x^2 + y^2]^{1/2}$).

in the superconducting region, $Z < 0$, Fig. 12.3. Here m_c is the boson mass in the direction of tunnelling along Z ($\hbar = c = k_B = 1$ in this section). Deep inside the superconductor the order parameter is a constant, $|\psi(Z)|^2 = n_s$ and $\mu = v n_s$, where the condensate density n_s is about $n_s \approx x/2$, if the temperature is well below T_c of the superconducting electrode. Here the in-plane lattice constant a and the unit cell volume are taken as unity, and x is the doping level as in $La_{2-x}Sr_xCuO_4$.

The normal barrier at $Z > 0$ is an underdoped cuprate above its transition temperature, $T_c' < T$ where the chemical potential μ lies below the bosonic band by some energy ϵ, Fig. 12.3, found from $\int dE N(E)[\exp((E + \epsilon)/T) - 1]^{-1} = x'/2$. Here $N(E)$ is the bipolaron density of states (DOS), and $x' < x$ is the doping level of the barrier. In-plane bipolarons are quasi-two dimensional repulsive bosons, propagating along the CuO_2 planes with the effective mass m several orders of magnitude smaller than their out-of-plane mass, $m_c \gg m$, [4]. Using bipolaron band dispersion, $E(\mathbf{K}) = K^2/2m + 2t_c[1 - \cos(K_\perp d)]$, the density of states is found as $N(E) = (m/2\pi^2) \arccos(1 - E/2t_c)$ for $0 < E < 4t_c$, and $N(E) = m/2\pi$ for $4t_c < E$. Here K and $K_\perp$ are the in-plane and out-of-plane center-of-mass momenta, respectively, $t_c = 1/2\, m_c d^2$, and d is the inter-plane distance. As a result one obtains

$$\epsilon(T) \leqslant -T \ln(1 - e^{-T_0/T}), \qquad (12.3)$$

which is exponentially small at $T_c' < T \ll T_0$ turning into zero at $T = T_c'$, where $T_c' \approx T_0/\ln(T_0/2t_c)$, and $T_0 = \pi x'/m \gg T_c' \gg t_c$.

It is important to note that $\epsilon(T)$ remains also small at $T_c'(2D) \leqslant T \ll T_0$ in the purely two-dimensional repulsive Bose-gas [36]. While in two dimensions Bose condensation does not occur in either the ideal or the inter-acting system, there is a phase transition to a superfluid state at $T'_c(2D) = T_0/\ln(1/f_0) \ll T_0$, where $f_0 \ll 1$ depends on the density of hard-core dilute bosons and their repulsion [36, 37]. The superfluid transition takes place only if there is a residual repulsion between bosons, i.e. $T'_c(2D) = 0$ for the ideal 2D Bose-gas. Actually $T'_c(2D)$ gives a very good estimate for the exact Berezinski-Kosterlitz-Thouless (BKT) critical temperature in the dilute Bose gas, where the BKT contribution of vertices is important only very close to $T'_c(2D)$ [37].

The GP equation in the barrier can be written as

$$\frac{1}{2m_c}\frac{d^2\psi(Z)}{dZ^2} = [v|\psi(Z)|^2 + \epsilon]\psi(Z). \tag{12.4}$$

Introducing the bulk coherence length, $\xi = 1/(2m_c n_s v)^{1/2}$ and dimensionless $f(z) = \psi(Z)/n_s^{1/2}$, $\tilde{\mu} = \epsilon/n_s v$, and $z = Z/\xi$ one obtains for a real $f(z)$

$$\frac{d^2 f}{dz^2} = f^3 - f, \tag{12.5}$$

if $z < 0$, and

$$\frac{d^2 f}{dz^2} = f^3 + \tilde{\mu}f, \tag{12.6}$$

if $z > 0$. These equations can be readily solved using first integrals of motion respecting the boundary conditions, $f(-\infty) = 1$, and $f(\infty) = 0$,

$$\frac{df}{dz} = -(1/2 + f^4/2 - f^2)^{1/2}, \tag{12.7}$$

and

$$\frac{df}{dz} = -(\tilde{\mu}f^2 + f^4/2)^{1/2}, \tag{12.8}$$

for $z < 0$ and $z > 0$, respectively. The solution in the superconducting electrode is given by

$$f(z) = \tanh\left[-2^{-1/2}z + 0.5\ln\frac{2^{1/2}(1+\tilde{\mu})^{1/2}+1}{2^{1/2}(1+\tilde{\mu})^{1/2}-1}\right]. \tag{12.9}$$

It decays in the close vicinity of the barrier from 1 to $f(0) = [2(1+\tilde{\mu})]^{-1/2}$ in the interval about the coherence length ξ. On the other side of the boundary, $z > 0$, it is given by

$$f(z) = \frac{(2\tilde{\mu})^{1/2}}{\sinh\{z\tilde{\mu}^{1/2} + \ln[2(\tilde{\mu}(1+\tilde{\mu}))^{1/2} + (1+4\tilde{\mu}(1+\tilde{\mu}))^{1/2}]\}}. \tag{12.10}$$

Its profile is shown in Fig. 12.3. Remarkably, the order parameter penetrates the normal layer up to the length $Z^* \approx (\tilde{\mu})^{-1/2}\xi$, which could be larger than ξ

by many orders of magnitude, if $\tilde{\mu}$ is small. It is indeed the case, if the barrier layer is sufficiently doped. For example, taking $x' = 0.1$, c-axis $m_c = 2,000$ m_e, in-plane $m = 10\,m_e$, $a = 0.4$ nm, and $\xi = 0.6$ nm, yields $T_0 \approx 140$ K and $(\tilde{\mu})^{-1/2} \gtrsim 50$ at $T = 25$ K. Hence the order parameter could penetrate the normal cuprate semiconductor up to a hundred coherence lengths or even more as observed (GPE) [30]. If the thickness of the barrier L is small compared with Z^*, and $(\tilde{\mu})^{1/2} \ll 1$, the order parameter decays following the power law, rather than exponentially,

$$f(z) = \frac{\sqrt{2}}{z + 2}. \tag{12.11}$$

Hence, for $L \lesssim Z^*$, the critical current should also decay following the power law rayher than exponentially. On the other hand, for the *undoped* barrier $\tilde{\mu}$ becomes larger than unity, $\tilde{\mu} \propto \ln(mT/\pi x') \to \infty$ for any finite temperature T when $x' \to 0$, and the current should exponentially decay with the characteristic length smaller that ξ, which is experimentally observed as well (NPE) [32].

12.3 Quantum magneto-oscillations, d-wave symmetry and checkerboard modulations

Until recently no convincing signatures of quantum magneto-oscillations have been found in the normal state of cuprate superconductors despite significant experimental efforts. There are no normal state oscillations even in high quality single crystals of overdoped cuprates like $Tl_2Ba_2CuO_6$, where conditions for de Haas-van Alphen (dHvA) and Shubnikov-de Haas (SdH) oscillations seem to be perfectly satisfied [38] and a large Fermi surface is identified in the angle-resolved photoemission spectra (ARPES) [39]. The recent observations of magneto-oscillations in kinetic [40, 41] and magnetic [42, 43] response functions of underdoped $YBa_2Cu_3O_{6.5}$ and $YBa_2Cu_4O_8$ are perhaps even more striking since many probes of underdoped cuprates including ARPES [44] clearly point to a non Fermi-liquid normal state. Their description in the framework of the standard theory for a metal [45] has led to a very small Fermi-surface area of a few percent of the first Brillouin zone [40–43], and to a low Fermi energy of only about *the room temperature* [42]. Clearly such oscillations are incompatible with the first-principle (LDA) band structures of cuprates, but might be compatible with a a low Fermi energy and non-adiabatic polaronic normal state of charge-transfer Mott insulators as discussed in Section 12.1.

Nevertheless one can raise a doubt concerning their normal state origin. The magnetic length, $\lambda \equiv (\pi\hbar/eB)^{1/2} \gtrsim 5$ nm, remains larger than the zero-temperature in-plane coherence length, $\xi \lesssim 2$ nm, measured independently, in any field reached in Ref. [40–43]. Hence the magneto-oscillations are observed in the vortex (mixed) state well below the upper critical field, rather than in the normal state, as also confirmed by the *negative* sign of the Hall resistance [40]. It is well known, that in "YBCO" the Hall conductivities of vortexes and quasiparticles have opposite sign causing the sign change in the Hall effect in the mixed state [46]. Also there is a substantial magnetoresistance [41], which is a signature of the flux flow regime rather than of the normal state. Hence it would be rather implausible if such oscillations have a normal-state origin due to small electron Fermi surface pockets [43] with the characteristic wave-length of electrons larger than the widely accepted coherence length.

Here I propose an alternative explanation of the magneto-oscillations [40–43] as emerging from the quantum interference of the vortex lattice and the checkerboard or lattice modulations of the order parameter observed by STM with atomic resolution [47]. The checkerboard effectively pins the vortex lattice, when the period of the latter, λ is commensurate with the period of the checkerboard lattice, a. The condition $\lambda = Na$, where N is a large integer, yields $1/B^{1/2}$ periodicity of the response functions, rather than $1/B$ periodicity of conventional normal state magneto-oscillations.

The integro-differential equation (12.1) in the continuum approximation, $E(\mathbf{K}) = \hbar^2 K^2/2m$, with the long-range Coulomb repulsion between double charged bosons, $V(\mathbf{r}) = V_c(\mathbf{r}) = 4e^2/\epsilon_0 r$, describes a single vortex with a *charged* core, Fig. 12.4, when the magnetic field, B is applied. The coherence length in this case is roughly the same as the screening radius, $\xi = (\hbar/2^{1/2}m\omega_p)^{1/2}$. Here $\omega_p = (16\pi n_s e^2/\epsilon_0 m)^{1/2}$ is the CBL plasma frequency, ϵ_0 the static dielectric constant of the host lattice, m is the (in-plane) boson mass, and n_s is the average condensate density. The chemical potential is zero, $\mu = 0$, if one takes into account the Coulomb interaction alone due to a neutralizing homogeneous charge background. Each vortex carries one flux

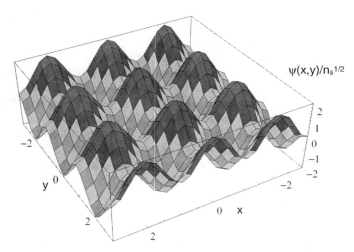

Fig. 12.4. The checkerboard d-wave order parameter of CBL [51] on the square lattice in zero magnetic field (coordinates x, y are measured in units of a).

quantum, $\phi_0 = \pi\hbar/e$, but it has an unusual core, Fig. 12.3a, Ref. [33], due to a local charge redistribution caused by the magnetic field, different from the conventional vortex [48], Fig. 12.3b. Remarkably, the coherence length turns out very small, $\xi \approx 0.5$ nm with the material parameters typical for underdoped cuprates, $m = 10\,m_e$, $n_s = 10^{21}$ cm^{-3} and $\epsilon_0 = 100$.

The coherence length ξ is so small at low temperatures, that the distance between two vortices remains large compared with the vortex size, $\lambda \gg \xi$, in any laboratory field reached so far [40–43]. It allows us to write down the vortex-lattice order parameter, $\psi(\mathbf{r}) = \psi_{vl}(\mathbf{r})$, as

$$\psi_{vl}(\mathbf{r}) \approx n_s^{1/2} \left[1 - \sum_j \phi(\mathbf{r} - \mathbf{r}_j) \right], \qquad (12.12)$$

where $\phi(\mathbf{r}) = 1 - f(\rho)$, and $\mathbf{r}_j = \lambda\{n_x, n_y\}$ with $n_{x,y} = 0, \pm 1, \pm 2, \ldots$ (if, for simplicity, we take the square vortex lattice [50]). The function $\phi(\rho)$ is linear well inside the core, $\phi(\rho) \approx 1 - 1.52\rho/\xi$ ($\rho \ll \xi$), and it has a small negative tail, $\phi(\rho) \approx -4\xi^4/\rho^4$ outside the core when $\rho \gg \xi$, Fig. 12.3 [33].

In the continuum approximation with the Coulomb interaction alone the magnetization of CBL follows the standard logarithmic law, $M(B) \propto \ln 1/B$ without any oscillations since the magnetic field profile is the same as in the conventional vortex lattice [49]. However, more often than not the center-of-mass Bloch band of preformed pairs, $E(\mathbf{K})$, has its minima at some finite wave vectors $\mathbf{K} = \mathbf{G}$ of their center-of-mass Brillouin zone [3, 4]. Near the minima the GP equation (12.1) is written as

$$\left[\frac{(-i\hbar\nabla - \hbar\mathbf{G} + 2e\mathbf{A})^2}{2m^{**}} - \mu \right] \psi(\mathbf{r}) +$$
$$\int d\mathbf{r}' V(\mathbf{r} - \mathbf{r}') |\psi(\mathbf{r}')|^2 \psi(\mathbf{r}) = 0, \qquad (12.13)$$

with the solution $\psi(\mathbf{r}) = \psi_{\mathbf{G}}(\mathbf{r}) \equiv e^{i\mathbf{G}\cdot\mathbf{r}} \psi_{vl}(\mathbf{r})$, if the interaction is the long-range Coulomb one, $V(\mathbf{r}) = V_c(\mathbf{r})$.

In particular, a nearest-neighbor (nn) approximation for the hopping of intersite bipolarons between oxygen p-orbitals on the CuO$_2$ 2D lattice yields four generate states $\psi_{\mathbf{G}}$ with $\mathbf{G}_i = \{\pm 2\pi/a_0, \pm 2\pi/a_0\}$, where a_0 is the lattice period [8]. Their positions in the Brillouin zone move towards Γ point beyond the nn approximation. The true ground state is a superposition of four degenerate states, respecting time-reversal and parity symmetries [51],

$$\psi(\mathbf{r}) = A n_s^{1/2} \left[\cos(\pi x/a) \pm \cos(\pi y/a) \right] \psi_{vl}(\mathbf{r}). \qquad (12.14)$$

Here we use the reference frame with x and y axes along the nodal directions and $a = 2^{-3/2} a_0$. Two "plus/minus" coherent states, Eq. (12.14), are physically identical since they are related via a translation transformation,

$y \Rightarrow y + a$. Normalizing the order parameter by its average value $\langle \psi(\mathbf{r})^2 \rangle = n_s$ and using $(\xi/\lambda)^2 \ll 1$ as a small parameter yield the following "minus" state amplitude, $A \approx 1 - N \sum_{n=0}^{\infty} 2[\tilde{\phi}_1(2^{1/2}\pi/a) + \tilde{\phi}_2(2^{1/2}\pi/a)]\delta_{n,R/2} + [\tilde{\phi}_1(2\pi/a) + \tilde{\phi}_2(2\pi/a)]\delta_{n,R}$ for the square vortex lattice [50] with the reciprocal vectors $\mathbf{g} = (2\pi/\lambda)\{n_x, n_y\}$. Here $\delta_{n,R}$ is the Kroneker symbol, $R = \lambda/a$ is the ratio of the vortex lattice period to the checkerboard period ($n = 0, 1, 2, ...$), $N = BS/\phi_0$ is the number of flux quanta in the area S of the sample, and $\tilde{\phi}_k(q) = (2\pi/S) \int_0^{\infty} d\rho \rho J_0(\rho q)\phi^k(\rho)$ is the Fourier transform of k's power of $\phi(\rho)$, where $J_0(x)$ is the zero-order Bessel function.

The order parameter $\psi(\mathbf{r})$, Eq. (12.14) has the d-wave symmetry changing sign in real space, when the lattice is rotated by $\pi/2$. This symmetry is due to the pair center-of-mass energy dispersion with the four minima at $\mathbf{K} \neq 0$, rather than due to a specific symmetry of the pairing potential. It also reveals itself as a *checkerboard* modulation of the carrier density with two-dimensional patterns in zero magnetic field, Fig. 12.4, as predicted by us [51] prior to their observations [47]. Solving the Bogoliubov-de Gennes equations with the order parameter, Eq. (12.14), yields the real-space checkerboard modulations of the single-particle density of states [51], similar to those observed by STM in cuprate superconductors.

Now we take into account that the interaction between composed pairs includes a short-range repulsion along with the long-range Coulomb one, $V(\mathbf{r}) = V_c(\mathbf{r}) + v\delta(\mathbf{r})$ [4]. At sufficiently low carrier density the short-range repulsion can be treated as a perturbation to the ground state, Eq. (12.14). Importantly the short-range repulsion energy of CBL, $U = (v/2)\langle \psi(\mathbf{r})^4 \rangle$, has a part, ΔU, oscillating with the magnetic field as

$$\frac{\Delta U}{U_0} \approx N \sum_{n=0}^{\infty} \left[A_1 \delta_{n,R/2} + A_2 \delta_{n,R} + A_3 \delta_{n,2R} \right], \qquad (12.15)$$

where $U_0 = v n_s^2/2$ is the hard-core energy of a homogeneous CBL, and the amplitudes are proportional to the Fourier transforms of $\phi(\rho)$ as

$$
\begin{aligned}
A_1 = {} & 15\tilde{\phi}_1\left(2^{1/2}\pi/a\right) - 45\tilde{\phi}_2\left(2^{1/2}\pi/a\right) + 24\tilde{\phi}_3\left(2^{1/2}\pi/a\right) \\
& - 6\tilde{\phi}_4\left(2^{1/2}\pi/a\right) + 8\tilde{\phi}_1\left(10^{1/2}\pi/a\right) - 12\tilde{\phi}_2\left(10^{1/2}\pi/a\right) \\
& + 8\tilde{\phi}_3\left(10^{1/2}\pi/a\right) - 2\tilde{\phi}_4\left(10^{1/2}\pi/a\right),
\end{aligned} \qquad (12.16)
$$

$$
\begin{aligned}
A_2 = {} & -(23/2)\tilde{\phi}_1\left(2\pi/a\right) + (57/2)\tilde{\phi}_2\left(2\pi/a\right) - 16\tilde{\phi}_3\left(2\pi/a\right) + 4\tilde{\phi}_4\left(2\pi/a\right) \\
& - 12\tilde{\phi}_1\left(2^{3/2}\pi/a\right) + 9\tilde{\phi}_2\left(2^{3/2}\pi/a\right) - 6\tilde{\phi}_3\left(2^{3/2}\pi/a\right) + 3\tilde{\phi}_4\left(2^{3/2}\pi/a\right)
\end{aligned}
$$
$$(12.17)$$

$$A_3 = -\tilde{\phi}_1\left(4\pi/a\right) + (3/2)\tilde{\phi}_2\left(4\pi/a\right) - \tilde{\phi}_3\left(4\pi/a\right) + (1/4)\tilde{\phi}_4\left(4\pi/a\right). \quad (12.18)$$

Fluctuations of the pulsed magnetic field and unavoidable disorder in cuprates induce some random distribution of the vortex-lattice period, λ. Hence one has to average ΔU over R with the Gaussian distribution, $G(R) = \exp[-(R - \bar{R})^2/\gamma^2]/\gamma\pi^{1/2}$ around an average $\bar{R}$ with the width $\gamma \ll \bar{R}$. Then using the Poisson summation formula yields

$$\frac{\Delta U}{U_0} = N \sum_{k=0}^{\infty} A_1 e^{-\pi^2 k^2 \gamma^2/16} \cos(\pi k \bar{R})$$
$$+ A_2 e^{-\pi^2 k^2 \gamma^2/4} \cos(2\pi k \bar{R}) + A_3 e^{-\pi^2 k^2 \gamma^2} \cos(4\pi k \bar{R}). \qquad (12.19)$$

The oscillating correction to the magnetic susceptibility, $\Delta\chi(B) = -\partial^2 \tilde{\Omega}/\partial B^2$, is strongly enhanced due to high oscillating frequencies in Eq. (12.19). Since the superfluid has no entropy we can use ΔU as the quantum correction to the thermodynamic potential $\tilde{\Omega}$ even at finite temperatures below $T_c(B)$. Differentiating twice the first harmonic ($k = 1$) of the first lesser damped term in Eq. (12.19) we obtain

$$\Delta\chi(B) \approx \chi_0 e^{-\delta^2 B_0/16B} \left(\frac{B_0}{B}\right)^2 \cos(B_0/B)^{1/2}, \qquad (12.20)$$

where $\chi_0 = U_0 S A_1 e^2 a^2/4\pi^4 \hbar^2$ is a temperature-dependent amplitude, proportional to the condensate density squared, $B_0 = \pi^3 \hbar/ea^2 = 8\pi^3 \hbar/ea_0^2$ is a characteristic magnetic field, which is approximately $1.1 \cdot 10^6$ Tesla for $a_0 \approx 0.38$ nm, and γ is replaced by $\gamma \equiv \delta\bar{R}$ with the relative distribution

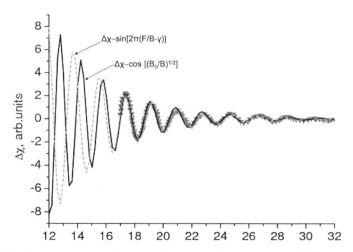

Fig. 12.5. Quantum corrections to the vortex-lattice susceptibility versus $1/B$, Eq. (12.20) (solid line, $B_0 = 1.000 \cdot 10^6$ Tesla, $\delta = 0.06$) compared with oscillating susceptibility of $YBa_2Cu_3O_{6.5}$ (symbols) and with the conventional normal state oscillations (dashed line) [43] at $T = 0.4$ K.

width δ. Assuming that $\xi \gtrsim a$, so that the amplitude A_1 is roughly a^2/S, the quantum correction $\Delta\chi$, Eq. (12.20), is of the order of wx^2/B^2, where x is the density of holes per unit cell. It is smaller than the conventional normal state (de Haas-van Alphen) correction, $\Delta\chi_{dHvA} \sim \mu/B^2$ [45], for a comparable Fermi-energy scale $\mu = wx$, since $x \ll 1$ in the underdoped cuprates.

Different from normal state dHvA oscillations, which are periodic versus $1/B$, the vortex-lattice oscillations, Eq. (12.20) are periodic versus $1/B^{1/2}$. They are quasi-periodic versus $1/B$ with a field-dependent frequency $F = B_0(B/B_0)^{1/2}/2\pi$, which is strongly reduced relative to the conventional-metal frequency ($\approx B_0/2\pi$) since $B \ll B_0$, as observed in the experiments [40–43]. The quantum correction to the susceptibility, Eq. (12.20) fits well the oscillations in $YBa_2Cu_3O_{6.5}$ [43], Fig. 12.5. The oscillations amplitudes, proportional to $n_s^2 \exp(-\delta^2 B_0/16B)$ decay with increasing temperature since the randomness of the vortex lattice, δ, increases, and the Bose-condensate evaporates.

12.4 Summary

A possibility of real-space pairing, proposed originally by Ogg [52] and later on by Schafroth and Blatt and Butler [53] has been the subject of many discussions as opposed to the Cooper pairing, particularly heated over the last 20 years after the discovery of high temperature superconductivity in cuprates. Our extension of the BCS theory towards the strong interaction between electrons and ion vibrations [4] proved that BCS and Ogg-Schafroth pictures are two extreme limits of the same problem. Cuprates are characterized by poor screening of high-frequency optical phonons, which allows the long-range Fröhlich EPI bound holes in superlight small bipolarons [3], which are several orders of magnitude lighter than small Holstein (bi)polarons.

The bipolaron theory accounts for GPE and NPE in slightly doped semiconducting and undoped insulating cuprates, respectively. It predicts the occurrence of a new length scale, $\hbar/\sqrt{2m_c\epsilon(T)}$, which turns out much larger than the zero-temperature coherence length in a wide temperature range above the transition point of the normal barrier, if bosons are almost two-dimensional (2D). The physical reason, why the quasi-2D bosons display a large normal-state coherence length, whereas 3D Bose-systems (or any-D Fermi-systems) at the same values of parameters do not, originates in the large DOS near the band edge of two-dimensional bosons compared with 3D DOS. Since DOS is large, the chemical potential is pinned near the edge with the magnitude, $\epsilon(T)$, which is exponentially small.

Here I also propose that the magneto-oscillations in underdoped cuprate superconductors result from the quantum interference of the vortex lattice and the lattice modulations of the d-wave order parameter, Fig. 12.3, which play the role of a periodic pinning grid. Our expression 12.20 describes the oscillations as well as the standard Lifshitz-Kosevich formula of dHvA and SdH effects [40–43]. The difference of these two dependencies could be resolved in ultrahigh magnetic fields as shown in Fig. 12.2. While our theory utilizes GP-type equation for composed charged bosons [33], the quantum interference of vortex and crystal lattice modulations of the order parameter is quite universal extending well beyond Eq. (12.1) independent of a particular pairing mechanism. It can also take place in the standard BCS superconductivity at $B < H_{c2}$, but hardly be observed because of much lower value of H_{c2} in conventional superconductors resulting in a very small damping factor, $\propto \exp(-\delta^2 B_0/16B) \ll 1$.

I appreciate valuable discussions with A.F. Andreev, I. Bozovic, L.P. Gor'kov, V.V. Kabanov, and support of this work by EPSRC (UK) (grant Nos. EP/D035589, EP/C518365).

References

1. A. S. Alexandrov, Phys. Rev. B **38**, 925 (1988).
2. A. S. Alexandrov and N. F. Mott, Rep. Prog. Phys. **57**, 1197 (1994).
3. A. S. Alexandrov, Phys. Rev. B **53**, 2863 (1996).
4. A. S. Alexandrov, *Theory of Superconductivity: From Weak to Strong Coupling* (IoP Publishing, Bristol/Philadelphia, 2003).
5. P. P. Edwards, C. N. R. Rao, N. Kumar, and A. S. Alexandrov, Chem. Phys. Chem. **7**, 2015 (2006).
6. A. S. Alexandrov, in *Studies in High Temperature Superconductors*, ed. A. V. Narlikar (Nova Science Pub., NY, 2006), v. **50**, pp. 1–69.
7. A. S. Alexandrov, J. Phys.: Condens. Matter **19**, 125216 (2007).
8. A. S. Alexandrov and P. E. Kornilovitch, J. Phys. Cond. Matt. **14**, 5337 (2002).
9. G. M. Zhao and D. E. Morris, Phys. Rev. B **51**, 16487 (1995); G. -M. Zhao, M. B. Hunt, H. Keller, and K. A. Müller, Nature (London) **385**, 236 (1997); R. Khasanov, D. G. Eshchenko, H. Luetkens, E. Morenzoni, T. Prokscha, A. Suter, N. Garifianov, M. Mali, J. Roos, K. Conder, and H. Keller, Phys. Rev. Lett. **92**, 057602 (2004).
10. A. Lanzara, P. V. Bogdanov, X. J. Zhou, S. A. Kellar, D. L. Feng, E. D. Lu, T. Yoshida, H. Eisaki, A. Fujimori, K. Kishio, J. I. Shimoyana, T. Noda, S. Uchida, Z. Hussain, and Z. X. Shen, Nature (London) **412**, 510 (2001); G. -H. Gweon, T. Sasagawa, S. Y. Zhou, J. Craf, H. Takagi, D. -H. Lee, and A. Lanzara, Nature (London) **430**, 187 (2004); X. J. Zhou, J. Shi, T. Yoshida, T. Cuk, W. L. Yang, V. Brouet, J. Nakamura, N. Mannella, S. Komiya, Y. Ando, F. Zhou, W. X. Ti, J. W. Xiong, Z. X. Zhao, T. Sasagawa, T. Kakeshita, H. Eisaki, S. Uchida, A. Fujimori, Z. -Y. Zhang, E. W. Plummer, R. B. Laughlin, Z. Hussain, and Z. -X. Shen, Phys. Rev. Lett. **95**, 117001 (2005).
11. D. Mihailovic, C. M. Foster, K. Voss, and A. J. Heeger, Phys. Rev. B **42**, 7989 (1990).
12. P. Calvani, M. Capizzi, S. Lupi, P. Maselli, A. Paolone, P. Roy, S. W. Cheong, W. Sadowski, and E. Walker, Solid State Commun. **91**, 113 (1994).
13. R. Zamboni, G. Ruani, A. J. Pal, and C. Taliani, Solid St. Commun. **70**, 813 (1989).
14. T. Timusk, C. C. Homes, and W. Reichardt, in *Anharmonic properties of High Tc cuprates* (eds. D. Mihailovic, G. Ruani, E. Kaldis, and K. A. Müller, Singapore: World Scientific, p. 171 (1995)).

15. T. R. Sendyka, W. Dmowski, T. Egami, N. Seiji, H. Yamauchi, and S. Tanaka, Phys. Rev. B **51**, 6747 (1995); T. Egami, J. Low Temp. Phys. **105**, 791 (1996).
16. D. Reznik, L. Pintschovius, M. Ito, S. Iikubo, M. Sato, H. Goka, M. Fujita, K. Yamada, G. D. Gu, and J. M. Tranquada, Nature **440**, 1170 (2006).
17. P. Zhang, S. G. Louie and M. L. Cohen, Phys. Rev. Lett. **98**, 067005 (2007).
18. A. S. Alexandrov, Physica C (Amsterdam) **363**, 231 (2001).
19. D. Mihailovic, V. V. Kabanov, K. Zagar, and J. Demsar, Phys. Rev. B **60**, 6995 (1999) and references therein.
20. A. S. Alexandrov, Physica C (Amsterdam) **182**, 327 (1991).
21. V. N. Zavaritsky, V. V. Kabanov, and A. S. Alexandrov, Europhys. Lett. **60**, 127 (2002).
22. A. S. Alexandrov, Phys. Rev. B **48**, 10571 (1993).
23. A. S. Alexandrov and N. F. Mott, Phys. Rev. Lett. **71**, 1075 (1993).
24. Y. Zhang, N. P. Ong, Z. A. Xu, K. Krishana, R. Gagnon, and L. Taillefer, Phys. Rev. Lett. **84**, 2219 (2000), and unpublished.
25. K. K. Lee, A. S. Alexandrov, and W. Y. Liang, Phys. Rev. Lett. **90**, 217001 (2003); Eur. Phys. J. B **30**, 459 (2004).
26. A. S. Alexandrov, Phys. Rev. B **73**, 100501 (2006).
27. A. S. Alexandrov, Phys. Rev. Lett. **96**, 147003 (2006).
28. H. Fehske and S. A. Trugman, in *Polarons in Advanced Materials*, ed. A. S. Alexandrov (Springer/Canopus, Bristol 2007), pp. 393–461; A. S. Mishchenko and N. Nagaosa, ibid, pp. 503–544.
29. A. S. Alexandrov, Phys. Rev. B **75**, 132501 (2007).
30. I. Bozovic, G. Logvenov, M. A. J. Verhoeven, P. Caputo, E. Goldobin, and M. R. Beasley, Phys. Rev. Lett. **93**, 157002 (2004), and references therein.
31. P. Abbamonte, L. Venema, A. Rusydi, G. A. Sawatsky, G. Logvenov, and I. Bozovic, Science **297**, 581 (2002).
32. I. Bozovic, G. Logvenov, M. A. J. Verhoeven, P. Caputo, E. Goldobin, and T. H. Geballe, Nature (London) **422**, 873 (2003).
33. A. S. Alexandrov, Phys. Rev. B **60**, 14573 (1999).
34. V. L. Ginzburg and L. D. Landau, Zh. Eksp. Teor. Fiz. **20**, 1064 (1950).
35. E. P. Gross, Nuovo Cimento **20**, 454 (1961); L. P. Pitaevskii, Zh. Eksp. Teor. Fiz. **40**, 646 (1961) (Soviet Phys. JETP **13**, 451 (1961)).
36. M. Yu. Kagan and D. V. Efremov, Phys. Rev. B **65**, 195103 (2002).
37. V. N. Popov, Theor. Math. Phys. **11**, 565 (1972); D. S. Fisher and P. C. Hohenberg, Phys. Rev. B **37**, 4936 (1988).
38. A. P. Mackenzie et al., Phys. Rev. Lett. **71**, 1238 (1993).
39. M. Plate et al., Phys. Rev. Lett. 95, 077001 (2005).
40. N. Doiron-Leyraud et al., Nature **447**, 565 (2007).
41. A. F. Bangura et al., arXiv:0707.4461.
42. E. A. Yelland et al., arXiv:0707.0057.
43. C. Jaudet et al., arXiv:0711.3559.
44. A. Damascelli, Z. Hussain, and Zhi-Xun Shen, Rev. Mod. Phys. **75** 473 (2003).
45. D. Schoenberg, *Magnetic Oscillations in Metals* (Cambridge University Press, Cambridge, 1984).
46. J. M. Harris, K. Krishana, N. P. Ong, R. Cagnon, and L. Taillefer, J. Low Temp. Phys. **105**, 877 (1996).
47. J. E. Hoffman et al. Science **295**, 466 (2002); C. Howald et al., Phys. Rev. B **67**, 014533 (2003); M. Vershinin et al. Science **303**, 1995 (2004).

48. A. A. Abrikosov, Zh. Eksp. Teor. Fiz. **32**, 1442 (1957); Soviet Phys. JETP **5**, 1174 (1957).
49. V. V. Kabanov and A. S. Alexandrov, Phys. Rev. B **71**, 132511 (2005).
50. Results for the square vortex lattice are also applied to the triangular lattice. Moreover there is a crossover from triangular to square coordination of vortices with increasing magnetic field in the mixed phase of cuprate superconductors (R. Gilardi et al., Phys. Rev. Lett. **88**, 217003 (2002); S. P. Brown et al., Phys. Rev. Lett. 92, 067004 (2004)).
51. A. S. Alexandrov, Physica C **305**, 46 (1998); Int. J. Mod. Phys. B **21**, 2301 (2007).
52. R. A. Ogg Jr., Phys. Rev. **69**, 243 (1946).
53. M. R. Schafroth, Phys. Rev. **100**, 463 (1955); J. M. Blatt and S. T. Butler, Phys. Rev. **100**, 476 (1955).

INTERLAYER TUNNELING IN STACKED JUNCTIONS OF HIGH TEMPERATURE SUPERCONDUCTORS, CDW MATERIALS AND GRAPHITE

Yu.I. LATYSHEV

Institute of Radio-Egineering and Electronics, Russian Academy of Sciences, Mokhovaya 11-7, 125009 Moscow, Russia.
`lat@mail.cplire.ru`

Abstract. A short review on interlayer tunneling studies of the stacked junctions of layered high temperature superconductors (HTSC), charge density wave (CDW) materials and graphite is presented. We specify individual features of each class of these materials and common features of interlayer tunneling. They are characterized by the layered crystalline structure of those materials. We emphasize the importance of the phase interlayer coupling in superconducting or CDW electron condensed states that provides interlayer coherent transport. We found that breaking of the phase coherence often happens via formation of phase topological defects (phase vortices) in one elementary junction of the stack that gives an opportunity for interlayer tunneling spectroscopy.

Keywords: interlayer tunneling, $Bi_2Sr_2CaCu_2O_{8+x}$, $NbSe_3$, TaS_{+3}, KMo_6O_{17}, graphite, energy gap, pseudogap, conductance peak

13.1 Introduction

Many layered conducting materials of different classes as superconductors, CDW materials, semimetals (graphite) has a common microscopic arrangement where highly conducting elementary planes of atomic thickness are separated by isolating elementary layers of atomic thickness. Macroscopically, that is characterized by the enormously high anisotropy of the in-plane and the out-of-plane conductivity, achieving 10^3–10^4 in those materials. Microscopically, the interlayer tunneling is associated with interlayer tunneling over intrinsic barriers. The method first appeared in studies of layered HTS materials [1] and later has been adapted to layered manganites [2], CDW materials [3] and graphite [4].

13.2 Interlayer tunneling in HTS materials

Figure 13.1 shows crystallographic structure of typical layered HTS material $Bi_2Sr_2CaCu_2O_{8+x}$ (Bi-2212). That represents a stacked nanostructure of intrinsic tunneling junctions S/I/S... S- superconnducting CuO bi-layers, I- isolating BiO + SrO bi-layers. The thickness of superconducting layer is 0.3 nm with their spacing s of 1.5 nm. The nanoscale value of the spacing provides a possibility of interlayer Josephson tunneling. The amplitude of the superconducting order parameter is modulated along the c-axis while the phases are coupled. That leads to the phase and amplitude effects in the interlayer tunneling [5].

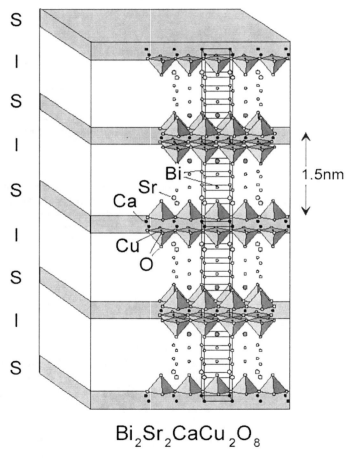

Fig. 13.1. Crystalline structure of $Bi_2Sr_2CaCu_2O_8$. Elementary superconducting layers are denoted as S-layers, while elementary isolating layers as I-layers.

Phase interference provides intrinsic DC Josephson effect, that has been predicted in early 90s for short stacked structures with in-plane sizes L less than Josephson penetration depth λ_J [6],

$$L \leqslant \lambda_J = s\lambda_c/\lambda_{ab} \qquad (13.1)$$

In that case the critical superconducting current across the layers is modulated by magnetic field oriented parallel to the layers with a periodicity of one flux quantum Φ_0 per elementary junction:

$$I_c(H) = I_{c0}\left|\frac{\sin(\frac{\pi sLH}{\Phi_0})}{\frac{\pi sLH}{\Phi_0}}\right| \qquad \Phi_0 = \frac{hc}{2e} \qquad (13.2)$$

The value of λ_J is small in Bi-2212 being about 1 micron and that makes it rather difficult to achieve condition (13.1) experimentally. The first indications of intrinsic Josephson effect have been obtained on rather big crystals with $L \sim 30\,\mu\mathrm{m}$ [1]. The micron sized stacked structures have been fabricated much later [7] and they, indeed, demonstrated good Fraunhofer patterns described by Eq. (13.2) (Fig. 13.2).

The $I_c(H)$ dependence of Fig. 13.2 resembles Fraunhofer patterns of conventional Josephson junction, however, the period of modulation, 1T, is about four orders of magnitude higher then in conventional junctions. That is a signature of the atomic size of those nanodevices.

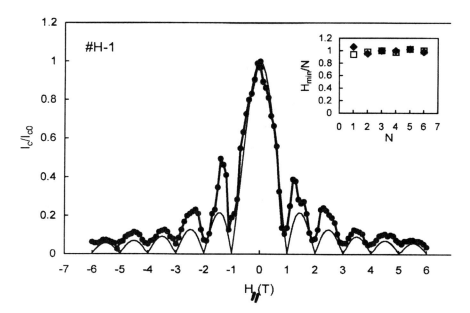

Fig. 13.2. Dependence of critical current across the layers of the stacked junction on magnetic field H oriented parallel to the layers. Thin solid line is a fit to Eq. (13.2).

a)

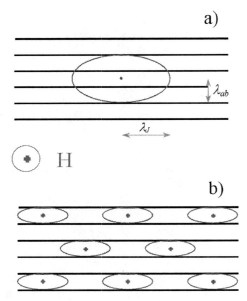

b)

Fig. 13.3. Schematic view of Josephson vortex (**a**) and Josephson vortex lattice (**b**) in the layered superconductor induced by magnetic field H.

In the opposite limit of long junctions $L > \lambda_J$ parallel magnetic field above some threshold value H_{c1} decouples phases between neighbour superconducting layers by creation of Josephson vortices (JVs) [8] (Fig. 13.3a). JVs are centered at the insulating layers and thus have no normal core. Therefore, they can move very fast along the layers while being driven by Lorentz force induced by steady current across the layers. Every Josephson vortex contains flux quantum Φ_0.

A Josephson vortex represents a phase topological defect. The circulation around its nonlinear core gives a phase variation of 2π. At higher fields the inter-vortex interaction leads to their ordering into the Josephson vortex lattice [9] (Fig. 13.3b). The Josephson vortex lattice motion induced by a current applied across the layers is accompanied by Josephson emission and may be identified by the Shapiro step response to the external microwave field (Fig. 13.4). Experiments of that type [10] revealed the existence of Josephson vortices and Josephson vortex lattice as well as the high coherence of the driven lattice.

Figure 13.3. Schematic view of Josephson vortex (a) and Josephson vortex lattice (b) in the layered superconductor induced by magnetic field H.

Another direction of studies was related with the quasiparticle interlayer tunneling [7,11–13]. The low energy quasiparticle conductivity across the layers was shown to have finite and universal value at low temperatures and does not depend on the scattering rate [12]. This and some other unusual properties of low energy quasiparticle tunneling have been attributed to the d-wave char-

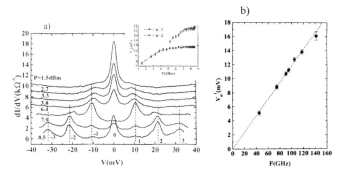

Fig. 13.4. (a) Shapiro step response of moving Joephson vortex lattice to the external microwawe radiation of frequency f = 90.4 GHz [10]. Peak position corresponds to the condition $V_i = iNhf/2e$ with i the integer, N the number of the synchronized junctions in the stack, N = 60. Insert shows a saturation of the step position with microwave power. (b) Frequency dependence of the first Shapiro step voltage position.

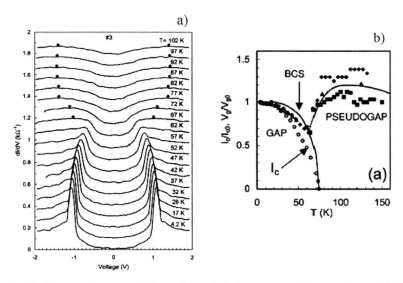

Fig. 13.5. Temperature evolution of the interlayer tunneling spectra of Bi-2212 stacked junction (a) [7] and temperature dependences of critical current I_c/I_{co}, superconducting gap V_g/V_{g0} and pseudogap V_{pg}/V_{g0}.

acter of supercoductivity in HTS materials [12, 14]. The stacked HTS structures also well demonstrates very clear the gap features related with quasiparticle tunneling over a gap (Fig. 13.5a). That was used for a spectroscopy of superconducting gap and a pseudogap above Tc (Fig. 13.5b) [7, 13]. These studies revealed that a pseudogap has another origin than superconducting one and can not be attributed to the superconducting fluctuations [7, 13].

13.3 Interlayer tunneling in the CDW state

Recently, the method of interlayer tunneling has been extended to other types of layered materials like layered manganites [2] and layered CDW materials [3, 5]. The CDW condensed state is in many respects similar to the superconducting state. Below the Peierls transition temperature an energy gap in electronic spectra is opened at the Fermi surface as well. The difference is as follows. The superconducting state is formed by the coupling of electrons on the opposite sites of the Fermi surface. The total momentum of this pair (Cooper pair) is equal to zero as well as a total spin. On the contrary, in the CDW condensed state the coupling occurs between electrons and holes at the opposite parts of the Fermi surface. Because of the negative effective mass of a hole, the pair of that type has a total momentum $2p_F$ or a wave vector $2k_F$. As a result the density of condensed carriers is spatially modulated forming a charge density wave [17]. The spins of electron and hole in the CDW pair are parallel. In zero magnetic field the CDW state is degenerated with respect to the spin orientation up ↑↑ or down ↓↓. In the case when the CDW is incommensurate with the underlying lattice its order parameter can be expressed as $\Delta_0 = A\cos(Qx + \varphi)$ with $Q = 2k_F$ the CDW wave vector and φ an arbitrary phase. A Peierls transition is preferable for compounds possessing considerably flat parts of the Fermi surface separated by $2p_F$ (nesting condition). That is typical for low dimensional (quasi 1D or quasi 2D) crystals containing conducting elementary chains or conducting planes.

For a number of typical CDW materials of MX_3-type (M- transition metal, X- chalcogen) the conducting chains are assembled in elementary conducting layers isolated from each other by a double barrier of insulating prism bases [3] (Fig. 13.6).

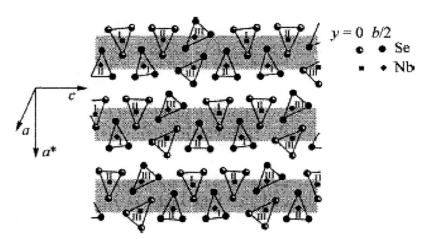

Fig. 13.6. Layered crystalline structure of NBSe3 in the a*c-plane. Highly conducting planes are shaded.

For NbSe$_3$ that results in a very high interlayer conductivity anisotropy $\sigma_{a*}/\sigma_b \sim 10^{-3}$ at low temperatures compared with the intralayer anisotropy $\sigma_c/\sigma_b \sim 10^{-1}$. In the Peierls state, similarly to the layered superconductors, the amplitude of the OP is modulated across the layers while phases remain coupled. That provides a ground for CDW gap spectroscopy by interlayer tunneling.

The CDW gaps and the zero bias conductance peak (ZBCP) have been identified in NbSe$_3$ using this method [3,5]. Here we present a short review of our recent studies on interlayer tunneling spectroscopy including observation and studies of new collective states with their energy lying inside the CDW gap.

13.4 Fabrication of the stacked nanostructures

For the stacked junction fabrication we used double sided focused ion beam (FIB) technique we first applied for fabrication of HTS stacked structures [16], and later modified for the CDW materials [3,5].

The FIB technique provides resolution of the etched patterns as small as 10 nm at low, pA scale ion currents. This resolution is, however, limited by the damaged region at the boundary of the pattern. The depth of damaged region depends on the ion energy and the ion incident angle. For standard ion beam energy, 15–30 kV, that is about 30–50 nm being maximal for the normal incidence. Even this reduced resolution is still high enough and competitive with electron lithography technique.

The stages of fabrication are shown in Figs. 13.7 and 13.8. At the first stage (Fig. 13.7b) a square trench of typical size $6 \times 6\,\mu$m is etched on one side of a thin single crystal by FIB to the depth of $d/2 + \delta$ where d is the thickness of the crystal and δ is the small excess depth. Typically, for d = 1 μm we selected $\delta = (0.05-0.1)$ d. The depth of etching is calculated from the time of etching the same area in the same etching conditions through the whole thickness of the crystal (Fig. 13.7a). The position of the trench is marked by small spots at the corners etched through the rest of the crystal thickness (Fig. 13.7b).

At the second stage (Fig. 13.8a) the crystal is turned over and a second trench is etched on the opposite side of the crystal near the first one to the same depth. The stacked junction with thickness of 2δ is thus formed between the trenches. At the third stage the stacked structure is trimmed, first fine (Fig. 13.8b) and then roughly, with high ion current, to the whole width of the crystal (Fig. 13.8c).

The typical junction is shown at Figs. 13.9a, b.

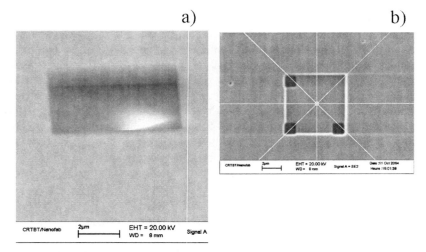

Fig. 13.7. One side FIB processing of NbSe$_3$ single crystal: (a) FIB etching pattern to the whole thickness of the crystal, (b) etching through the one half of the crystal thickness and putting markers at the corners.

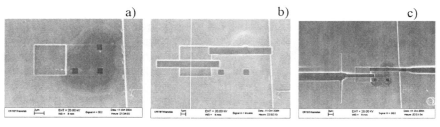

Fig. 13.8. Opposite side FIB processing: (a) half thickness etching, (b) fine trimming, (c) rough trimming to the whole width of the crystal.

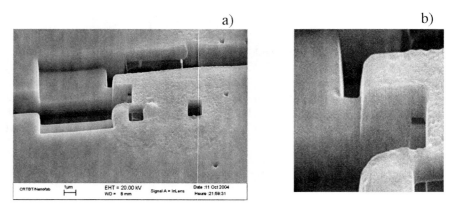

Fig. 13.9. SEM image of the stacked structure obtained: (a) general view, (b) enlarged central part. The stack size is $1 \times 1 \times 0.2\,\mu$m.

13.5 CDW gap features on interlayer tunneling spectra of NbSe₃

Figure 13.10a shows temperature evolution of interlayer tunneling spectra measured on NbSe₃ stacked junction [5] and by point contact NbSe₃-NbSe₃ [3] along the a^*-axis.

A remarkable feature of the spectra obtained on the mesa structure is very sharp and distinctive both CDW gap features at low temperatures marked by arrow as $2\Delta_2$ (low temperature CDW) and (upper temperature CDW) and zero bias conductance peak (ZPCP). In spite of big variation of spectra with temperature decrease (dynamic conductivity changes by many times) the integral value $s = \int_{-V_0}^{V_0} (dI/dV)dV$ with $V_0 = 200$ mV remains nearly constant (insert to Fig. 13.10) indicating that $dI/dV(V)$ spectra, indeed, present tunneling density of states and the total number of states expressed as an integral s does not depend on temperature. A small variation of s observed may be related with temperature variation of the barrier transparency with temperature.

Comparison of the spectra obtained on the stacked junction containing few tens of elementary tunnel junctions and the spectra obtained on point contact NbSe₃-NbSe₃ along the a*-axis [3] (Fig. 13.10b) containing only one CDW/I/CDW junction shows the same value for the gap feature $2\Delta_2 \approx 60$ mV. That implies that only one junction in the stack is effectively working. We consider this junction as a weakest junction at the stack.

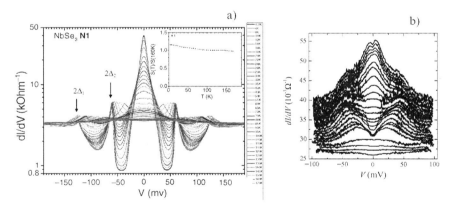

Fig. 13.10. A temperature evolution of interlayer tunneling spectra of NbSe3 [5] (a). The CDW gaps at law temperatures are marked by arrows. High temperature CDW gap $2\Delta_1 = 140$ mV, low temperature CDW gap $2\Delta_2 = 65$ mV. The insert shows weak temperature variation of the integral number of states s defined as $s = \int_{-V_0}^{V_0} (dI/dV)dV$ with $V_0 = 200$ mV. (b) Interlayer tunneling spectra of micro-contact NbSe₃-NbSe₃ measured along the a^*-axis. The temperature has been varied from 4.2 K (top spectrum) to 77 K (bottom spectrum).

With an increase of bias voltage across the stack the voltage drops on the weakest elementary junction [5] thus providing a possibility of interlayer tunneling spectroscopy on a single elementary tunnel junction. This process is accompanied by the CDW phase decoupling in the weakest junction [5].

Theoretical model developed by Brazovskii [9] describes this process by successive entering of phase CDW dislocations or dislocation lines (DLs) in the weakest junction at voltages exceeding some threshold value, $V_t \approx 0.1\Delta$ [17, 18]. The DL has a charge $2e$ localized mostly in its core. The core of dislocation line has a very short size in transverse direction across the layers, which is about the spacing between elementary conducting layers $s = 1-2$ nm, while its in-plane size l is 20–50 times bigger, $l \approx 2s\omega_p/T_p$ [17], where ω_p is a plasma frequency of the layered material. At low bias voltage $V < V_t$ the potential is uniformly distributed over a stack. Entering of charged DLs leads to a redistribution of the potential with the most of voltage applied to a stack drops on a set of dislocation lines located along the weakest junction [17]. That means that the voltage mostly drops on this junction.

From geometrical considerations this junction should be located at the place where the maximum current density is achieved. For thin enough stacks, containing few tens of elementary junctions that is likely to be the central junction. For thick stacks of few hundred junctions the more likely configuration includes two weak junctions located near the vertical ends of the stack where current concentration occurs (Fig. 13.11). Experimentally we regularly observed that for thin mesas of thickness ≈50 nm the voltage position of the main peak of interlayer tunneling spectra corresponds to the overgap tunneling of single junction, while for the "thick" stack of thickness more then 200 nm this peak is located at twice higher bias voltage indicating that voltage drops on two junctions connected in series. That observation supports our geometrical considerations.

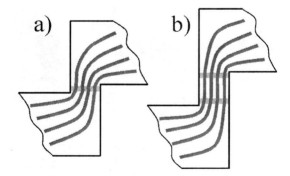

Fig. 13.11. Schematic view of formation of weak junction in thin stack (a) and a couple of weak junctions in a "thick" stack (b) in regions of highest current density.

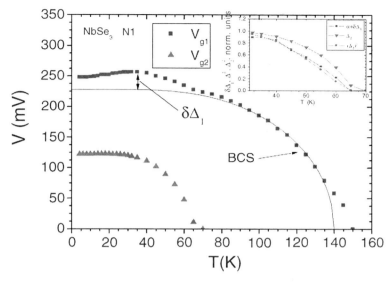

Fig. 13.12. Temperature dependences of the CDW gap peaks V_{g1} and V_{g2} of the interlayer tunneling spectra [23]. Solid line is a fit to the BCS dependence at $T > 65$ K. The insert shows comparison of the temperature dependences of the enhancement of V_{g1} below T_{p2} and normalized Δ_2 and Δ_2^2.

The extracted CDW gap values at low temperatures have the following values $2\Delta_{1,2} = 140$ mV, 60 mV. These values are consistent with the data of STM [19], ARPES [20], infrared [21] and point contact [22] spectroscopies.

Figure 13.12 shows temperature dependences of both CDW gaps in $NbSe_3$. They generally follow BCS type dependence. Some deviations happen above T_p in the region of CDW fluctuations and also below the T_{p2}, where both CDWs coexist. As our experiment showed they are not independent. Appearance of low temperature CDW leads to a small enhancement of the energy gap of high temperature CDW (Fig. 13.12). This effect was theoretically predicted more than 20 years ago [24] and is associated with joint commensurability of both CDWs.

Namely, the wave vectors characterizing the two CDWs at low temperatures, at the range of their coexisting [25] $q_1 = (0, 0241, 0)$ and $q_2 = (0.5, 0.260, 05)$ in units of the reciprocal unit lengths satisfy the approximate relation:

$$2(q_1 + q_2) \cong (1, 1, 1) \tag{13.3}$$

i.e. twice their sum is nearly a reciprocal lattice vector. That has been considered [24] as an evidence of phase coupling between the two CDW's below T_{p2}. The analysis based on simple Ginzburg-Landau theory [24] pointed out that phase-locking effect can be accompanied by small enhancement of CDW1 energy gap below T_{p2}. The observed gap enhancement [23] is small,

$\delta\Delta1/\Delta1 \approx 10\%$, however, that is reproducibly observed on interlayer tunneling spectra. That points out to the high resolution of this technique. Note that this effect has been searched for by X-ray diffraction technique, but has not been found [26].

13.6 The intragap states in NbSe₃

Careful and systematic studies of interlayer tunneling in NbSe₃ revealed an existence of the states localized inside the CDW gap [17, 27] (Fig. 13.13). The difficulties to record these states was related with the existence of relatively broad zero bias conductance peak. However, this peak is significantly suppressed at high temperatures and also may be considerably narrowed by parallel magnetic field at low temperatures. Under those conditions the intragap states may be clearly identified. Figure 13.13 shows two types of intragap features on interlayer tunneling spectrum at high temperatures near T_{P1}. One is related with a broad peak at the bias voltage V close to $V_s = 2\Delta/3$, another one appears at even lower energies of about $V \approx 0.1\Delta$. That manifests itself as a threshold of sharp increase of interlayer dynamic conductivity.

Figure 13.14 shows scaling of the intragap peak voltage position, V_s with a gap voltage $V = 2\Delta$. At various temperatures (Fig. 13.14a) and magnetic fields (Fig. 13.14b) its position is very close to the vertical line corresponding to the condition $V_s = 2\Delta/3$.

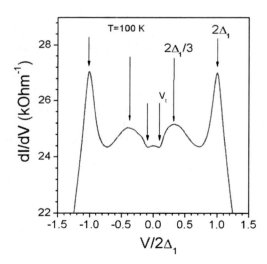

Fig. 13.13. Interlayer tunneling spectrum of NbSe₃ stacked junction in the upper CDW state. The arrows mark CDW gap peaks at $V = 2\Delta_1$, the intra gap peaks at $V = 2\Delta_1/3$ and a threshold voltage for sharp increase of interlayer tunneling $V = V_t \approx 0.1\Delta_1$.

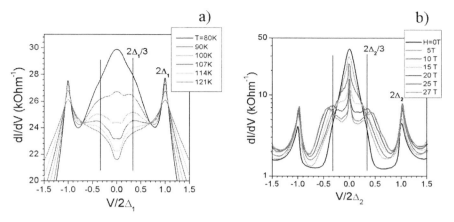

Fig. 13.14. Interlayer tunneling spectra of NbSe₃ at various temperatures (a) and at 4, 2 K at various magnetic fields oriented parallel to the layers H// c-axis [27] (b). The vertical lines indicate voltage position $V/(2\Delta) = 1/3$.

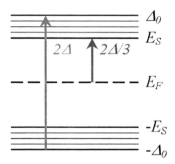

Fig. 13.15. The scheme of tunnel transitions which produce peaks on tunneling spectra at $V = 2\Delta$ (red line) and at $V = 2\Delta/3$ (blue line).

We consider the intragap peak at $V_s = 2\Delta/3$ to be related with the existence of the intragap states at the energy $E_s = 2\Delta/3$. Correspondingly, the peak is associated with tunneling transitions from Fermi energy, where the ungapped carriers are localized, to these states (Fig. 13.15). We attribute these states to the amplitude CDW solitons (π-solitons) [28].

As it was mentioned above the incommensurate CDW (ICDW) is characterized by the CDW order parameter Δ_0, $\Delta_0 = A\cos(Qx + \varphi_0)$. The uniform ground state in the ICDW state with $A(x) = const$ is degenerated with respect to the phase shift by π that changes A by - A. That leads to the possibility of non-uniform ground state that can be realized by local change of the phase by π and simultaneous acceptance of one electron from free band to conserve electro-neutrality. The resulting state with $A = \tanh(x/\xi_0)$ is known as the amplitude soliton [28] (Fig. 13.16). The amplitude soliton has an energy $E_s = 2\Delta/\pi$ that is lower by $\sim\Delta/3$ than the lowest energy of electron in a free

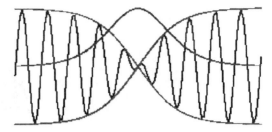

Fig. 13.16. Spatial distribution of charge (black curve) and spin (blue curve) near the point of the formation of amplitude soliton.

band Δ. Therefore, free electrons near the band edge tend to be self-localized into amplitude soliton states. Experimentally the existence of the amplitude soliton states has been reliably demonstrated only for dimeric compounds like polyacetylene. Using interlayer tunneling technique we found the amplitude soliton states in inorganic MX_3 compounds with commensurability parameter m close to $m = 4$ [27].

13.7 Amplitude solitons in o-TaS$_3$

We have studied also interlayer tunneling spectra of another CDW material o-TaS$_3$. In contrast to NbSe$_3$ the gap in o-TaS$_3$ is opened on the full Fermi surface and below Peierls transition temperature ($T_p = 220$ K) that material becomes Peierls nsulator. Interlayer tunneling spectra are shown at Fig. 13.17. The CDW gap is clearly seen, however, that is much more smearing to compare with NbSe$_3$ [29].

The CDW gap value at low temperatures below 120 K is $2\Delta \approx 200$. That is consistent with Hall measurements [30] and the photoconductivity data [31]. A remarkable feature of the spectra is the absence of zero bias conductivity peak, we observed in NbSe$_3$. That points out that the origin of this peak associated with a presence of the ungapped carriers in NbSe$_3$.

To extract CDW gap value at high temperatures we sabstracted parabolic background as shown at the Fig. 13.18a. Parabolic type of background has been confirmed by measurements at the temperatures above Peierls transition.

The extracted temperature dependence of the CDW gap in o-TaS$_3$ is shown at the Fig. 13.18b. There are two sharp increases of the CDW gap with a temperature decrease, one below Peierls transition temperature and another below 130 K, where transition from incommensurate CDW phase (ICDW) to commensurate phase (CCDW) occurs.

We explain this remarkable gap increase below 130 K in the following way. At the ICDW state the edge of the free band is characterized by the energy of the CDW amplitude solitons (Fig. 13.19a).

In the CCDW phase the CDW phase φ is fixed by the commensurability potential and π-solitons can not exist. Their energy is pushed out to the end

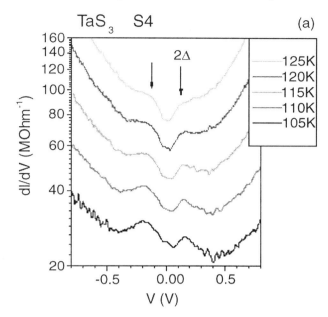

Fig. 13.17. Interlayer tunneling spectra of o-TaS$_3$ at low temperatures below 125 K [29].

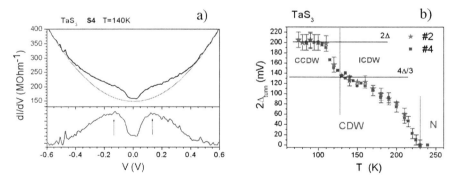

Fig. 13.18. A procedure of subtraction of parabolic background for a determination of the CDW gap in o-TaS$_3$ stacked junction at high temperatures (a). Temperature dependence of the CDW gap in o-TaS$_3$ (b).

of the free band (Fig. 13.19b). In the experiment that manifests itself as an effective increase of the CDW gap. The value of increase is as follows $(2\Delta - 4\Delta/3)/(4\Delta/3) = 0.5$ [29], in a good agreement with experiment Fig. 13.18b. Thus, our studies of interlayer tunneling point out to the existence of the amplitude solitons in the ICDW phase of another CDW material, o-TaS$_3$.

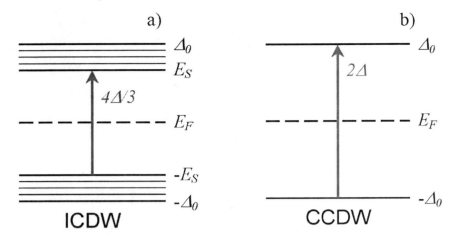

Fig. 13.19. The scheme of tunnel transitions determining the edge of the energy gap at the incommensurate CDW (ICDW) state and at the commensurate CDW (CCDW) state in o-TaS$_3$.

13.8 Phase decoupling effects on interlayer tunneling in layered CDW materials

Another remarkable feature that appears at lower energies within the CDW gap is a sharp voltage threshold V_t for onset of the interlayer tunneling conductivity (Fig. 13.20). The value of V_t at low temperatures was found to be close to the energy of 3D CDW ordering kT_p (T_p is the Peierls transition temperature) for both CDWs in NbSe$_3$ and for o-TaS$_3$. Therefore, V_t was attributed to the energy of the CDW phase decoupling between the neigbour layers.

Brazovskii developed a model that describes phase decoupling via the subsequent formation in the weakest junction of an array of dislocation lines (DLs) [17, 19], the CDW phase topological defects [29]. The DLs are oriented across the chains. They appear as a result of the shear stress of the CDW induced by the electric field. As in a case of Josephson vortices the circulation around the DL core gives a phase variation of 2π. That corresponds to the CDW elementary charge $2e$ per chain. The excess charge of the DL accumulates the electric field within the DL core. In the vertical direction the DL core has atomic size, i.e. inside the core, the field is concentrated within one elementary junction (Fig. 13.21). The in-plane size of the DL core is much bigger, about 50 nm [17]. That means that for a junction of 1 μm, about 10 DLs can completely overlap its area and the voltage then drops on one elementary junction as it was observed experimentally.

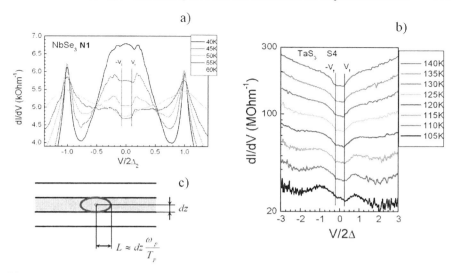

Fig. 13.20. Interlayer tunneling spectra of NbSe₃ (a) and o-TaS₃ (b) demonstrating threshold behaviour for interlayer dynamic conductivity. Position of the threshold is marked by vertical lines. Schematic view of the core of CDW dislocation line (c).

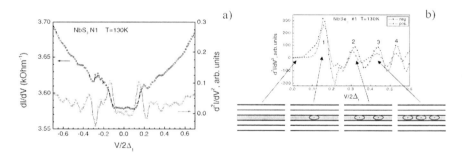

Fig. 13.21. Interlayer tunneling spectra of NbSe₃ stacked junction (a) and its derivative (b) demonstrating multiple character of the threshold voltage. Insert to Fig. 13.21b schematically shows entering of new dislocation lines in the weakest junction.

13.9 Interlayer tunneling of CDW spectroscopy above T_p in NbSe₃

As that was mentioned in previous section electric field across the layers creates CDW dislocation lines which contribute to the interlayer conductivity. Above Peierls transition temperature the CDW dilocation lines can appear due to thermal fluctuations even at zero electric field.

Therefore, one can expect to observe excess zero bias conductivity above transition temperature. That feature can be used for studies of CDW

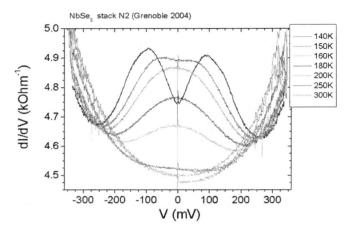

Fig. 13.22. Interlayer tunneling spectra of NbSe3 stacked junction slightly below and above the upper Peierls transition temperature.

fluctuations above T_p. Figure 13.22 shows the interlayer tunneling spectrum of NbSe3 below and above transition temperature.

At T = 140 K the CDW gap maxima are clearly seen on interlayer tunneling spectrum. At T = 150 K they are nearly merged into each other, remaining broad maximum at zero voltage. This maximum (or excess zero bias conductivity) has been observed up to very high temperatures about 290 K twice exceeding Peierls transition temperature $T_{p1} = 145$ K. The CDW fluctuations at so high temperatures has also been observed by X-ray diffraction techniques [32]. Our observation shows that interlayer tunneling technique is very sensitive method for studies of CDW fluctuations above Peierls transition temperature.

13.10 Interlayer tunneling spectroscopy in NbSe3 at high magnetic fields

As the next step of our studies we used interlayer tunneling for CDW gap spectroscopy at high magnetic fields. The interest to these studies was related with enormously high magnetoresistance at low temperatures [33]. To explain that it was predicted [34] that magnetic field can abolish Fermi surface pockets responsible for ungapped carriers and in this way contribute to the excess magneto-resistance. In more recent papers [35] that was predicted the possibility to increase Peierls transition temperature by high magnetic field. However, until recently there were no experiments on tunneling studies of the CDW gap at high magnetic fields.

Here we report on interlayer tunneling CDW gap spectroscopy near the lower Peierls transition temperature at high pulsed magnetic fields [36].

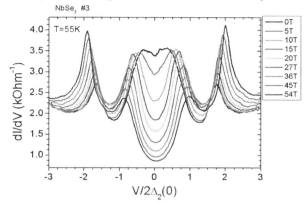

Fig. 13.23. Interlayer tunneling spectra of NbSe₃ stacked junction under high magnetic fields H//a*- axis below (a) and above (b) Peierls transition temperature.

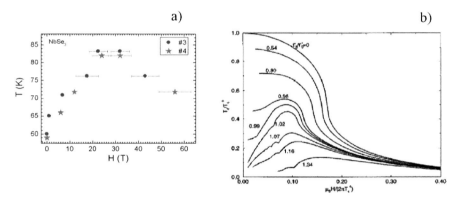

Fig. 13.24. Magnetic field dependence of Peierls transition temperature: (a) extracted from interlayer tunneling experiment, (b) calculation results from Ref. [35].

The experiments have been done in Toulouse at the National Laboratory of Pulsed Magnetic Fields. WE used pulsed fields up to 55T. That has been provided by discharge of a battery of yigh voltage capacitors through the coil. The duration of the field sweap was about 300 ms. We developed a system of fast data acquisition that allowed us to collect about 1,000 interlayer tunneling spectra within one puls.

The results are shown at the Fig. 13.23. Below T_p (55 K) magnetic field significantly increases the CDW gap. The magnetic field also induces the CDW gap at temperatures above T_p [37]. That means the increase of T_p by magnetic field. The phase diagram of T_p – H is shown in the Fig. 13.24.

One can see that $T_p(H)$ dependence found from our experiments has non-monotonic character, T_p first increases, reaches maximum at about 30 T and then decreases. We explain this behaviour following theoretical paper [35] by the interplay between orbital and Pauli effects on CDW ordering. Low field interacts with ungapped carriers increasing their energy and, as a result, improving the nesting condition. That leads to the increase of the CDW gap and T_p. At fields above 30 T the Zeeman splitting of ground state becomes important. The CDW sate with spin up configuration $\uparrow\uparrow$ increases its energy in field while the state with spin down confuguration $\downarrow\downarrow$ decrease energy. As a result, the CDW wave vector $Q_{\uparrow\uparrow}$ increases whereas $Q_{\downarrow\downarrow}$ decreases with field. Hence, a CDW state with a fixed CDW wave vector at zero field Q_0 tends to be destroyed with field. We found that Pauli contribution becomes important at H > 30T. That corresponds to a condition $2\ \mu_B H > k\ T_p$. Theoretical dependences $T_p(H)$ are shown at Fig. 13.24b. For systems with non-perfect nesting parameter theory predicts non-monotonic behaviour $T_p(H)$ similar to what we found experimentally a crossover field corresponds to the parameter $\mu_B H/(2\pi T_p)$ to be about 0.08. That gives for $\mu_B H$ a value $1/2k\ T_p$ very close to what we found experimentally.

13.11 Interlayer tunneling spectroscopy of KMo_6O_1 and graphite

For comparison with $NbSe_3$ and $o\text{-}TaS_3$ compounds, which are quasi-one-dimensional materials, we studied interlayer tunneling spectra on two quasi-two-dimensional compounds KMo_6O_{17} (purpe bronze) and graphite. The interlayer tunneling spectra on KMo_6O_{17} is shown on Fig. 13.25. Along with sharp CDW gap peaks we found on that compound existence of pseudogap at low temperatures below 25K [38] (Fig. 13.25).

Similar pseudogap was observed on interlayer tunneling spectra of graphite below 30 K [4] (Fig. 13.26). Both pseudogaps in KMo_6O_{17} and graphite are characterized by a zero bias drop of density of states linearly dependent on voltage. We observed a pseudogap only on qusi-two-dimensional materials. That was observed non-regularly. That points out that its origin related with local desorder. Preliminarly we associate a pseudogap with a 2D Coulomb gap with a presence of a disorder. In a 2D-case pseudogap should be linearly dependent on energy [40].

In a summary, we have fabricated stacked structures for interlayer tunneling studies in quasi-one-dimensional and quasi-two-dimensional HTS, CDW materials and graphite. Studies of unterlayer tunneling on those structures let us to define the energy gaps and pseudogaps in those materials. We identified the states inside the CDW gap related with the amplitude and phase CDW excitations. Using interlayer tunneling technique we found the enhancement of the CDW gap and Peierls transition temperature by magnetic field. We characterize interlayer tunneling technique as a powerful method for studies of condensed electron states in the layered materials.

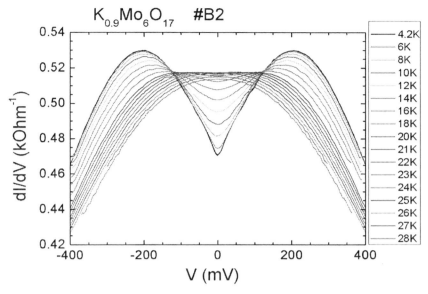

Fig. 13.25. Interlayer tunneling spectra on KMo_6O_{17} stacked junction. A pseudogap appears at zero bias voltage with temperature decrease below 25 K.

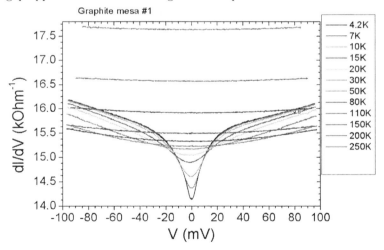

Fig. 13.26. Interlayer tunneling spectra obtained on graphite stacked junction. A pseudogap appears with a temperature decrease below 30 K [4].

Acknowledgements

I would like to express acknowledgements to my coauthors, A.P. Orlov, A.A. Sinchenko, S.A. Brazovakii, P. Monceau, Th. Fournier, J. Marcus, E. Mossang and D. Vignolles. The work has been supported by grants of RFBR (No 05-02-17578, No 06-02-72551_a), INTAS Program, grant No 05-7972, and programs of the presidium RAS "Quantum nanostructures" and "Strongly correlated electron systems and quantum critical phenomena".

References

1. Kleiner R, Steinmeyer F, Kunkel G and Mueller P 1992 *Phys. Rev. Lett.* **68**, 2394 as a recent review, see Yurgens A 2000 *Supercond. Sci. Technol.* **13**, R85.
2. Nachtrab P, Helm S, Moessle M, Kleiner R, Waldmann O, Koch R, Mueller P, Kimura T and Tokura Y 2002 *Phys. Rev.* **B 65** 012410.
3. Nachtrab Yu I, Monceau P, Sinchenko A A, Bulaevskii L N, Brazovskii S A, Kawae T and Yamashita T 2003 *J. Phys. A: Math. Gen.* **36**, 9323.
4. Yu.I. Latyshev, Z.Ya. Kosakovskaya, A.P. Orlov, A.Yu. Latyshev, V.V. Kolesov, P. Monceau, J. Marcus, Th. Fournier, "Nonlinear interlayer transport in the aligned carbon nanotube films and graphite" to appear in the Journal "Fullerene, Nanotubes and Carbon Nanostructures".
5. Yu.I. Latyshev, P. Monceau, A.P. Orlov, S.A. Brazovskii, and Th. Fournier, "Interlayer tunneling spectroscopy of charge density waves", Supercond. Sci. Technol., **20**, S87–S92 (2007).
6. Bulaevskii L N, Clem J R and Glazman L I 1992 *Phys. Rev B* **46**, 350.
7. Yu.I. Latyshev, S.-J. Kim, V.N. Pavlenko, T. Yamashita, and L.N. Bulaevskii, "Interlayer tunneling of quasiparticles and Cooper pairs in Bi-2212 from experiments on small stacks", Physica C **362**, 156–163 (2001).
8. Clem J R and Coffey M W 1989 *Phys. Rev. B* **42**, 6209.
9. Bulaevskii L N and Clem J R 1991 *Phys. Rev. B* **44**, 10234.
10. Yu.I. Latyshev, M.B. Gaifullin, T. Yamashita, M. Machida, and Yuji Matsuda, "Shapiro Step Response in the Coherent Josephson Flux Flow State of $Bi_2Sr_2CaCu_2O_{8+x}$", Phys. Rev. Lett., **87**, (2001) 247007(4).
11. Tanabe K, Yidaka Y, Karimoto S and Suzuki M 1996 *Phys. Rev. B* **53**, 9348.
12. Yu.I. Latyshev, T. Yamashita, L.N. Bulaevskii, M.J. Graf, A.V. Balatsky, and M.P. Maley, "Interlayer transport of quasiparticles and Cooper pairs in $Bi_2Sr_2CaCu_2O_{8+x}$ superconductors", Phys. Rev. Lett., **82**, (1999) 5345–5348.
13. V.M. Krasnov et al., Phys. Rev. Lett., **84**, 5860 (2000).
14. Yu.I. Latyshev,. "Evidence for *d*-wave order parameter symmetry in Bi-2212 from experiments on interlayer tunneling." in a book Symmetry and Heterogeneity in High Temperature Superconductors ed. by A. Bianconi, Dordrecht: Kluwer Academic Publisher (2006) Chapter **7**, 101–118.
15. G. Gruner, *Density Waves in Solids*, 1994 (Reading, MA, Addison-Wesley).
16. Yu.I. Latyshev, S.-J. Kim, and T. Yamashita, IEEE Trans. on Appl. Supercond., **9**, 4312 (1999).
17. Yu.I. Latyshev, P. Monceau, S. Brazovskii, A.P. Orlov and T. Fournier, "Subgap collective Tunneling and Its Staircase Structure in Charge Density Waves", Phys. Rev. Lett, **96**, 116402 (2006).
18. S. Brazovskii, Yu.I. Latyshev, S.I. Matveenko, P. Monceau "Recent views on solitons in Density Waves", J. Physique, France IV, **131**, (2005) 77–80.
19. Z. Dai, c.G. Slough, R.V. Coleman, Phys. Rev. B **45**, R9469 (1992).
20. J. Schafer et al. Phys. Rev. Lett., **91**, 066401 (2003).
21. A. Perucchi, L. Degiorgy, R.E. Thorne, Phys. Rev. B **69**, 195114 (2004).

22. T. Ekino, J. Akimitsu, Jpn. J. Appl. Phys., 26, 625 (1987); A.A. Sinchenko, P. Monceau, Phys. Rev. B 67, 125117 (2003).
23. A.P. Orlov, Yu.I. Latyshev, A.M. Smolovich, P. Monceau, "Interaction of both charge density waves in NbSe3 from interlayer tunneling exsperiments", JETP Lett. 84, 89–92, 2006.
24. R. Bruinsma, S.E. Trullinger, Phys. Rev. B 22, 4543 (1980).
25. R.M. Fleming, D.E. Moncton, D.B. McWhan, Phys. Rev. B 18, 5560 (1978).
26. A.H. Moudden, J.D. Axe, P. Monceau, F. Levy, Phys Rev. Lett., 65, 223 (1990).
27. Yu.I. Latyshev, P. Monceau, S. Brazovskii, A.P. Orlov, and T. Fournier, "Observation of charge density wave solitons in overlapping tunnel junctions", Phys. Rev. Lett., 95, (2005) 266402.
28. S.A. Brazovskii Sov. Phys. JETP, 51, 342 (1980).
29. Yu.I. Latyshev, P. Monceau, S.A. Brazovskii, A.P. Orlov, A.A. Sinchenko, Th. Fournier, E. Mossang, "Interlayer tunneling spectroscopy of layered CDW materials", J. Physique, France IV, 131, (2005) 197–202.
30. Yu.I. Latyshev, Ya.S. Savitskaya and V.V. Frolov, "Hall effect accompanying a Peierls transition in TaS3", Pis'ma Zh. Exsp. Teor. Fiz. 38 (1983) 446–449; JETP Lett. 38, (1983) 541–545.
31. S.V. Zaitsev-Zotov and V.E. Minakova, Phys. Rev. Lett. 97, 266404 (2006).
32. J.P. Pouget, R. Moret, A. Meerschaut, L. Guemas and J. Rouxel, J.Physique, 44, C3-1729 (1983).
33. R.V. Coleman, G. Eiserman, M.P. Everson, A. Johnson, and A.M. Falikov, Phys. Rev. Lett., 55, 863 (1985).
34. C.A. Baltseiro, L.M. Falikov, Phys. Rev, B 34, 863 (1985).
35. D. Zanchi, A. Bjelis, G. Montabeaux, Phys. Rev., B 53, 1240 (1996).
36. Yu.I. Latyshev, A.P. Orlov, P. Monceau, Th. Fournier, E. Mossang, D. Vignolles, "Enhancement of Peierls Transition Temperature in NbSe3 by High Magnetic Field" Abstract book of International School "Magnetic Fields for Science" August 27 – September 8, 2007, Cargese, p. Fri07-4.
37. A.P. Orlov, Yu.I. Latyshev, P. Monceau, D. Vignolles, "Probing of the CDW ordering in NbSe3 by high magnetic field" to appear elsewhere.
38. Yu.I. Latyshev, P. Monceau, A.P. Orlov, A.A. Sinchenko, S.A. Brazovskii, L.N. Bulaevskii, Th. Fournier, T. Yamashita, T. Hatano, J. Marcus, J. Dumas, C. Schlenker, "Interlayer tunneling spectroscopy of layered high temperature superconductors and charge density wave materials", Extended abstracts of the conference on Recent Developments in Low Dimensional Charge Density Wave Conductors, Skradin, Croatia, June 29-July 3, 2006, pp. 4–5.
39. Yu.I. Latyshev, Z.Ya. Kosakovskaya, A.P. Orlov, A.Yu. Latyshev, V.V. Kolesov, P. Monceau, J. Marcus, Th. Fournier, "Nonlinear interlayer transport in the aligned carbon nanotube films and graphite" to appear in the Journal "Fullerens and Carbon Nanostructures".
40. V.F. Gandmakher, *Electrons in disordered media*, Fizmatgiz, 2nd edition, 2005 (in Russian).

14

MULTIBAND DESCRIPTION OF THE ELECTRON-DOPED CUPRATE GAPS ON THE DOPING SCALE

N. KRISTOFFEL[1] AND P. RUBIN[2]

[1]*Institute of Theoretical Physics, University of Tartu, Tähe 4, 51010 Tartu, Estonia.* kolja@fi.tartu.ee

[2]*Institute of Physics, University of Tartu, Riia 142, 51014 Tartu, Estonia*

Abstract. A model for the description of electron-doped cuprate superconductor excitation gaps on the doping scale is developed. Interband pairing channel is supposed between the defect states (two subbands) created by doping and the upper Hubbard band. Bare gaps between these systems leading to the appearance of pseudogaps close with extended doping. Extincting Hubbard bands supply the defect subsystem. Illustrative calculations for T_c, superconducting gaps, pseudogaps, supercarrier density, thermodynamical critical field and the coherence length are performed on the whole electron-doping scale. Critical dopings correspond to achieved band overlaps. Gross features of the phase diagram of electron-doped materials are reproduced in some analogy with the hole-doping case but in different realization.

Keywords: cuprates, electron doping, multiband model, gaps

14.1 Introduction

Parent oxycuprate materials of high-temperature superconductors belong to strongly correlated charge-transfer insulators. Antiferromagnetic ordering present in pure materials is destroyed by doping being necessary to initiate superconductivity. The electronic spectrum remains in no means rigid at this. The carriers created by doping can be of electron or hole nature. The phase diagrams of hole- and electron-doped cuprate superconductors are comparable in gross features. Various experimental methods have been elaborated and improved to reach reliable results on the quasiparticle excitations represented by the spectral gaps [1]. The class of hole-doped systems is larger

J. Bonča, S. Kruchinin (eds.), *Electron Transport in Nanosystems.*
© Springer Science + Business Media B.V. 2008

and is also more widely investigated. It repays attention that direct measurement of two coexisting superconducting gaps [2] and demonstration of the presence of two gaps in the same spectral region [3] have been reached quite recently, as also the extraction of the true superconducting gap from the background [4] and the demonstration (quantum oscillations of resistance) of the presence of a well defined Fermi surface in the underdoped copper oxide [5].

A long time ago there have been various appointments that superconducting cuprates are two-component systems with functioning itinerant and "defect" carriers [6]. The latter are associated with the new electron states created by the perturbative action of doping. Electronic phase separation [7] and spatial inhomogeneity [8,9] entered essentially the physics of cuprates. A huge amount of theoretical approaches including also various two-component scenarios have not given decisive conclusions about the superconductivity mechanism in cuprate superconductors until the recent time. However, promising results have been obtained by the multiband approaches. For complex multiband systems where various bands resonate with the Fermi energy the interband pairing channel becomes opened which can support high transition temperatures by repulsive interactions [10–13]. Multiband approaches have been approved for cuprate superconductors long time ago (reviews [10, 11]). The nature of the combining bands has remained often unclear. A first approach taking account of the presence of "defect" carriers has been seemingly [14]. Later the background of minibands created by the striped phase separation has been investigated [9]. The interest to multiband models has been especially grown in connection with the two-gap superconductor MgB_2 [15, 16]. It continues in applications to cuprates [17–23, 40] and other novel superconducting natural and artificial systems.

The authors of the present contribution have developed a simple multiband model for the description of two-component hole-doped cuprate superconductors [21–23]. For both types of cuprates a "defect" band near the Fermi level is created by doping [24–26]. In the case of hole doping it borns near the top of the oxygen band between the Cu-Hubbard band components. It has been postulated that the effective pairing channel acts between the itinerant and defect subsystem states [18], associated with the hole-poor and hole-rich parts of the material. It means that doping not only metallizes the sample but also prepares the specific electronic background of the CuO_2 planes in which interband pairing is realized. Bare gaps are supposed between the itinerant and defect bands, which become closed by extended doping leading to bands overlap. This suggests a novel source of the phase diagram critical points. In the normal phase insulator to metal transitions are expected here and the pseudogaps to transform smoothly into the corresponding superconducting gaps. The self-consistent complex of data obtained without inner discrepancies by this model describes qualitatively correctly the observed properties of hole-doped cuprates on the doping scale [21–23].

14.2 Model scenario for electron-doped cuprates

The present work is devoted to electron-doped cuprate superconductors [1,27–32]. The same general ideas as in the case of hole-doped systems will be used, however, the concrete realization is different. The defect states are now built up near the bottom of the UHB [24] (see also [33–35]) and are supplied by both extincting Hubbard components without a phase separation. It is supposed that the leading pairing interaction consists in the pair transfer between the defect states and the UHB. The supposed bare gaps between them will vanish with extended doping. Besides the UHB (β) we introduce antinodal ($\alpha 1$) and nodal ($\alpha 2$) defect subbands. These latter two distinguish the functionality of the $(\pi, 0)$ and $(\frac{\pi}{2}, \frac{\pi}{2})$ centered regions of the momentum space. The total number of states is normalized to one (including the LHB) and the doping concentration for one Cu site and one spin is c. At half-filling ($c = 0$) the spectral weight of Hubbard bands (of width Δd) is $1/2$. Under extended electron doping it diminishes to $(1/2 - c)$ [24]. The loosed weight $2c$ goes to the defect subbands equally.

The energy zero is taken at the $\alpha 2$ band bottom, the $\alpha 1$ band bottom lies higher by d. The β-band bottom is fixed at d_1. At extremely small doping the bare gap between the UHB and the defect states energy domain reads $(d_1 - d)$. However, the defect band tops must be supposed to be shifted towards the UHB bottom with extended doping. First, it is widely known that progressive doping leads to formation of a common carrier family which resembles the Fermi liquid behaviour. And second, pseudogap energies diminish with doping [27, 31, 32]. Correspondingly the $\alpha 1$ and $\alpha 2$ top energies evolve as $(d + \alpha c^2)$ and αc^2. This quadratic dependence is suggested by the pseudogap curves in [27, 32]. The defect subbands densities of states $\rho_\alpha(1, 2) = (\alpha c)^{-1}$ rise with doping and of the Hubbard components $\rho_{\beta, \gamma} = (\frac{1}{2} - c)(\Delta d)^{-1}$ diminish.

There are two critical points in the model. At $c_0 = [(d_1 - d)\alpha^{-1}]^{1/2}$ the $\alpha 1 - \beta$ gap closes. For $c < c_0$ the chemical potential $\mu_1 = d + \alpha c^2$ remains connected with the $\alpha 1$, so at low doping electrons occupy the antinodal region [28, 29]. For $c \geq c_0$ ($c < c_x$) the $\alpha 1$ and β bands overlap, being both intersected by μ_2. Experimentally one sees here a steep and a flat energy dispersion [29]. The second critical point appears when $\mu_2(c_x) = \alpha c_x^2$ and all three actual band components overlap and are intersected by μ_3. Additional spectral intensity in the nodal region appears and a common Fermi surface will be formed. Experimentally one has revealed the growing participation of the hole carriers in the electron doped cuprates [36]. In our model it becomes explained by the chemical potential in the upper parts of defect subbands. The optimal interband pairing conditions near c_x become destroyed by further doping.

14.3 Superconducting characteristics and excitations

The superconducting system under consideration will be described by the mean-field Hamiltonian analogous to used in [21] which can be simply diagonalized. The interband pairing interaction (constant W) will be taken the same over the whole momentum space and intraband pairing channels are not introduced. The quasiparticle energies

$$E_\sigma(\mathbf{k}) = \pm[\epsilon_\sigma^2(\mathbf{k}) + \Delta_\sigma^2(\mathbf{k})]^{1/2} \qquad (14.1)$$

preserve its usual form in this approach. Here $\epsilon_\sigma = \xi_\sigma - \mu$ with band energies ξ_σ, and Δ_σ stands for superconducting gaps.

The observable spectral gaps correspond to the minimal quasiparticle energies. In the presence of bare normal state gaps the band components not containing μ create pseudogaps type excitations [18]. At heavy underdoping $c < c_0$ this is the case for β-band and $\alpha2$-band with pseudogap energies

$$\Delta_{p\beta} = [(d_1 - d - \alpha c^2)^2 + \Delta_\beta]^{1/2} \,, \qquad (14.2)$$

$$\Delta_{p\alpha} = [d^2 + \Delta_\alpha^2]^{1/2} \,. \qquad (14.3)$$

The normal state gap corresponding at $\Delta_{p\beta}$ closes at c_0 and the separation of occupied and empty states vanishes indicating the vanishing antiferromagnetic order. The smaller pseudogap $\Delta_{p\alpha}$ is attributed to excitations in the nodal spectral window and at $c > c_0$ it diminishes until c_x as

$$\Delta_{p\alpha} = [(\mu_2 - \alpha c^2)^2 + \Delta_\alpha]^{1/2} \,. \qquad (14.4)$$

In this region the Δ_α superconducting gap is expected to be observable in the antinodal region. Beyond c_x the excitation spectrum is expected to be determined by both superconducting gaps Δ_β and Δ_α. The latter must now be detectable in the whole momentum window. With vanishing normal state gap $|\mu - \alpha c^2|$ at c_x the tracks of AF order become quenched also in the $\left(\frac{\pi}{2}, \frac{\pi}{2}\right)$ window.

The illustrative calculations have been made by calculating T_c and superconducting gaps numerically according to the system ($\Theta = k_B T$)

$$\Delta_\alpha = W \Delta_\beta \sum_k E_\beta^{-1}(\mathbf{k}) th \frac{E_\beta(\mathbf{k})}{2\Theta}$$

$$\Delta_\beta = W \Delta_\alpha \sum_k {}^{1,2} E_\alpha^{-1}(\mathbf{k}) th \frac{E_\alpha(\mathbf{k})}{2\Theta} \,. \qquad (14.5)$$

Here the indices at $\sum$ point to the integration in energy intervals occupied by $\alpha1$ and $\alpha2$ subbands. The paired carrier density is expressed as

$$n_s = \frac{1}{2} \left\{ \Delta_\beta^2 \sum_k E_\beta^{-2}(\mathbf{k}) th^2 \frac{E_\beta(\mathbf{k})}{2\Theta} + \Delta_\alpha^2 \sum_k {}^{1,2} E_\alpha^{-2}(\mathbf{k}) th^2 \frac{E_\alpha(\mathbf{k})}{2\Theta} \right\} \,. \qquad (14.6)$$

The thermodynamic critical field

$$H_{c0} = [4\pi(2\rho_\alpha \Delta_\alpha^2 + \rho_\beta \Delta_\beta^2)]^{1/2} \qquad (14.7)$$

represents the condensation energy as $H_{c0}/8\pi$.

The expression for the electronic critical coherence length can be obtained analogously to [22, 39] with the result $(T = 0)$

$$\xi_0^2 = \frac{[\eta_\alpha(1) + \eta_\alpha(2)]B_\beta m_\beta^{-1} + \eta_\beta[B_\alpha(1) + B_\alpha(2)]m_\alpha^{-1}}{[\eta_\alpha(1) + \eta_\alpha(2)]A_\beta + \eta_\beta[A_\alpha(1) + A_\alpha(2)]} \; . \qquad (14.8)$$

The quantities in (14.8) for μ between integration limits $\Gamma_{0\sigma}$ and $\Gamma_{c\sigma}$ are estimated as

$$\eta_\sigma = W\rho_\alpha \ln\left\{ |\Gamma_{0\sigma} - \mu||\Gamma_{c\sigma} - \mu|\left(\frac{2\gamma}{\pi\Theta_c}\right)^2 \right\}, \qquad (14.9)$$

$$A_\sigma = 2W\rho_\alpha \; ; \quad B_\sigma = \frac{7\zeta(3)}{2}W\rho_\alpha\frac{|\mu - \Gamma_{0\sigma}|}{(\pi\Theta_c)^2} \qquad (14.10)$$

with $\gamma = 1.78$ and $\zeta(3) = 1.2$. For μ out of the band $A_\sigma = B_\sigma = 0$ and

$$\eta_\sigma = W\rho_\alpha \ln\left|\frac{\Gamma_{c\sigma} - \mu}{\Gamma_{0\sigma} - \mu}\right| . \qquad (14.11)$$

The bare effective masses in bands are defined by the densities of states as $m_\sigma = 2\pi\hbar^2\rho_\sigma V^{-1}$, where $V = a^2$ is the CuO_2 plane plaquette area.

14.4 Illustrative results

For illustrative calculations concerning a "typical" electron doped cuprate with T_c round 30 K near $c = 0.15$ in the superconductivity domain between $c = 0.07$ and 0.3 the following parameter set has been chosen: $d = 0.03$; $d_1 = 0.1$; $\Delta d = 1.0$ and $\alpha = 10$ (eV). The interband coupling channel with $W = 0.175$ eV leads then to $T_{cm} = 28$ K at $c = 0.15$. The critical doping concentrations read $c_0 = 0.08$ and $c_x = 0.13$.

Figure 14.1 illustrates the calculated T_c, μ and both pseudogaps on the doping scale. T_c exhibits the "usual" bell like curve and the density of paired carriers shown in the inset of Fig. 14.1 follows the same route. Note that n_s diminishes behind $T_c(max)$ where the number of doped carriers grows. The chemical potential rises with doping as observed [30]. The present model contains the second pseudogap which can enter the superconducting domain. Experimental paper [27] points to a comparable event under a question mark.

The calculated coherence length (seemingly for the first time) in Fig. 14.2 is nonmonotonic and of opposite behaviour with the condensation energy shown for comparison also in Fig. 14.2. One concludes that the strength of the pairing

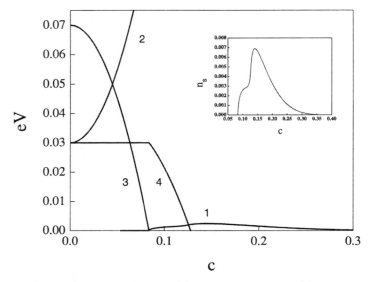

Fig. 14.1. The transition temperature (1), chemical potential (2), the large (3) and the small (4) normal state (pseudo)gaps on the electron doping (c) scale. The inset shows the paired carrier density.

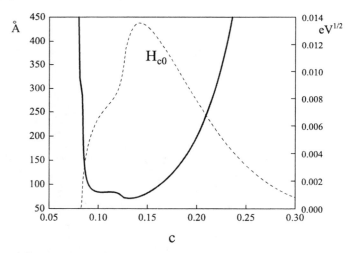

Fig. 14.2. The thermodynamic critical field and the $T = 0$ coherence length on the doping scale.

and the phase coherence develop jointly in the present approach. The same conclusion has been made in the hole-doping case [22] in agreement with the recent experimental result for the whole doping scale [37]. Our parameter set serves T_c of observed magnitude, however, ξ_0 is obtained without any further fittings. Comparing T_c-s of hole and electron doped materials one

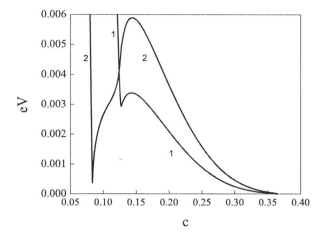

Fig. 14.3. The excitation energies represented by pseudogap and superconducting components ($T = 0$). Curve $1 - \Delta_{p\alpha}$ and Δ_{α}; $2 - \Delta_{p\beta}$ and Δ_{β}.

expects greater ξ_0-s for the latter case. Indeed, the present calculation has given roughly an order of magnitude greater coherence lengths for the electron doping case.

In Fig. 14.3 the behaviour of minimal quasiparticle excitation energies at $T = 0$ are shown. The curve 1 shows the smaller pseudogap and the super-conducting gap Δ_{α} created by the $\alpha 2$ defect subsystem. For $c > c_x$, Δ_{α} must appear in both nodal and antinodal windows by the cooperation of $\alpha 1$ and $\alpha 2$. At $c < c_x$, Δ_{α} manifests in the $(\pi, 0)$ window (not shown in Fig. 14.3). The expected insulator-metal crossover at c_x in the normal state can be attributed to the effect observed in [38]. The experimental resistivity curve is roughly comparable with $|\mu_2 - \alpha c^2|$, see Fig. 14.1.

The larger pseudogap transforms into the UHB superconducting gap at the lower critical point c_0, as the curve 2 shows. At $c > c_x$ the spectral response is determined by both superconducting gaps. In the manifestation of two gaps the states associated with the smaller gap can mask the edge of the larger one. The peak associated with the larger gap represents a spectral hump cf. [29].

The characteristic gap relations $2\Delta_{\alpha,\beta}/kT_c$ violate the BCS universality being approximately 4.8 and 2.6 for α- and β-bands, correspondingly.

14.5 Conclusion

It seems that the present model is approaching to describe some properties of electron-doped cuprate superconductors on the doping phase diagram. Experimental investigations of excitation gaps over the whole doping scale could be useful for more thorough comparison of the present theoretical conclusions with observations.

If accepting the positive contribution of the modelling of hole-doped cuprates [21,22,39,40] and of the present work one arrives to a general logistic scheme concerning the functioning of cuprate high-temperature superconductors. Doping metallizes the basic antiferromagnetic materials and builds up a new complex electron spectrum where an interband pairing channel opens. However, the detailed realization of these processes follows in different manner and on different sublattices for hole- and electron-doped materials. The characteristic functioning of the interband pairing mechanism on the multiband background leads to comparable phase diagrams of both types of cuprate high-temperature superconductors.

This work was supported by Estonian Science Foundation grant No. 6540.

References

1. A. Damascelli and Z. Hussain and Z. -H. Shen, Angle-resolved photoemission studies of the cuprate superconductors, Rev. Mod. Phys., **75**, 0, 473–541, 2003
2. R. Khasanov and A. Shengelaya and A. Maisuradze and F. La Mattina and A. Bussmann-Holder and H. Keller and K. A. Muller, Experimental evicende for two gaps in the high-temperature $La_{1.83}Sr_{0.17}CuO_4$ superconductor, Phys. Rev. Lett., **98**, 057007–4, 2007
3. K. Tanaka and W. S. Lee and H. Lu and A. Fujimori and T. Fujii and A. Risdiana and I. Terasaki and D. J. Scalapino and T. P. Deveraux and Z. Hussain and Z. -H. Shen, Distinct Fermi-momentum dependent energy gaps in deeply underdoped Bi 2212, Science, **314**, 1910–1913, 2006
4. M. C. Boyer and W. D. Wise and K. Chatterjee and M. Yi, T. Kondo and T. Takeychi and H. Ikuta and E. W. Hudson, Imaging the two gaps of the high-T_c superconductor $Pb-Bi_2Sr_2CuO_{6+x}$,cond.-mat. supr.-com., 0705.1731, 0, 1–6, 2007
5. N. Dairon-Leyraud and C. Proust and D. Le Boeuf and J. Levallois and J. -B. Bonnemaison and R. Liang and D. A. Bonn and W. N. Hardy and L. Taillefer, Quantum oscillationsn and the Fermi surface in an underdoped high-T_c superconductor, Nature, **447**, 0, 565–568, 2007
6. K. A. Müller, Recent experimental insights into HTSC materials, Physica C, **341–348**, 0, 11–18, 2000
7. K. A. Müller and G. Benedek, Phase separation in cuprate superconductors, publisher World Sci., Singapore, 1993
8. J. C. Phillips, Self-organized networks and lattice effects in high-temperature superconductors, Phys. Rev. B, **75**, 0, 214503–23, 2007
9. A. Bianconi and A. Valletta and A. Perali and N. L. Saini, Superconductivity of a striped phase at the atomic limit, Physica C, **296**, 0, 269–280, 1998,
10. V. A. Moskalenko and M. E. Palistrant and V. M. Vakalyuk, High-temperature superconductivity basing on the account of electron spectrum peculiarities, Uspekhi Fiz. Nauk, **161**, 155–178, 1991
11. N. Kristoffel and P. Konsin and T. Örd,Two-band model for high-temperature superconductivity, Riv. Nuovo Cimento, **17**, 0, 1–41, 1994
12. H. Nagao and S. P. Kruchinin and A. M. Yaremko and K. Yamaguchi, Multiband superconductivity, Intern. J. Modern. Phys. B, **16**, 0, 3419–3428, 2002

13. A. Bianconi, Feshbach shape resonance in multiband superconductivity in heterostructures, J. Supercond., **18**, 0, 25–36, 2005
14. L. Gorkov and A. V. Sokol, Phase separation of electronic liquid in novel superconductors, Pis'ma ZETF, **46**, 0, 333–336, 1987
15. A. Liu and Y. Y. Mazin and Y. Kortus, Beyound Eliashberg superconductivity in MgB$_2$: anharmonicity, two-phonon scattering and multiple gaps, Phys. Rev. Lett., **87**, 0, 087005–4, 2001
16. N. Kristoffel and T. Örd and K. Rägo, MgB$_2$ two-gaps superconductivity with intra- and interband couplings, Europhys. Lett., **61**, 0, 104–115, 2003
17. A. Perali and C. Castellani and C. Di Castro and M. Grilli and E. Piegari and A. A. Varlamov, Two-gap model for underdoped cuprate superconductors, Phys. Rev. B, **62**, 0, R9295–R9298, 2000
18. N. Kristoffel and P. Rubin, Pseudogap and superconductivity gaps in a two-band model with the doping determined components, Solid State Commun., **122**, 0, 265–268, 2002
19. R. Micnas and A. Robaszkiewicz and A. Bussmann-Holder, On the superconductivity in the induced pairing model, Physica C, **387**, 0, 58–64, 2003
20. H. Kamimura and H. Ushio and S. Matsuno and S. Hamada, Theory of Copper Oxide Superconductors, Springer, Berlin, 2005
21. N. Kristoffel and P. Rubin, Superconducting gaps and pseudogaps in a composite model of two-component cuprate, Physica C, **402**, 0, 257–262, 2004
22. N. Kristoffel and T. Örd and P. Rubin, Doping dependence of cuprate coherence length, paired carrier effective mass and penetratinon depth in a two-component model, Physica C, **437/438**, 0, 168–170, 2006
23. N. Kristoffel and P. Rubin, Cuprate interband superconducting density for doping driven spectral overlaps, J. Supercond., **18**, 0, 705–708, 2005
24. M. B. J. Meinders and H. Eskes and G. A. Sawatzky, Breakdown of low-energy-scale sum rules in correlated systems, Phys. Rev. B, **48**, 0, 3916–3926, 1993
25. T. Takahashi and H. Matsuyama and H. Yoshida-Katayama and K. Feni and K. Kamiya and H. Inokuchi, Angle resolved study of nonsuperconductive Br$_2$Sr$_2$Ca$_{0.4}$Y$_{0.6}$Cu$_2$O$_8$, Physica C, **170**, 0, 416–418, 1990
26. S. Uchida and T. Ido and H. Takanagi and T. Arima and Y. Tokura and S. Tajima, Optical spectra of La$_{2-x}$Sr$_x$CuO$_y$: Effect of carrier doping on the electronic structure of the CuO$_2$ plane, Phys. Rev. B, **43**, 0, 7942–7954, 1991
27. Y. Alff and Y. Krockenberger and B. Welter and M. Schonecke and R. Gross and D. Manske and M. Naito, A hidden pseudogap under the dome of superconductivity in electron-doped high-temperature superconductors, Nature, **422**, 0, 698–701, 2003
28. S. R. Park and Y. S. Roh and Y. K. Yoon and C. S. Leem and B. J. Kim and H. Koh and H. Eisaki and N. P. Armitage and C. Kim, Electronic structure of electronic-doped Sr$_{1.86}$Ce$_{0.14}$CuO$_4$, Phys. Rev. B, **75**, 0, 060501(R)–4, 2007
29. H. Matsui and K. Terashima and T. Sato and T. Takahashi and S. -C. Wang and H. -B. Yang and H. Ding and T. Uefuji and K. Yamada, Angle resolved photoemission spectroscopy of the AF superconductors Nd$_{1.87}$Ce$_{0.13}$CuO$_4$, Phys. Rev. Lett., **94**, 0, 047005–9, 2005
30. N. Harima and J. Matsuno and A. Fujimori and Y. Onose and Y. Taguchi and Y. Tokura, Chemical potential shift in Nd$_{2-x}$Ce$_x$CuO$_4$, Phys. Rev. B, **64**, 0, 22507(R)–4, 2001

31. N. P. Armitage and D. H. Lu and C. Kim and A. Damascelli and K. M. Shen and F. Bonning and D. L. Feng and R. Bogdanov and Z. -X. Shen and Y. Onoso and Y. Taguchi and Y. Tokura and P. K. Mang and N. Kaneko and M. Greken, Anomalous electronic structure and pseudogap effects in $Nd_{1.85}Ce_{0.15}CuO_4$, Phys. Rev. Lett., **87**, 0, 147003–4, 2001

32. Y. Onose and Y. Taguchi and K. Ishizaka and Y. Tokura, Doping dependence of pseudogap and related charge dynamics in $Nd_{2-x}Ce_xCuO_4$, Phys. Rev., **87**, 0, 217001–4, 2001

33. Q. Yuan and X. -Z. Yan and H. Tings-wave like excitation in the superconducting state of electron-doped cuprates with d-wave pairing, Phys. Rev. B, **74**, 0, 214503–9, 2006

34. C. Kusko and R. S. Markiewicz and M. Lindroos and A. Bansil, Fermi surface evolution and collapse of the Mott pseudogap in $Nd_{2-x}Ce_xCuO_{4+\delta}$, Phys. Rev. B, **66**, 0, 140513(R)–4, 2002

35. B. Kyung and V. Hankevych and A. -M. Dare and A. -M. S. Tremblay, Pseudogap and spin fluctuations in the normal state of the electron-doped cuprates, Phys. Rev. Lett., **93**, 0,147004–4, 2004

36. Y. Dagan and R. L. Greene, Hole superconductivity in the electron-doped superconductor $Pr_{2-x}Ce_xCuO_4$, Phys. Rev. B, **76**, 0, 024506–4, 2007

37. H. H. Wen and H. P. Yang and S. L. Li and X. H. Zeng and A. A. Sankissian and W. D. Si and X. X. Xi, Hole doping dependence of the coherence length in $La_{2-x}Sr_xCuO_4$ thin films, Europhys. Lett., **64**, 0, 790–796, 2003

38. P. Fournier and P. Mohanty and E. Maiser and S. Darzens and T. Venkatesan and C. J. Lobb and G. Czjezek and R. A. Webb and R. L. Greene, Insulator-metal crossover near optimal doping in $Pr_{2-x}Ce_xCuO_4$, Phys. Rev. Lett., **81**, 0, 4720–4723, 1998

39. N. Kristoffel and T. Örd and P. Rubin, in: Electron correlation in new materials and nanosystems, edited by K. Scharnberg and S. Kruchinin, pp. 275–282, Springer, Dordrecht, 2007

40. V. J. Belyavski and I. V. Kopaev and S. V. Shevtsov, Mirror nesting of the Fermi contour and superconducting pairing from the repulsive interaction, J. Supercond, **17**, 0, 303–315, 2004

15

ANTIADIABATIC STATE – GROUND STATE OF SUPERCONDUCTORS: STUDY OF YBCO

P. BAŇACKÝ

Faculty of Natural Science, Institute of Chemistry, Chemical Physics division, Comenius University, Mlynská dolina CH2, 84215 Bratislava, Slovakia and S-Tech a.s., Dubravská cesta 9, 84105 Bratislava, Slovakia. banacky@fns.uniba.sk

Abstract. It has been shown that el-ph coupling to A_g, B_{2g} and B_{3g} modes in $YBa_2Cu_3O_7$ induces T-dependent electronic structure instability which is related to fluctuation of analytic critical point of (d-pσ) band across Fermi level. It results in considerable reduction of chemical potential and to breakdown of the Born-Oppenheimer approximation. At critical temperature T_c, superconducting system undergoes transition from adiabatic electronic ground state into anti-adiabatic state at broken symmetry that is stabilized due to the effect of nuclear dynamics. This effect is absent in non-superconducting $YBa_2Cu_3O_6$. Formation of asymmetric gaps in **a** and **b** direction of $YBa_2Cu_3O_7$ has been shown. In a good agreement with experimental T_c of superconducting state transition, critical temperature has been calculated - $Tc \approx 92.8\,K$. Present study has also revealed that in c direction there should be identified the next gap that is considerably smaller than gaps in **a**, **b** directions.

Keywords: non-adiabatic electron-phonon interactions, anti-adiabatic state, superconducting state transition

15.1 Introduction

The ARPES study of high-T_c cuprates [1–3] and theoretical results of low-Fermi energy band structure fluctuation [4–7] indicate that electron coupling to pertinent phonon modes drive system from adiabatic ($\omega \ll E_F$) into antiadiabatic state ($\omega > E_F$). At these circumstances, not only Migdal-Eliashberg approximation is not valid, but basic adiabatic Born-Oppenheimer approximation (BOA) does not hold. The *ab initio* nonadiabatic electron-vibration theory of the complex electronic ground state that respects this fact has been elaborated in [8]. From the theory follows that due to EP interactions that drive system from adiabatic to antiadiabatic state, symmetry breaking is induced and system is stabilized in antiadiabatic state at distorted geometry with respect to adiabatic equilibrium high symmetry structure. Stabilization effect in antiadiabatic state is due to strong dependence of the electronic motion on the instantaneous nuclear kinetic energy, i.e. on the effect that is neglected on the adiabatic level within the Born-Oppenheimer approximation (BOA). Antiadiabatic ground state at distorted geometry is geometrically degenerate with fluxional structure of nuclear positions in the phonon modes that drive system into this state. It has been shown that while system remains in antiadiabatic state, nonadiabatic polaron – renormalized phonon interactions are zero in well defined k region of reciprocal lattice. Along with geometric degeneracy of the antiadiabatic state it enables formation of mobile bipolarons that can move on the lattice as supercarriers without dissipation in a form of polarized inter-site charge density distribution. More over, it has been shown that due to EP interactions at transition to antiadiabatic state, k-dependent gap in one-electron spectrum is opened. Gap opening is related to shift of the original adiabatic Hartree-Fock orbital energies and to the k-dependent change of density of states of particular band(s) at Fermi level. The shift of orbital energies determines in a unique way one-particle spectrum and thermodynamic properties of system. Resulting one-particle spectrum yields all thermodynamic properties that are characteristic for system in superconducting state, i.e. temperature dependence of the gap, specific heat, entropy, free energy and critical magnetic field. The k-dependent change in the density of states close to Fermi level at transition from adiabatic (nonsuperconducting) to antiadiabatic state (superconducting) can be experimentally verified by ARPES or tunneling spectroscopy as spectral weight transfer at cooling superconductor from temperatures above T_c down to temperatures below T_c. This theory has been applied recently to the study of MgB_2 superconducting state transition [6] and obtained results have been in a good agreement with the experimental data. In the present paper, the results obtained at study of the transition to superconducting state for the high-T_c cuprate $YBCO$ – $YBa_2Cu_3O_7$, from the stand-point of the ab initio theory are presented.

15.2 Nonadiabatic effects in $YBa_2Cu_3O_7$

15.2.1 PRELIMINARIES

The study starts with the LCAO-based HF-SCF calculation of the electronic band structure of $YBa_2Cu_3O_7$ for clumped nuclei configuration at the high-symmetry experimental geometry [9, 10]. The band structure calculations have been performed by the computer code SOLID 2000 [11]. The code is based on the method of cyclic cluster [12] with the quasi-relativistic INDO Hamiltonian [13, 14]. Within the INDO calculation scheme, the one-electron off-diagonal inter site matrix elements $h_{\mu_A \nu_B}$ (β integrals - hopping terms) are not restricted only to nearest-neighbor or next nearest-neighbor terms but all terms among involved atoms and AOs are included. Two-electron coulomb repulsion is calculated for one and two-center terms (three and four-center terms are neglected) in the form $(\mu\mu|\nu\nu)$ over the Slater-Condon parameters. For two-electron one-center terms, also exchange repulsion in the form $(\mu\nu|\mu\nu)$ is involved in calculations. Incorporating INDO Hamiltonian within the cyclic cluster method (with Born-Karman boundary conditions) yields good results for properties related to electrons lying at Fermi level (frontier orbital properties) and for calculations of equilibrium structures [15–17]. The same method is used for study the effects of lattice distortion on the electronic band structure. Study of the nonadiabatic effects is performed as the post-SCF calculations based on the results derived from the band structures.

15.3 Results

The experimental [10] lattice parameters of $YBa_2Cu_3O_7$ (orthorhombic structure, space group Pmmm, oP14); fractional coordinates of the unit cell atoms: $Cu(1) = (0,0,0)$; $Cu(2) = (0,0,0.355)$; $Y = (1/2,1/2,1/2)$, $Ba = (1/2,1/2,0.186)$, $O(1) = (0,1/2,0)$, $O(2) = (1/2,0,0.380)$, $O(3) = (0,1/2,0.376)$, $O(4) = (0,0,0.156)$, and lattice constants $\mathbf{a} = 3.817A^o$, $\mathbf{b} = 3.882A^o$, $\mathbf{c} = 11.671A^o$ have been used for band structure (BS) study. The unit cell has 13 atoms as it corresponds to the formula unit with the chain oxygen $O(1)$ in $\mathbf{b}$-direction and vacancy in a direction. The basic cluster of the dimension $5 \times 5 \times 5$, has been generated by the corresponding translations of the unit cell in the directions of crystallographic axes, $\mathbf{a}$ (5), $\mathbf{b}$ (5), $\mathbf{c}$ (5).The cluster of this size (1625 atoms) generates a grid of 125 points in k-space. The HF-SCF procedure is performed for each k-point of this grid with the INDO Hamiltonian matrix elements that obey the boundary conditions of the cyclic cluster. Pyykko-Lohr quasi-relativistic basis set of the valence electron atomic orbitals (s, p-AO for Ba, O, and s, p, d-AO for Cu and Y) has been used, i.e., 72 AO/unit cell and the total number of STO type functions in the basic cluster has been 9,000.

15.3.1 BAND STRUCTURES OF $YBA_2CU_3O_7$

15.3.1.1 Experimental - undistorted geometry

Calculated, LCAO-based HF-SCF band structure of $YBa_2Cu_3O_7$ is presented in Fig. 15.1a. Character of the bands has been discussed in details elsewhere - see e.g. [18–21] and it is not necessary to be analyzed here. From the standpoint of the present study, the important is the topology of the couple of $Cu(2)$-$O(2)/O(3)$ planes – derived (d-pσ) bands with the maximum in antibonding region at the S point. This topology feature is common to the present and DFT – based band structures. The difference is in the chain oxygen $O(1)$-

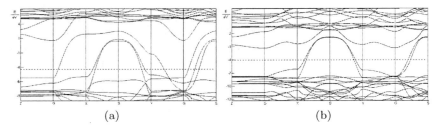

(a)   (b)

Fig. 15.1. Part of the band structure of $YBa_2Cu_3O_7$ (a) and $YBa_2Cu_3O_6$ (b) calculated at the experimental - undistorted geometry. The fractional coordinates of the selected high symmetry points of the Brillouin zone are: G($\equiv \Gamma$) (0, 0, 0), X(1/2, 0, 0), Y(0, 1/2, 0), S(1/2, 1/2, 0), Z(0, 0, 1/2).

derived band topology and it will be discussed on the other place of this paper in relation to opening the gaps in one-electron spectrum on the -Y and -X lines with respect to the recent ARPES results [22, 23] on the untwined $YBa_2Cu_3O_7$.

The band structure of nonsuperconducting – deoxygenated YBCO, YBa_2 Cu_3O_6, which is without the chain oxygen $O(1)$ is in Fig. 15.1b. As it can be expected, comparing to the band structure of $YBa_2Cu_3O_7$ – Fig. 15.1a, the CuO planes-derived (d-pσ) bands are with the same topology but the chain oxygen $O(1)$-derived p band is absent. Recent ARPES experiments [1, 2] of high-T_c cuprates have shown an abrupt change of the electron velocity in low-Fermi energy region, 50–80 meV below the Fermi level, to be universal common feature of the high-T_c cuprate superconductors. From the character of the calculated band structure of $YBa_2Cu_3O_7$, Fig. 15.1a, as well from the other published DFT-based band structures of other high-T_c cuprates, it is clear that in the low-Fermi energy region below the Fermi level of these compounds, there is not any abrupt change in the slope ($\partial\varepsilon/\partial k$ – velocity (quasi-momentum) of electrons) of bands that cross Fermi level at the basal plane (Γ, X, Y, S). Expressed explicitly, electronic band structures of the high-T_c cuprates calculated for respective high-symmetry experimental structures do not indicate any low-Fermi energy velocity decrease of electrons or importance of non-adiabatic EP coupling at superconducting state transition.

15.3.1.2 Distorted geometries

Early theoretical studies of EP coupling in $YBa_2Cu_3O_7$, based on rigid-muffin-tin or related rigid-atom approximations [24, 25] have indicated EP coupling strength to be very small. Results [26] obtained by LAPW method, using the frozen-phonon technique, are different. For A_g phonon modes strong EP coupling with coupling constant $\lambda \cong 1.7$–1.9 have been obtained. Much smaller EP coupling has been calculated for B_{2g} and B_{3g} phonon modes – $\lambda < 0.6$. In spite of strong EP coupling, mainly of A_g phonon modes, dramatic

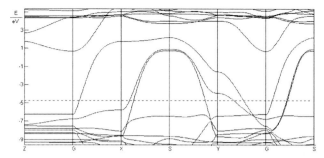

Fig. 15.2. Band structure at the coupling to A_g, B_{2g} and B_{3g} modes with displacements of apical O4 ($\Delta f_c = -0.0027$), planar O2 ($\Delta f_a = 0.0057$) and O3 ($\Delta f_b = -0.0057$) atoms out of the equilibrium - experimental positions. At this distortion, there is a shift of the SP of one of the $Cu(2)$-$O(2)/O(3)$ plane - derived (d-pσ) band on the Γ-Y line at Y point, from below to above-Fermi level position.

changes in the topology of bands, e.g. shift of analytic critical point of some band above or below Fermi level, like in MgB_2 [4–7], has not be reported for $YBa_2Cu_3O_7$.

Present HF-SCF band structure calculations confirm these results as far as a single isolated mode distortion is concerned. Situation is changed substantially when combination of A_g, B_{2g} and B_{3g} modes (for notation see e.g. [27–29]) is studied. For the **c**-direction displacement $\Delta f_c = -0.0027(-0.03151A^o)$ of the apical O4, i.e. change of the inner coordinate $0.156 \rightarrow 0.1533$, and **a, b** - direction displacements of O2 and O3 atoms from 0.5 by $\Delta f_a = 0.0057(0.02175A^o)$ for O2, and $\Delta f_b = -0.0057(-0.02213A^o)$ for O3, the band structure significantly changes the topology – Fig. 15.2.

For this combination of displacements, the ground state energy with respect to undistorted structure has been increased – destabilized by $+170$ meV/unit cell. For single modes displacements destabilization is smaller, but without the shift of the SP across the Fermi level.

Continuation in displacement of the apical O4 yields again the topology change. For $\Delta f_c = -0.0025$, the SP in Y point is shifted back across the Fermi level to below-Fermi level position and the band structure of the topology of Fig. 15.1a is recovered. Expressed explicitly, present results of HF-SCF calculations, show that combination of O4, O2, O3 atoms vibration (A_g, B_{2g} and B_{3g} modes) shifts periodically up and down the SP of one of the CuO plane derived (d-pσ) band on the Γ-Y line at Y point across the Fermi level. The same effect is reached if the displaced position of O4 is fixed and displacement of O3 (O2) atoms is increased.

The distortion resulting in Fermi level crossing, as reported above, corresponds to regular oxygen atoms displacements with respect to nuclear vibration in corresponding modes. It can be shown that fictitious forced cooperative displacements of O2 and O3 atoms that preserve orthogonal direction of the displacement vectors with the same $|\Delta f| = 0.0057$ (O4 remains in the displaced position with $\Delta f_c = -0.0027$), generate an infinite number of distorted structures with nearly the same ground state energy destabilization and with the band structure characterized by the SP shift above the Fermi level in the Y point (Fig. 15.2). The distorted structures are characteristic by O2, O3 atoms placed on perimeters of circles with the same radius $\Delta f = 0.0057$. The circles are centered at the undistorted coordinates of these atoms. Basically, keeping orthogonal position of the displacement vectors, an infinite number of different distorted structures can be generated by the cooperative motion of O2, O3 atoms along the perimeters of the specified circles. For the distorted structures characteristic by the SP shift across the Fermi level, there is a significant change in the one-particle spectrum with respect to undistorted, high-symmetry experimental structure. The most significant is formation of the new Fermi level with down-energy shift by -0.35 eV with respect to the original Fermi level of the undistorted structure.

Results of the distortions for non-superconducting $YBa_2Cu_3O_6$ are different. The studied distortions do not yield the SP shift of (d-pσ) band on the Γ -Y line at Y point across the Fermi level over the investigated wide range of A_g, B_{2g} and B_{3g} modes displacements. The distortion that in the case of $YBa_2Cu_3O_7$ results in Fermi level crossing leaves the band structure of $YBa_2Cu_3O_6$ – Fig. 15.1b, without significant change of the bands topology and without significant shift of the Fermi level. Also the destabilization energy due to distortion, on the BOA level, is in this case smaller, $+129$ meV/unit cell.

15.3.2 NONADIABATIC EP INTERACTIONS IN $YBA_2CU_3O_7$

15.3.2.1 Nonadiabatic correction to zero-particle term of the fermionic Hamiltonian; Correction to the fermionic ground state energy

Results of band structure calculations have shown that at vibration motion the EP coupling induces fluctuation of analytic critical point of (d-pσ) band across Fermi level. It is related to decrease of chemical potential of electrons in this band to very small value, $\mu_{dp\sigma} \to 0$, i.e. to transition from adiabatic to antiadiabatic state. At these circumstances, standard adiabatic BOA is not valid and electronic motion has to be studied as a function of nuclear coordinates Q as well of nuclear momenta P.

Fermionic ground state energy correction due to EP interactions on the non-adiabatic level is determined by two terms [8],

$$\Delta E^0_{(na)} = \sum_{rAI} \hbar\omega_r \left(|c^r_{AI}|^2 - |\hat{c}^r_{AI}|^2 \right) =$$

$$= \sum_A^{unocc} \sum_I^{occ} \sum_r |u^r_{AI}|^2 \left[\hbar\omega_r \Big/ \left((\varepsilon^0_A - \varepsilon^0_I)^2 - (\hbar\omega_r)^2 \right) \right] \qquad (15.1)$$

The first term, related to $\{c^r_{PQ}\}$ – coefficients of coordinate Q-dependent canonical transformation matrix C,

$$c^r_{PQ} = u^r_{PQ} \left[\left(\varepsilon^0_P - \varepsilon^0_Q \right) \Big/ \left((\hbar\omega_r)^2 - \left(\varepsilon^0_P - \varepsilon^0_Q \right)^2 \right) \right] ; P \neq Q \qquad (15.2)$$

represents the adiabatic diagonal Born-Oppenheimer correction (DBOC).

The second term, related to $\left\{ \hat{c}^r_{PR} \right\}$ – coefficients of momentum P-dependent canonical transformation matrix $\hat{C}$,

$$\hat{c}^r_{PQ} = u^r_{PQ} \left[\hbar\omega_r \Big/ \left((\hbar\omega_r)^2 - \left(\varepsilon^0_P - \varepsilon^0_Q \right)^2 \right) \right] ; P \neq Q \qquad (15.3)$$

is pure nonadiabatic contribution that express influence of the electronic motion on the nuclear kinetic energy and vice-versa. It has to be stressed that on the adiabatic - BOA level, the second term is absent and hence-forth, ground state energy correction is always positive. Negative value of the ground state energy correction (stabilization contribution to the ground state energy) is possible only for antiadiabatic state due to participation of the nuclear kinetic effect. Correction to electronic ground-state energy in **k**-space representation due to interaction of pair of states mediated by phonon mode r can be written as,

$$\Delta E^0_{(na)} = 2 \sum_{\varphi_{Rk}} \sum_{\varphi_{Sk'}} \int_0^{\varepsilon_{k',\max}} n_{\varepsilon_{k'}} \left(1 - f_{\varepsilon^0 k'} \right) d\varepsilon^0_{k'} \times$$

$$\int_{\varepsilon_{k,\min}}^{\varepsilon_{k\max}} f_{\varepsilon^0 k} \left| u^r_{k-k'} \right|^2 n_{\varepsilon_k} \frac{\hbar\omega_r}{\left(\varepsilon^0_k - \varepsilon^0_{k'} \right)^2 - (\hbar\omega_r)^2} d\varepsilon^0_k,$$

$$\varphi_{Rk} \neq \varphi_{Sk'} \qquad (15.4)$$

In general, all bands of 1st BZ of multi-band system are covered, including intra-band terms, i.e., $\varphi_{Rk}, \varphi_{Rk'}$, $k \neq k'$ whil $\varepsilon^0_k < \varepsilon_F$e, $\varepsilon^0_{k'} > \varepsilon_F$.

Fermi-Dirac populations $f_{\varepsilon^0 k}, f_{\varepsilon^0 k'}$ make correction (15.4) temperature dependent. Term $u^r_{k-k'}$ stands for matrix element of EP coupling and n_{ε_k}, $n_{\varepsilon_{k'}}$ are DOS of interacting bands at $\varepsilon^0_{k'}$ and ε^0_k. For adiabatic systems, e.g. metals, this correction is positive and negligibly small (DBOC). Only for systems in antiadiabatic state the correction is negative and absolute value depends on the magnitudes of $u^r_{k-k'}$ and n_{ε_k}, $n_{\varepsilon_{k'}}$ at displacement for FL crossing. At the moment when ACP approach FL, system not only undergoes transition to antiadiabatic state but DOS of fluctuating band is considerably increased at FL.

Inspection of Figs. 15.1a and 15.2., indicates that for basal-**a**, **b** plain, the main correction to the ground state energy can be expected from EP interactions of occupied states ε_k of fluctuating (d-pσ) band with unoccupied states $\varepsilon_{k'}$ of O1-derived p band in Γ-Y and Γ-X directions. Interaction of two (d-pσ) bands that correspond to different $Cu - O$ layers which are separated

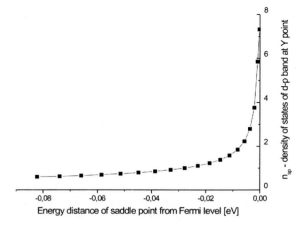

Fig. 15.3. Density of states of fluctuating d-pσ band as a function of the energy for situation when the SP touches Fermi level at Y point.

nearly 8.3 A^o, can be expected to be negligibly small. Stabilization contribution (negative value) of EP interactions starts as soon as the SP of the fluctuating (d-pσ) band approaches Fermi level from the bonding region on the energy distance $-\hbar\omega$ and it continues until the SP is not more than $+\hbar\omega$ above Fermi level. In principle, Eq. (15.4) can be solved exactly providing that functional dependences of corresponding density of states **n** and matrix elements $u_{k-k'}$ of EP interactions on orbital energies are known.

For density of states these functional dependences can be derived from the band structure. Calculated density of states for fluctuating d-pσ band as a function of energy for situation when the SP touches Fermi level at Y point is shown in Fig. 15.3. As it can be seen from this figure, shift of the d-pσ band considerably increases density of states close to Y point at Fermi level as soon as the SP approaches Fermi level or crosses it at fluctuation. Density of states of the chain oxygen O1-derived p band at **k**-point(s) where the band intersects Fermi level is constant over energy interval $\pm\hbar\omega$ at Fermi level. For Γ-Y direction it is $n_{\Gamma Y} = 0.04$, and for direction Γ-X corresponding value is even smaller, $n_{\Gamma X} = 0.03$ states/eV.

Functional dependence $u_{k-k'} \equiv f(\varepsilon_{k'} - \varepsilon_k)$ of the EP interaction matrix elements is not available on the corresponding level. However, from the SCF-HF calculation the overall value of EP interactions at displacement of vibrating atoms that results to Fermi level crossing can be extracted, i.e. change of one-electron core term $\Delta h(Q_{cros\sin g}) = u^{(1)}(Q_{cros\sin g}) \equiv \bar{u}$. The SCF-HF calculations yield $\bar{u} = 2.5$ eV. In what follows, dependence of EP coupling on orbital energy distance is approximated by functional form,

$$(u_z)^2 = (\bar{u})^2.(0,99 + 0,12.z - 0,31.z^2) \qquad (15.5)$$

It holds for $\Delta\varepsilon_{kk'} \leq 2\hbar\omega$, and $u_z = 0$ for, $\Delta\varepsilon_{kk'} > 2\hbar\omega$, $z = \frac{\Delta\varepsilon_{kk'}}{\hbar\omega}$.

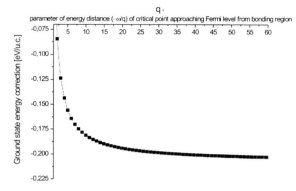

Fig. 15.4. The nonadiabatic correction to the fermionic ground state energy, $\Delta E^0_{(na)}$, of the $YBa_2Cu_3O_7$ for the distorted structure due to EP coupling to A_g, B_{2g}, B_{3g} modes as function of the parameter q – see text.

The functional dependence $u(z)$ is an important factor that determines not only the ground state energy correction but also the form of the final density of states. In this respect, reported absolute values of calculated physical parameters should be understood as theoretical simulation rather than the exact numerals.

At the moment when SP approach FL, system not only undergoes transition to antiadiabatic state but DOS of fluctuating band at FL is considerably increased – Fig. 15.3. At the calculation of nonadiabatic ground state energy correction, density of states of fluctuating d-pσ band has been approximated by mean value $\bar{n}_{SP} = 2$ states/eV (value for nearly degenerate states, $\Delta\varepsilon_{kk'} \approx 0.014$ eV – see Fig. 15.3), and phonon energy of $B_{2g}(B_{3g})$ phonon mode corresponds to the experimental value $\hbar\omega = 0.072$eV. Calculated dependence of nonadiabatic ground state energy correction as function of the parameter q that "regulates" energy distance of the SP from FL is shown in Fig. 15.4.

The maximal stabilization effect is reached when the SP of $Cu2 - O2/O3$-derived d-pσ band in Y point just approaches Fermi level from the bonding side (q$\to \infty$ in $\hbar\omega/q$). The nonadiabatic correction reaches the value -0.204 eV/unit cell. An equivalent result can be obtained in case of back-crossing. In such a case, the SP would approach Fermi level from the antibonding side and the maximal stabilization effect is reached when the SP is just $+\hbar\omega$ above Fermi level (then q $= 1$).

The ground state energy correction in antiadiabatic state at 0 K, $\Delta E^0_{(na)} \approx -204$ meV/u.c., prevails in absolute value the electronic energy increase $\Delta E_{cr} = E_{d,cr} - E_{eq} \approx 170$ meV/u.c. at nuclear displacements d when SP crosses FL. At these circumstances, system is stabilized (-34 meV/u.c.) in antiadiabatic electronic ground state at broken symmetry with respect to adiabatic equilibrium high-symmetry structure. Antiadiabatic ground state is geometrically (quasi-)degenerate. There are an infinite number of the $O2, O3$ atoms in-plane displacements, i.e. different nuclear configurations with the same ground state energy. On the lattice scale, geometric degeneracy of the fermionic ground state energy for distorted structure, i.e. existence of an infinite number of $O2, O3$ atoms displacements (fluxional structure of $Cu2 - O2/O3$ plane), enables cooperative and dissipationless motion of displaced $O2, O3$ atoms along the perimeters of circles centered at the undistorted positions of $O2, O3$ atoms, with the same radii equal to the fractional displacement $|\Delta f_{a,b}| \approx 0.0056 - 0.0057(\approx 0.022 A^o)$. This is a new, the coherent macroscopic quantum state.

The effect detectable by the ARPES experiments should be mentioned. For the lattice distortion when the SP of d-pσ band approaches Fermi level at Y point and the ground state energy is stabilized by nonadiabatic EP interactions, the dispersion of the d-pσ band at Fermi level down to -250 meV along the Γ-Y direction calculated from the band structure is shown in Fig. 15.5.

As it can be seen, the deviation of the dispersion curve from the straight-line direction starts at 75–80 meV below the Fermi level. This Figure resembles the results obtained by the ARPES experiments [1, 2] for high-T_c cuprates. The authors [1, 2] has described it as the low-Fermi energy "electron velocity", $(\partial|\varepsilon_k^0|/\partial k)$, decrease. In case of $YBa_2Cu_3O_7$, as it is evident from the present study, this effect should be T-dependent and it appears in the off-nodal Γ-Y direction close to Y point at about 75–80 meV below the Fermi level. Recently, beside the presence of T-independent small kink in the nodal direction as

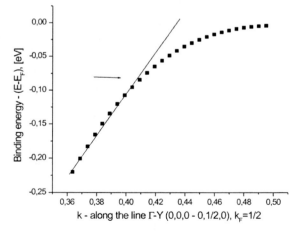

Fig. 15.5. Calculated dispersion of the fluctuating d-pσ band in Γ-Y direction at the lattice distortion due to electron coupling with A_g, B_{2g}, B_{3g} phonon modes, for situation when the SP of the band touches Fermi level (0.0 eV) at Y point. As a guide for eyes, the straight line and arrow are drawn to indicate the sudden change of low-Fermi energy "electron velocity" that start at about 75–80 meV below Fermi level.

reported in [1,2], formation of the T-dependent giant kink has been measured in off-nodal direction at the Fermi level for $Bi2223$ [3]. This kink is present below T_c and disappears above T_c at about 70 meV from the Fermi level, like the result presented in Fig. 15.5 that has been calculated for $YBa_2Cu_3O_7$.

As it has been mentioned, studied distortions do not change the band structure of non-superconducting $YBa_2Cu_3O_6$. It means that the density of states of d-pσ band remains small (≈ 0.04) over the relevant energy interval $\pm\hbar\omega_{B_{l'g,l_g}}$, and due to the adiabatic character of the bands topology (Fig. 15.1b), the EP interactions do not stabilize distorted structure.

15.3.2.2 Nonadiabatic correction to the one-particle term of the fermionic Hamiltonian: Corrections to the orbital energies and gap opening in a metallic one-particle spectrum

According to nonadiabatic theory of EP interactions [8,9], for nonadiabatic correction $\Delta\varepsilon_k$ to unoccupied state $\varepsilon_{k'}^0$ for system in intrinsic nonadiabatic state follows,

$$\Delta\varepsilon(Pk') = \sum_{Rk_1'>k_F} \left|u^{k'-k_1'}\right|^2 \left(1 - f_{\varepsilon^0 k_1'}\right) \frac{\hbar\omega_{k'-k_1'}}{\left(\varepsilon_{k'}^0 - \varepsilon_{k_1'}^0\right)^2 - \left(\hbar\omega_{k'-k_1'}\right)^2} -$$

$$- \sum_{Sk<k_F} \left|u^{k-k'}\right|^2 f_{\varepsilon^0 k} \frac{\hbar\omega_{k-k'}}{\left(\varepsilon_{k'}^0 - \varepsilon_k^0\right)^2 - \left(\hbar\omega_{k-k'}\right)^2}, \qquad (15.6)$$

for $k' > k_F$. For correction to occupied state ε_k^0 holds

$$\Delta\varepsilon(Pk) = \sum_{Rk_1'>k_F} \left|u^{k-k_1'}\right|^2 \left(1 - f_{\varepsilon^0 k_1'}\right) \frac{\hbar\omega_{k-k_1'}}{\left(\varepsilon_k^0 - \varepsilon_{k_1'}^0\right)^2 - \left(\hbar\omega_{k-k_1'}\right)^2} -$$

$$- \sum_{Sk_1<k_F} \left|u^{k-k_1}\right|^2 f_{\varepsilon^0 k} \frac{\hbar\omega_{k-k_1}}{\left(\varepsilon_k^0 - \varepsilon_{k_1}^0\right)^2 - \left(\hbar\omega_{k-k_1}\right)^2}, \qquad (15.7)$$

for $k \leq k_F$.

Replacement of discrete summation by integration, $\sum_k \ldots \rightarrow \int n(\varepsilon_k)$, introduces DOS $n(\varepsilon_k)$ in (7, 8), which is of crucial importance in relation to fluctuating band - see Fig. 15.1. The original HF states ε_k^0 is by this correction shifted on the energy scale to a new position,

$$\varepsilon_k = \varepsilon_k^0 + \Delta\varepsilon_k \qquad (15.8)$$

Nonadiabatic shift of the HF orbital energies induces change of the original density of state at Fermi level for the band where the gap in one-particle spectrum has been opened. For corrected DOS $n(\varepsilon_k)$, that is the consequence of shift $\Delta\varepsilon_k$ of orbital energies, the following relation can be derived;

$$n(\varepsilon_k) = \left|1 + \left(\partial\left(\Delta\varepsilon_k\right)/\partial\varepsilon_k^0\right)\right|^{-1} n^0\left(\varepsilon_k^0\right) \qquad (15.9)$$

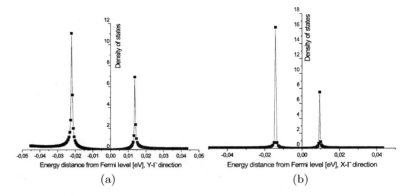

Fig. 15.6. Corrected density of states of the O1- derived pσ band in **b** (Γ-Y) direction (a) and **a** (Γ-X) direction (b) at k point where the band intersects Fermi level.

Term $n^0\left(\varepsilon_k^0\right)$ stands for uncorrected DOS of the original adiabatic states of particular band, $n^0\left(\varepsilon_k^0\right) = \left|\left(\partial\varepsilon_k^0/\partial k\right)\right|^{-1}$.

Close to k-point where the original band, which interacts with fluctuating band, intersects FL, the occupied states near FL are shifted downward below FL and unoccupied states are shifted upward - above FL. The gap is identified as an energy distance between created peaks in corrected DOS above FL (half-gap) and below FL. Formation of peaks is related to spectral weight transfer that is observed by ARPES or tunneling spectroscopy at cooling below T_c. Calculated results are in Figs. 15.6a, b.

While the gap (energy distance of the peaks at Fermi level) opened in **b** direction is $\Delta_b(0) = 35.7$ meV, gap in **a** direction is smaller, $\Delta_a(0) = 24.2$ meV. Calculated density of states simulates the **a**, **b** asymmetry of the ARPES spectra as it has been recorded [22] for untwined $YBa_2Cu_3O_7$. Experimental ratio of the gaps in the **a** and **b** direction is $(\Delta_a(0)/\Delta_b(0))_{exp} \approx 0.66$. The present calculations yield for this ratio, $(\Delta_a(0)/\Delta_b(0))_{theor} \approx 0.68$. Also the ratio of peaks intensities in **a** and **b** direction (higher intensity in **a** direction) is in good qualitative agreement with the experimental results [22].

In this connection, existence of the gap in the **a** direction (X-Γ line) should be mentioned. According to the present results, the gap in one-electron spectrum is opened in the chain oxygen O1- derived pσ band at the k point where the band intersects X-Γ line (Figs. 15.1a, 15.2). Existence of this gap can hardly be expected from the band structures based on DFT calculations since according to the corresponding published results, pσ band intersects Fermi level on X-S line and there is no band at all that intersects Fermi level on X-Γ line.

Critical temperature T_c of the antiadiabatic-adiabatic state transition that is conected to the gap extinction $\Delta(T_c) \to 0$, can be derived from the gap equation [8],

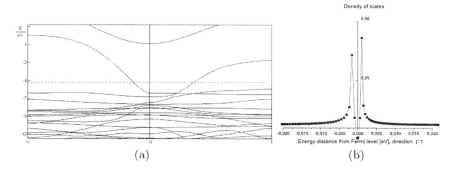

Fig. 15.7. Dispersion of the bands; $Cu - O$ plane (d-pσ) band in Γ-T and chain oxygen O1-derived pσ band in Γ-U direction (a) and resulting corrected DOS at FL(b)

$$\Delta(T) = \Delta(0).tgh(\Delta(T)/4k_BT), \Delta(T_c) \to 0, T_c = \Delta(0)/4k_B \qquad (15.10)$$

Calculated value of the critical temperature that corresponds to the larger gap is $T_c = 92.8$ K.

Related to the mechanism of gap formation in one-electron spectrum of $YBa_2Cu_3O_7$, unexpected new result should be mentioned. Calculated band structure has no bands crossing FL in vertical directions (Γ-Z, X-U, Y-T, S-R lines), that indicates strong 2D character of this material. Closer inspection of the band structure reveals, however, that 2D character is disturbed by strong dispersions of two bands; the $Cu - O$ plane (d-pσ) band with fluctuating SP and chain oxygen $O1$-derived pσ band. These bands interests Fermi level in c-direction. In particular, the d-pσ band intersects Fermi level in **b-c** plane on Γ-T line and O1-derived pσ band intersects Fermi level in **a-c** plane on Γ-U line – see Fig. 15.7.

Also in this case it can be expected that for intrinsic nonadiabatic state when the SP at Y point approaches Fermi level, nonadiabatic EP interactions open gaps in one-electron spectrum. For (d-pσ) intraband interactions, gap should appear in Γ-T direction and interbands interactions should give rise to a gap on pσ band in Γ-U direction. By simple rescaling of with respect to Γ-Y and Γ-T (or Γ-U) distances, mean value of EP interaction in Γ-T direction is approximated by the value $\bar{u}_{\Gamma T} \approx 0.6$ eV. Calculated gap in this direction is $\Delta_{\Gamma T}(0) \approx 5$ meV (Fig. 15.7b) that is considerably smaller comparing to the gaps that are opened in the basal plane.

15.3.2.3 Formation of van hove singularity in DOS of the $Cu - O$ plane d-pσ band

One of the reasons that at the very beginning was proposed to be responsible for high T_c in cuprates, has been possibility of the existence of Van Hove singularities in the density of states (DOS) at Fermi level [30, 31]. For a long time there were only indirect evidences for existence of such peak(s) in the DOS at Fermi level, e.g. discontinuity in the specific heat at T_c, and studies of thermopower and quasiparticle lifetime broadening [32–34]. For the first time, direct experimental evidence of the Van Hove singularity in DOS of high-T_c cuprates has been reported by Gofron et al. [35] in 1994. The existence of the Van Hove singularity in DOS of CuO d-pσ band at 19 meV below the Fermi level has been measured at the study of superconducting $YBa_2Cu_4O_8$. While, the peak is present below T_c at Y point of the Brillouin zone along Y-Γ direction, it is absent above T_c. This effect has been found to be a common feature of the DOS of CuO plane d-pσ band of all high-T_c cuprate superconductors.

Formation of the Van Hove singularity follows straightforwardly from the nonadiabatic EP interaction mechanism. From the band structure at high-symmetry equilibrium nuclear configuration (Fig. 15.1a) one can see that analytic critical points of the d-pσ bands of $Cu-O$ plane at Y point that are related to singularities in DOS, are far below the Fermi level. This situation corresponds to adiabatic regime, and Van Hove singularities at the Fermi level do not exist, as it is experimentally detected above T_c. However, electron coupling to the phonon modes generates formation of antiadiabatic state which is stabilized at distorted nuclear configuration. At these circumstances, the SP of fluctuating d-pσ band of $Cu-O$ plane at Y point has approached Fermi level. It has been shown above that as a consequence of the band shift, the ground state is stabilized at distorted nuclear configuration and the gap in the one-particle spectrum of the chain oxygen $O1$-derived pσ band has been opened due to effective nonadiabatic EP interactions.

The nonadiabatic EP interactions which result in opening of the gaps in the one-particle spectrum (Fig. 15.6) of the chain oxygen $O1$-derived p band influences also DOS of the d-pσ band of $Cu-O$ plane at Y point in Y-Γ direction. Without the account for nonadiabatic EP interactions the DOS of the fluctuating d-pσ band, at the moment when the SP of this band touches Fermi level, has the maximum at 0 eV (Fig. 15.3 with rescaled energy value of the Fermi level to 0 eV). The analytic expression of the uncorrected DOS corresponding to this situation is $n^0\left(\varepsilon^0_{k,d-p\sigma}\right) = \left|\left(\partial\varepsilon^0_{k,d=p\sigma}/\partial k\right)^{-1}\right|$.

The nonadiabatic interactions shift not only the orbital energies of the chain oxygen $O1$-derived pσ band (gap opening as described above), but the EP interactins shift also orbital energies of the fluctuating d-pσ band in the downward direction away from Fermi level, to the new positions,

$$\varepsilon_{k,d-p\sigma} = \varepsilon^0_{k,d-p\sigma} + \Delta\varepsilon_{k,d-p\sigma} \qquad (15.11)$$

For corrected DOS, that reflects the shift due to nonadiabatic EP interactions in the intrinsic nonadiabatic state, holds Eq. (15.9). Calculation of final corrected DOS of the d-pσ band of $Cu-O$ plane at the Y point according to Eq. (10) yields result presented in Fig. 15.8.

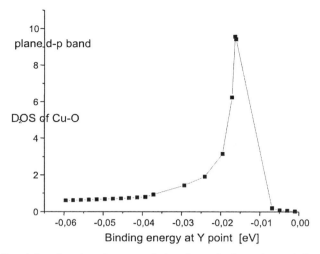

Fig. 15.8. Resulting density of states of the $Cu - O$ plane derived d-pσ band at Y point with the account for nonadiabatic EP interactions. The energy of the Fermi level is rescaled to 0 eV.

As it can be seen from this figure, the peak in the DOS is at ≈ 17 meV below the Fermi level, which is in a good agreement with the experimental results measured below T_c [35].

15.4 Conclusion

Results of the first principles study of $YBa_2Cu_3O_7$ obtained by application of the nonadiabatic theory of EP interactions offer substantially different scenario of SC state transition than BCS or BCS-like pairing theories.

For antiadiabatic conditions with broken translation symmetry, the Bose-Einstein condensation can be related to fermionic ground state energy stabilization due to nonadiabatic EP interactions in antiadiabatic state at broken translation symmetry. In this respect, "condensate" is represented by "charged bosons"- real space singlet pairs (polarized intersite charge density - bipolarons), rather than by BCS superfluid with Cooper's pairs at adiabatic BOA based high-symmetry equilibrium structure. This picture is consistent with the experimental results on cuprates [36, 37], where it has been concluded that charged bosons are bipolarons. In order to stabilize distorted structure, the energy gain due to nonadiabatic EP interactions has to be greater than energy loss due to symmetry lowering. Meaning of the gap is also different here. Within the antiadiabatic picture, the gap has its usual meaning, i.e. quasiparticle ("non-adiabatic polaron") excitation energy over the one-particle spectrum.

At finite temperatures, with increasing temperature from 0 K, due to the Fermi statistics of the one-particle state populations, the nonadiabatic EP corrections become smaller (T-dependent contribution of antiadiabatic state at distorted nuclear geometry), and at crossing critical temperature T_c, the electronic energy loss due to symmetry lowering (T-independent contribution) becomes greater than the energy gain due to nonadiabatic EP interactions. The system, in order to minimize fermionic ground state energy, goes to normal adiabatic metal state with the equilibrium geometry of the higher symmetry. For cooling, the situation is opposite. At crossing T_c in downward direction, the distorted structure becomes more stable than the undistorted one. The reason is that at lowered symmetry, there is the proper structure of the one-particle spectrum at Fermi level. This structure, antiadiabatic state, effectively enables to switch-on and maximizes the nonadiabatic EP interactions. The nonadiabatic energy gain starts to prevail over the energy loss of the distortion.

With respect to presented results, the SC state transition in $YBa_2Cu_3O_7$ can be characterized as a nonadiabatic sudden increase of the cooperative kinetic effect at lattice energy stabilization. It is exactly participation of the nuclear kinetic energy term on the nonadiabatic level that stabilizes (negative contribution) fermionic ground state energy at a distorted structure. At the adiabatic conditions, nuclear kinetic effect is absent, and adiabatic correction (DBOC - effect of the nuclear positions) to the fermionic ground state energy is always positive for equilibrium as well as for distorted clumped nuclei structures.

Geometrical degeneracy of the fermionic ground state energy for antiadiabatic distorted structure (existence of an infinite number of the CuO plane $O2$, $O3$ atoms in-plane displacements, due to electron coupling to the A_g and, B_{2g}, B_{3g} modes) enables cooperative and dissipationless motion of displaced $O2$, $O3$ atoms along the perimeters of circles centered at the undistorted $O2$, $O3$ atoms positions, with the same radii equal to the fractional displacement $|\Delta f_{a,b}|$. In the case of $YBa_2Cu_3O_7$, calculated radius of circles is $|\Delta f_{a,b}| = 0.0057 (0.022 A^o$ in the absolute value). The cooperative nuclear motion, i.e. "fluxional microcirculations" of $O2$, $O3$ nuclei in the CuO layers, induces dynamic cooperative formation of shortened and elongated in-plane $Cu2 - O2(O3)$ and $O2 - O3$ "bond" distances on the lattice scale. This microcirculation is connected with a dynamic intersite charge density polarization, i.e. dynamic formation of increased and decreased interatomic charge densities in the **a-b** plane of CuO layers. It should be interpreted as dynamic formation of nonadiabatic bipolarons [8] as it indicates the form of spinorbital in antiadiabatic state that is corrected by contribution of momenta P_1, P_2 of pair of nuclei ($O2$, $O3$ atoms) in phonon mode(s) that drives transition to antiadiabatic state,

$$\varphi_k(x, Q, P) \propto l \left\{ 1 + \sum_q \left[u^{|q|} \frac{\hbar\omega_q}{(\hbar\omega_q)^2 - \left(\varepsilon_k^0 - \varepsilon_{k+q}^0\right)^2} \times \right. \right.$$

$$\left. \left. \times \left(P_1 e^{iq \cdot [x - (m_1 - d_1)]} + P_2 e^{iq \cdot [x - (m_2 + d_2)]} \right) \right] \right\} \varphi_k^0(x, 0, 0) \qquad (15.12)$$

In the spinorbital form, the site approximation for momentum has been used, i.e. $P_q \propto (sign.q) \sum_m P_m e^{iq \cdot m}$ and m_1 and m_2 are equilibrium site positions of involved nuclei on (crude)adiabatic level while d_1 and d_2 are nuclear displacements at which crossing to intrinsic nonadiabatic state occur.

Nonadiabatic bipolarons are charge supercarriers and theirs motion on the lattice in the **a**, **b** plane is coherent and dissipationless. It has been shown in [8] that at strong nonadiabatic limit, $(|\varepsilon_{Y,d-p\sigma} - \varepsilon_F|) \to \infty$, which is characteristic for antiadiabatic state at broken translation symmetry, interaction energy of new quasi-particles, i.e. new fermions (nonadiabatic polarons) and nonadiabatic phonons, goes to zero, $\Delta H'_{nFB} \to 0$. More over, due to geometrical degeneracy of the fermionic ground state energy (fluxional structure of CuO layers), there are not energy barriers for the motion of bipolarons on the lattice and tunneling mechanism is not necessary to be considered.

For adiabatic systems, limit $(|\varepsilon_{Y,d-p\sigma} - \varepsilon_F| / \hbar\omega) \to \infty$, which is characteristic for normal metal state of superconductors at equilibrium, high-symmetry nuclear geometry, interaction energy of the adiabatic quasiparticles (polarons and renormalized phonons) is $\Delta H'_{nFB} \approx (u_{k'k})^2 / (\varepsilon_k - \varepsilon_{k'})$. It is basically well known polaron energy./Possibility of bipolaron superconductivity has been proposed and discussed elsewhere; see e.g. Alexandrov [38] and references therein.

As it has been shown, the nonadiabatic effects, which are present in $YBa_2Cu_3O_7$, are absent in $YBa_2Cu_3O_6$. This is the reason, from the standpoint of the nonadiabatic theory, that deoxygenated YBCO is not superconductor.

Present study suggests possibility of the experimental verification of the described nonadiabatic mechanism at transition to SC state. The effect of sudden decrease of low-Fermi energy (50–80 meV) electron velocity by ARPES experiments for a wide group of high-T_c cuprates, has been reported [1,2] in the nodal, $(0, 0, 0)$–$(\pi, \pi, 0)$, direction. The results concerning $YBa_2Cu_3O_7$ had not been reported. According to present calculations (see Fig. 15.5) this effect should be registered also for $YBa_2Cu_3O_7$, however, not in the nodal direction but close to Y point on the Γ-Y line, i.e. in the off-nodal $(0, 0, 0)$–$(0, \pi, 0)$ direction at about 75–80 meV below the Fermi level, if the corresponding experiment is performed. Like in the case of $Bi2223$ [3], this effect should be T-dependent, i.e. it should appear only at $T \leq T_c$. The high precision ARPES or tunneling spectroscopy should also detect the existence of small gap in c-direction ($\Gamma - T(U)$ line – Fig. 15.7), the existence of which has not been reported so far.

Acknowledgements

The author is acknowledged to S-Tech a.s for financial research support, as well the partial support of the grant VEGA1/2465/05.

References

1. A. Lanzara, P.V. Bogdanov, X.J. Zhou et al., Nature **412**, 510 (2001)
2. X.J. Zhou, T. Yoshida, A. Lanzara et al., Nature **423**, 398 (2003)
3. T. Takahashi, T. Sato, H. Matsui and K. Terashima, New J. Phys. **7**, 105 (2005)
4. J.M. An and W.E. Picket, Phys. Rev. Lett. **86**, 4366 (2001)
5. T. Yilderim, O. Gulseren, J.W. Lynn et al., Phys. Rev. Lett. **87**, 037001 (2001)
6. P. Banacky, Int. J. Quant. Chem. **101**, 131 (2005)
7. L. Boeri, E. Cappelluti and L. Pietronero, Phys. Rev. B **71**, 012501 (2005)
8. M. Svrcek, P. Banacky and A. Zajac, Int. J. Quant.Chem. **a/43**, 393 (1992); **b/43**, 415 (1992); **c/43**, 425 (1992); **d/43**, 551 (1992) P. Banacky, Phys. Rev. B, submitted (August 2006)
9. C.W. Chu, P.H. Hor, R.L. Meng et al., Phys. Rev. Lett. **58**, 405 (1987)
10. Y. Muto, N. Kobayashi and Y. Syono, in Novel Superconductivity (editors S. Wolf and V. Kresin, Plenum Press, New York (1987)
11. SOLID 2000, Computer code for electronic structure calculation of periodic systems. S-Tech a.s., Bratislava, Slovakia (www.stech.sk)
12. J. Noga, P. Baňacký, S. Biskupič et al., J. Comp. Chem. **20**, 253 (1999)
13. J.A. Pople, D.L. Beveridge and P.A. Dobosh, J. Chem. Phys. **47**, 2026 (1967)
14. J.A. Pople and D.L. Beveridge, in Approximate Molecular Orbital Theory, (McGraw-Hill, New York, 1970)
15. A. Zajac, P. Pelikán, J. Noga et al., J. Phys. Chem. B **104**, 1708 (2000)
16. A. Zajac, P. Pelikán, J. Minar et al., J. Solid. State Chem. **150**, 286 (2000)
17. P. Pelikán, M. Kosuth, S. Biskupič et al., Int. J. Quant. Chem. **84**, 157 (2001)
18. Picket, W.E. Rev. Mod. Phys. **61**, 433 (1989)
19. W.E. Picket, R.E. Cohen and H.A. Krakauer, Phys. Rev. B **42**, 8764 (1990)
20. O.K. Andersen, O. Jepsen, A.I. Liechtenstein and I.I. Mazin, Phys. Rev. B **49**, 4145 (1994)
21. R. Kouba, C. Ambrosch-Draxl and B. Zangger, Phys. Rev. B **60**, 9321 (1999)
22. D.H. Lu, D.L. Feng, N.P. Armitage et al., Phys. Rev. Lett . **86**, 4370 (2001)
23. A. Damascelli, Z. Hussain and Z.x. Shen, Rev. Mod. Phys. **75**, 473 (2003)
24. W. Weber and L.F. Mattheiss, Phys. Rev. B **37**, 599 (1988)
25. P.B. Allen, W.E. Picket and H. Krakauer, Phys. Rev. B **37**, 7482 (1988)
26. R.E. Cohen, W.E. Picket and H. Krakauer, Phys. Rev. Lett. **64**, 2575 (1990)
27. J.H. Chung, T. Egami, R.J. McQuinney et al., Phys. Rev. B **67**, 014517 (2003)
28. R. Lin, C. Thomsen, W. Kress et al., Phys. Rev.B **37**, 7971 (1988)
29. K.F. McCarty, J.Z. Lin, R.N. Shelton and M.B. Radomsky, Phys. Rev. B **41**, 8792 (1990)
30. J.E. Hirsh and D.J. Scalapino, Phys. Rev. Lett. **56**, 2735 (1986)

31. J. Labbe and J. Bok, Europhys. Lett. **3**, 1225 (1987)
32. C.C. Tsuei, C.C. Chi, D.M. Newns, Phys. Rev. Lett. **69**, 2134 (1992)
33. D.M. Newns, H.R. Krishnamurthy, P.C. Pattnaik, Phys. Rev. Lett. **69**, 1264 (1992)
34. P.C. Pattnaik, C.L. Kane, D.M. Newns, Phys. Rev. B **145**, 5714 (1992)
35. K. Gofron, J.C. Campuzano, A.A. Abrikosov et al., Phys. Rev. Lett. **73**, 3302 (1994)
36. G. Zhao, M.B. Hunt, H. Kellerand K.A. Muller, Nature (London) **385**, 236 (1997)
37. J.P. Franck, in Physical Properties of High Temperature superconductors IV, edited by D. Ginsberg (World Scientific, Singapore, 1994)
38. A.S. Alexandrov, Phys. Rev. B **61**, 12315 (2000)

ELEMENTS OF MODERN HIGH-TEMPERATURE SUPERCONDUCTIVITY

J.D. DOW[1] AND D.R. HARSHMAN[2,3]

[1] *Department of Physics, Arizona State University, Tempe, AZ 85287-1504 U.S.A.* cats@dancris.com
[2] *Physikon Research Corporation, Lynden, WA 98264 U.S.A.*
[3] *Department of Physics, Arizona State University, Tempe, AZ 85287-1504 U.S.A.*

Abstract. Evidence is presented that all high-temperature superconductors: (i) are p-type; (ii) involve holes located outside the host's cuprate-planes; (iii) contain superconducting charge carriers which are Cooper-paired holes; (iv) have holes with a nodeless, s-wave pairing function that is consistent with a gap function of s-wave symmetry; and (v) involve pairing of holes by Coulomb interactions (not by phonons). The p-type superconductivity in $YBa_2Cu_3O_7$ resides in the BaO layers and is accompanied by n-type normal conductivity in the cuprate-planes. Consequently the linear-in-temperature term observed in the specific heat and the non-zero offset found in the thermal conductivity divided by temperature are explained.

Keywords: high-temperature superconductivity

16.1 Introduction

High-temperature superconductivity has been studied by thousands of authors with the result being that twenty years after its discovery [1], and 50 years after the theory of low temperature superconductivity has been published [2], the phenomenology of high-temperature superconductivity is not completely understood.

The important unanswered (or incompletely answered) questions include: (i) is it possible to form both n-type (electron) and p-type (hole) high-temperature superconductors? (ii) what part of the crystal structure superconducts? (iii) are the superconducting carriers holes or electrons or both? (iv) are the paired carriers of s-wave or d-wave symmetry? and (v) what force acts between the holes or the electrons of a Cooper pair?

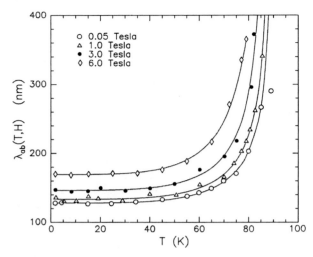

Fig. 16.1. Magnetic penetration depth (in nm) of single-crystal YBa$_2$Cu$_3$O$_7$ in the ab plane as a function of temperature (in Kelvin) for applied fields of 0.05, 1.0, 3.0, and 6.0 Tesla. As discussed in [15], these data collapse onto a single curve for H = 0.

Currently a majority of workers in the field would answer the above questions (incorrectly) as follows: (i) while most superconducting materials are p-type, some, such as Ce- or Am-doped PSYCO (Pb$_2$Sr$_2$YCu$_3$O$_8$), can be n-type; (ii) the cuprate-planes superconduct; (iii) the charge carriers are commonly holes, but electrons can produce n-type superconductivity in some high-T$_C$ superconductors (an example of which is Ce-doped PSYCO); (iv) the paired carriers have d-wave symmetry; and (v) the superconductivity is mediated by phonons, which pair the holes.

We have evidence that all of the above answers are incorrect. Our research indicates that: (i) currently there are no n-type high-temperature superconductors [3,4]; (ii) the superconductivity occurs in planes with holes: the BaO layers, the SrO layers, or the interstitial oxygen regions of various superconducting materials [5–7], or with holes associated with S in the organic material κ-[BEDT-TTF]$_2$Cu[NCS]$_2$ (which has T$_C$ ∼10 K) [8,9]; (iii) there are no high-temperature n-type superconducting materials (in which the superconducting carriers are electrons) [10–14]; (iv) the carriers of superconductivity are paired holes with s-wave symmetry in high-temperature superconductors [7,15–17], (Fig. 16.1); and (v) the hole-pairing is mediated by the screened Coulomb interaction [18–20].

16.2 Are there any n-type superconductors? No

Most high-temperature superconductors are obviously p-type, and there is no dispute about the sign of the charge carriers that superconduct. But some materials are thought to be n-type superconductors, most notably

$Nd_{2-z}Ce_zCuO_4$ and $Pb_2Sr_2YCu_3O_8$. The material $Nd_{2-z}Ce_zCuO_4$ and its related Ce-doped sisters, the compounds (rare-earth)$_{2-z}Ce_zCuO_4$, are thought by most workers to superconduct n-type. This belief is based on the ideas that Ce is the primary dopant of $Nd_2Ce_zCuO_4$, that Ce is isolated when it dopes, and that isolated-Ce dopes the material n-type. But it is possible that Ce pairs with interstitial oxygen when it dopes $Nd_{2-z}Ce_zCuO_4$, so that the dopant is a $(Ce,O_{interstitial})$ pair which can be a p-type dopant if the oxygen is not fully charged to O^{-2}. Moreover, the way that $Nd_{2-z}Ce_zCuO_4$ is prepared and made to superconduct suggests that a transition to $(Ce,O_{interstitial})$ doping might occur during preparation. Unfortunately, efforts to probe the character of Ce-doped $Nd_{2-z}Ce_zCuO_4$ failed [21], and we were unable to prove which charge state Ce was in, or if interstitial oxygen was present. So the nature of Ce-doping in $Nd_{2-z}Ce_zCuO_4$ is currently unresolved. The present state is that Tokura et al. [22] and Takagi et al. [23] have both claimed that $Nd_{2-z}Ce_zCuO_4$ is an n-type superconductor, although they did not eliminate the possibility of interstitial oxygen (which could have changed the net doping to p-type). Contradicting this, Brinkmann et al. [24] claimed that the material is p-type (without invoking interstitial oxygen). Hence the nature of the doping of $Nd_{2-z}Ce_zCuO_4$ and its sister compounds is currently unresolved, and so $Nd_{2-z}Ce_zCuO_4$ is neither clearly an n-type superconductor, nor a p-type one. To find a material that unquestionably can be doped n-type, we consider PSYCO ($Pb_2Sr_2YCu_3O_8$ with T_C up to 84 K [25–27]), which can be doped both p-type and n-type on the trivalent Y site. Ce or Am dopants create a quadrivalent ion, Ce^{+4} or Am^{+4}, making PSYCO an n-type material, while Ca^{+2} doping (in place of Y^{+3}) creates a p-type material that superconducts. But when PSYCO is doped n-type with Ce^{+4} or Am^{+4}, it does not superconduct [13,28], while the p-doped material does superconduct [4]. We conclude that, as with PSYCO, only p-type high-T_C materials superconduct [11,21].

16.3 Do high-temperature materials superconduct in their cuprate-planes? No

First, we ask the question, "Do all high-temperature materials superconduct in their cuprate-planes?" The best answer to this question is provided by Cu-doped Ba_2YRuO_6 [29,30], which has virtually no cuprate-planes and yet begins superconducting with an onset critical temperature T_C of 93 K [31]. Its sister compound, Cu-doped Sr_2YRuO_6, begins superconducting at 49 K, while the related compounds to doped Sr_2YRuO_6, $GdSr_2Cu_2RuO_8$ and $Gd_{2-z}Ce_zSr_2Cu_2RuO_{10}$, have onset critical temperatures near, but somewhat below, 49 K [32–35]. For those who are skeptical of the claimed superconductivity in doped Ba_2YRuO_6 and Sr_2YRuO_6, the data for $GdSr_2Cu_2RuO_8$ and $Gd_{2-z}Ce_zSr_2Cu_2RuO_{10}$ confirm the importance of the SrO layers. The cuprate-planes in these materials are either weakly ferromagnetic or (more likely) antiferromagnetic [36], an indication that those planes do not superconduct. Likewise the Gd layers, with no oxygen ions, are excluded from the superconductivity. Since the Ru ion has a large electronic moment, the magnetization of the RuO_2 plane almost certainly inhibits superconductivity in that layer. The only remaining plane for superconductivity in $GdSr_2Cu_2RuO_8$ is the SrO layer. The situation is similar for $Gd_{2-z}Ce_zSr_2Cu_2RuO_{10}$. Holes in the SrO layers superconduct in the three ruthenate compounds, doped Sr_2YRuO_6, $GdSr_2Cu_2RuO_8$, and $Gd_{2-z}Ce_zSr_2Cu_2RuO_{10}$; and the BaO layers superconduct in doped Ba_2YRuO_6. Clearly in the above ruthenates, the BaO or SrO layers host the superconductivity.

16.3.1 EVIDENCE THAT THE CUPRATE-PLANES DO NOT SUPERCONDUCT IN $YBA_2CU_3O_7$

One of the most widely circulated beliefs has been that the cuprate-planes superconduct. But the primary "proof" of such superconductivity comes from neutron spectroscopy data taken by Cava et al. [37] and by Jorgensen et al. which actually *disagreed with each other*. Jorgensen, in his Physics Today article [39], inexplicably claimed (contrary to the facts) to have confirmed the Cava measurements in Ref. [38]. There is only one datum among all of the neutron data of Cava et al. which is inconsistent with the data of Jorgensen et al.; and this single datum purportedly shows that the cuprate-planes superconduct. We believe that the single Cava et al. datum is incorrect, and that the Jorgensen data contain no evidence that the cuprate-planes superconduct. That brings us to muon spin rotation data of $YBa_2Cu_3O_7$, which show that the superconductivity is nodeless, *without any evidence of a Cu d-band signature*, consequently indicating that the superconductivity resides *not in the cuprate-planes, but in the BaO layers* [15]. The most interesting fact about positive muon spin rotation (μ^+SR) of $YBa_2Cu_3O_7$ is that, once fluxon de-pinning has been accounted for, the superconducting signal detected has *no detectable Cu signature* at a level of $\ll 1\%$. Since band structure calculations [40] find that $YBa_2Cu_3O_7$ has a 24% cuprate-plane signature and a 7% CuO-chain signature, but μ^+SR detected *no Cu signature*, neither the cuprate-planes nor the CuO-chains host the superconductivity. Hence, the *superconducting holes must reside in the BaO layers*. Consequently the μ^+SR data prove convincingly that the BaO layers *alone* are the host of the *p*-type superconductivity in $YBa_2Cu_3O_7$ [15].

16.4 What is the symmetry of the paired carriers?
s-wave

Analysis of the muon data requires that fluxon de-pinning be accounted for first, after which the bulk superconductivity of $YBa_2Cu_3O_7$ is found to be 100% nodeless, consistent with *s*-wave pairing, and best described by the two-fluid formula [15]. This two-fluid model is also consistent with old measurements [41–44] on heavily twinned crystals and sintered powders in which the fluxons are strongly pinned. (The authors that claim that the superconductivity of $YBa_2Cu_3O_7$ is d-wave make several mistakes: (i) they often analyze surface sensitive data, (ii) they normally do not account for fluxon de-pinning, and (iii) they overlook the fact that the fit of any d-wave model has a probability less than 4×10^{-6} achieving a fit as good as the successful two-fluid model fit [15,16].) In the absence of electronic moments, the μ^+SR technique only detects the superconducting bands, and finds no evidence of a Cu signature. These results support the arguments of Sec. III that the superconducting hole condensate does not reside in the cuprate-planes. There is no basis for claims that the Cooper pairing is d-wave. In the ruthenates, the superconductivity is also unquestionably s-wave (nodeless) [34].

16.5 Mechanism of Cooper pairing: The Coulomb force

Several years ago it was reported that Cooper pairing in high-temperature superconductors originates from the Coulomb interaction between holes [18–20]. This work supercedes recent attempts to assign the pairing to phonon assistance, and is confirmed by the recent analysis of oxygen isotope measurements [45]. The Cooper pairing is Coulombic.

16.6 Specific heat and thermal conductivity

Specific heat data for $YBa_2Cu_3O_7$ [46] exhibit a zero-field linear-in-temperature term at the lowest temperatures T. Interestingly, the thermal-conductivity data (κ) show a corresponding *non-zero* offset in κ/T extrapolated to zero temperature. Both of these observations point to the presence of a significant density of normal carriers below T_C. Taillefer et al. [47] have argued that the thermal conductivity of $YBa_2Cu_3O_{6.9}$ can be explained in terms of d-wave superconductivity. However, they have assumed that there is no normal conductivity. In particular, they have not accounted for the possibilities that (i) not all layers superconduct the same, or (ii) some layers are normal rather than superconducting. We have found that the cuprate-layers and the CuO-layers of $YBa_2Cu_3O_7$ *do not superconduct,* and that only the BaO layers superconduct. Consequently there exists at low temperatures a pool of normal electrons that do not superconduct, in the cuprate-layers [48]. These findings are consistent with the data in that they (unlike the Taillefer theory) explain why the thermal conductivity κ divided by temperature T *does not approach zero* as the temperature T approaches zero: the electrical conductivity of the planes and the chains is *normal conductivity,* not superconductivity. This is consistent with the fact that κ/T versus T^2 *approaches a constant,* 0.2 mW/K^2-cm, as T^2 approaches zero in $YBa_2Cu_3O_{6.9}$, instead of *approaching zero.* There is s-wave superconductivity of holes in *only* the BaO layers, and some thermal conductivity (divided by T) from the other layers is left over at T = 0 [48].

16.7 Summary

Evidence is presented that high-temperature superconductivity (i) is invariably p-type; (ii) does not reside in the host's cuprate-planes; (iii) involves holes as the only superconducting charge carriers; (iv) assigns s-wave symmetry to the Cooper-paired holes; (v) attributes the pairing force to Coulomb interactions, not to phonons; (vi) excludes n-type superconductivity; and (vii) does not involve superconductivity in the cuprate-planes. We hope that these constraints, which limit most existing theories, will be applied to future theories, so that a new theory of high-temperature superconductivity that is consistent with the data can be developed.

Acknowledgments

JDD and DRH thank the U.S. Army Research Office (Contract W911NF-05-1-0346) for their support; DRH is also supported by Physikon Research Corporation (Project No. PL-206). We thank A. T. Fiory for assistance with the preparation of figures.

References

1. J. G. Bednorz and K. A. Müller, Z. Phys. **B 64**, 189 (1986).
2. J. Bardeen, L. N. Cooper, and J. R. Schrieffer, Phys. Rev. **106**, 162 (1957); **108**, 1175 (1957).
3. J. D. Dow and M. Lehmann, Philos. Mag. **83**, 527 (2003).
4. M. Lehmann, J. D. Dow, and H. A. Blackstead, Physica **C 341–348**, 309 (2000).
5. J. D. Dow, D. R. Harshman, and A. T. Fiory, "High-T_C superconductivity of cuprates and ruthenates," in *Electron Correlations in New Materials and Nanosystems,* edited by K. Scharnberg and S. Kruchinin (Springer, Berlin, 2007) pp. 263–274 (2007) [NATO Workshop, Yalta, Crimea, Ukraine, of 19–23 September 2005.]
6. J. D. Dow and D. R. Harshman, *High-temperature superconductivity of Oxides,* in "New Challenges in Superconductivity: Mathematics, Physics, and Chemistry," **183**, 129 (2005), IOS Press, Amsterdam/Springer, P. O. Box 17, 3300 Dordrecht, The Netherlands, ed. by J. Ashkenazi, M. V. Eremin, J. L. Cohn, I. Eremin, D. Manske, D. Pavuna, and F. Zuo.
7. J. D. Dow and D. R. Harshman. Brazilian J. Phys. **33**, 681 (2003).
8. D. R. Harshman, A. T. Fiory, R. C. Haddon, M. L. Kaplan, T. Pfiz, E. Koster, I. Shinkoda, and D. Ll. Williams, Phys. Rev. **B 49**, 12990 (1994).
9. D. R. Harshman, A. T. Fiory, R. C. Haddon, M. L. Kaplan, T. Pfiz, E. Koster, I. Shinkoda, and D. Ll. Williams, Phys. Rev. **B 49**, 12990 (1994).
10. H. A. Blackstead and J. D. Dow, Phys. Rev. **B 62**, 9050 (2000).
11. M. Lehmann, J. D. Dow, and H. A. Blackstead, Physica **C 341–348**, 309 (2000).
12. M. A. Subramaniam, J. Gopalakrishnan, C. C. Torardi, P. L. Gai, E. D. Boyers, T. R. Askew, R. B. Flippen, W. E. Farveth, and A. W. Sleight, Physica **C 157**, 124 (1989).

13. S. Skanthakumar and L. Soderholm, Phys. Rev. **B 53**, 920 (1996).
14. L. Soderholm, S. Skanthakumar, U. Staub, M. R. Antonio, and C. W. Williams, J. Alloys and Compounds **250**, 623 (1997).
15. D. R. Harshman, W. J. Kossler, X. Wan, A. T. Fiory, A. J. Greer, D. R. Noakes, C. E. Stronach, E. Koster, and J. D. Dow, Phys. Rev. **B 69**, 174505 (2004).
16. For a discussion of our studies of $YBa_2Cu_3O_7$, and comparison with the claims of d-wave superconductivity by Sonier et al., see D. R. Harshman, W. J. Kossler, X. Wan, A. T. Fiory, A. J. Greer, D. R. Noakes, C. E. Stronach, E. Koster, and J. D. Dow, [Reply to Comment by J. E. Sonier, D. A. Bonn, J. H. Brewer, W. N. Hardy, R. F. Kiefl, and Ruixing Liang,] "Nodeless pairing state in single-crystal $YBa_2Cu_3O_7$." Phys. Rev. **B 72**, 146502 (2005).
17. J. D. Dow, J. Supercond. **18**, 63–65 (2005).
18. D. R. Harshman and A. P. Mills, Jr., Phys. Rev. **B 45**, 10684 (1992).
19. M. Gurvitch and A. T. Fiory, Phys. Rev. Lett. **59**, 1337 (1987).
20. M. Gurvitch, A. T. Fiory, L. S. Schneemeyer, R. T. Cava, G. P. Espinosa, and J. V. Waszczak, Physica **C 153–155**, 1369 (1988).
21. Before stating our position that n-type materials do not superconduct, we note that we attempted to show that $Nd_{2-z}Ce_zCuO_4$ is a p-type superconductor, although some authors feel that $Nd_{2-z}Ce_zCuO_4$ is n-type. The reason for this feeling is that the Hall coefficient of $Nd_{2-z}Ce_zCuO_4$ is *both* n-type and p-type, depending on the measurement conditions. This has produced a dispute over whether $Nd_{2-z}Ce_zCuO_4$ material is p-type or n-type [22][23]. Our explanation of this problem is that doping of $Nd_{2-z}Ce_zCuO_4$ with Ce is *not simple doping of the material with isolated Ce^{+4} (which would make the material n-type)*, but is doping with $(Ce,O_{interstitial})$ pairs, and the pairs can be p-type dopants, depending on the charges of the interstitial oxygen and the Ce. Our attempt to prove this point failed, however, and the issue is currently unresolved.
22. Y. Tokura, H. Takagi, and S. Uchida, Nature **337**, 345 (1989).
23. H. Takagi, S. Uchida, and Y. Tokura, Phys. Rev. Lett. **62**, 1197 (1989).
24. M. Brinkmann, T. Rex, M. Stief, H. Bach, and K. Westerhalt, Physica **C 269**, 76 (1996).
25. J. S. Xue, J. E. Greedan, and M. Maric, J. Solid State Chem. **102**, 501 (1993).
26. R. Prasad, N. C. Soni, K. Adikary, S. K. Malik, and C. C. Tomy, Solid State Commun. **76**, 667 (1990).
27. J. S. Xue, M. Reedyk, Y. P. Lin, C. V. Stager, and J. E. Greedan, Physica **C 166**, 29 (1990).
28. L. Soderholm, C. Williams, S. Skanthakumar, M. R. Antonio, and S. Conradson, Z. Physik **B 101**, 539 (1996).
29. D. R. Harshman, W. J. Kossler, A. J. Greer, C. E. Stronach, D. R. Noakes, E. Koster, M. K. Wu, F. Z. Chien, H. A. Blackstead, D. B. Pulling, and J. D. Dow, Physica **C 364–365**, 392 (2001).
30. S. M. Rao, J. K. Srivastava, H. Y. Tang, D. C. Ling, C. C. Chung, J. L. Yang, S. R. Sheen, and M. K. Wu, J. Crystal Growth **235**, 271 (2002).
31. The onset of superconductivity in doped Ba_2YRuO_6 is 93 K, but the actual T_C occurs below 30 K, when the Ru librations freeze out.
32. M. K. Wu, D. Y. Chen, D. C. Ling, and F. Z. Chien, Physica **B 284–288**, 477 (2000).
33. D. R. Harshman, H. A. Blackstead, W. J. Kossler, A. J. Greer, C. E. Stronach, E. Koster, B. Hitti, M. K. Wu, D. Y. Chen, F. Z. Chien, and J. D. Dow. Intl. J. Mod. Phys. **B 13**, 3670 (1999).

34. J. D. Dow and D. R. Harshman, J. Supercond. **15**, 455 (2002).
35. D. R. Harshman, J. D. Dow, W. J. Kossler, D. R. Noakes, C. E. Stronach, A. J. Greer, E. Koster, Z. F. Ren and D. Z. Wang, Philos. Mag. **83**, 3055 (2003).
36. H. A. Blackstead, W. B. Yelon, M. Kornecki, M. P. Smylie, Q. Cai, J. Lamsal, V. P. F. Awana, S. Valamurugan, and E. Takayama-Muromachi, Phys. Rev. **B 76,** page (2007).
37. R. J. Cava, A. W. Hewat, E. A. Hewat, B. Batlogg, M. Marezio, K. M. Rabe, J. J. Krajewski, W. F. Peck, Jr., and L. W. Rupp, Jr., Physica **C 165**, 419 (1990).
38. J. D. Jorgensen, B. W. Veal, A. P. Paulikas, L. J. Nowicki, G. W. Crabtree, H. Claus, and W. K. Kwok, Phys. Rev. **B 41**, 1863 (1990).
39. J. D. Jorgensen, Phys. Today, 34 (June, 1991).
40. P. de la Mora, J. D. Dow, D. R. Harshman, M. de Llano, and S. Ramírez, "Superconductivity in the BaO layers of $YBa_2Cu_3O_x$," to be published.
41. D. R. Harshman, G. Aeppli, E. J. Ansaldo, B. Batlogg, J. H. Brewer, J. F. Carolan, R. J. Cava, M. Celio, A. C. D. Chaklader, W. N. Hardy, S. R. Kreitzman, G. M. Luke, D. R. Noakes, and M. Senba, Phys. Rev. **B 36**, 2386 (1987).
42. B. Pümpin, H. Keller, W. Kündig, W. Odermatt, I. M. Savic, J. W. Schneider, H. Simmler, P. Zimmermann, J. G. Bednorz, Y. Maeno, K. A. Müller, C. Rossel, E. Kaldis, S. Rusiecki, W. Assmus, and J. Kowalewski, Physica **C 162–164**, 151 (1989).
43. D. R. Harshman, L. F. Schneemeyer, J. V. Waszczak, G. Aeppli, R. J. Cava, B. Batlogg, L. W. Rupp, Jr., E. J. Ansaldo, and D. Ll. Williams, Phys. Rev. **B 39**, 851 (1989).
44. B. Pümpin, H. Keller, W. Kündig, W. Odermatt, I. M. Savic, J. W. Schneider, H. Simmler, P. Zimmermann, E. Kaldis, S/Rusiecki, Y. Maeno, and C. Rossel, Phys. Rev. **B 42**, 8019 (1990).
45. D. R. Harshman, J. D. Dow, and A. T. Fiory, "Isotope effect in high-T_C superconductors," Phys. Rev. **B** , to be published (2008).
46. K. A. Moler, D. L. Sisson, J. S. Urbach, M. R. Beasley, A. Kapitulnik, D. J. Baar, R. Liang, and W. N. Hardy, Phys. Rev. **B 55**, 3954 (1997).
47. L. Taillefer, B. Lussier, R. Gagnon, K. Behnia, and H. Aubin, Phys. Rev. Lett. **79**, 483 (1997).
48. D. R. Harshman and J. D. Dow, Intl. J. Mod. Phys. **B 19,** 147 (2005).

Part III

Spintronics

17

NONEQUILIBRIUM DENSITY OF STATES AND DISTRIBUTION FUNCTIONS FOR STRONGLY CORRELATED MATERIALS ACROSS THE MOTT TRANSITION

J.K. FREERICKS AND A.V. JOURA
Department of Physics, Georgetown University, Washington, DC 20057, U.S.A. freericks@physics.georgetown.edu

Abstract. We examine the local density of states and the momentum-dependent distribution functions as they evolve in time for systems described by the Falicov-Kimball model initially in equilibrium, and then driven by a large uniform electric field turned on at time $t = 0$. We use exact dynamical mean-field theory, extended to nonequilibrium situations, to solve the problem. We focus on the accuracy of the numerics and on the interesting new features brought about by the strong correlations.

Keywords: nonequilibrium formalism, strong correlations, Mott transition

17.1 Introduction

There has been increasing interest in the behavior of quantum mechanical systems that are driven out of equilibrium due to the presence of large external electrical fields, motivated in part by the miniaturization of electronics and nanotechnology, which routinely have large electric fields placed over structures with small feature sizes. The general formalism for the nonequilibrium many-body problem was worked out by Kadanoff and Baym [1] and by Keldysh [2] in the 1960s. Unfortunately, at that time there was no known way to solve the resulting equations in cases with strong electron correlations, and most analysis was based on perturbative approaches in the interaction strength (because the noninteracting Green's functions in a field were known exactly [3], the theory treated all electric field effects to all orders in the field). In 1989, dynamical mean-field theory was invented [4], and it has allowed us to solve nearly all equilibrium many-body problems in solid-state physics [5].

J. Bonča, S. Kruchinin (eds.), *Electron Transport in Nanosystems.*
© Springer Science + Business Media B.V. 2008

It has recently been generalized to the nonequilibrium case [6–10], and the work we report on here describes some of the results emerging from these calculations.

The model Hamiltonian we will consider is the Falicov-Kimball model [11], which involves two sets of spinless electrons—conduction electrons (which hop between neighboring lattice sites and are denoted by c) and localized electrons (which do not hop and are denoted by f). The two electrons have a mutual on-site Coulomb repulsion of strength U when two electrons are in the same unit cell. The Hamiltonian (in the absence of an external field) is

$$\mathcal{H} = -\frac{t^*}{2\sqrt{d}}\sum_{\langle ij\rangle}(c_i^\dagger c_j + c_j^\dagger c_i) + U\sum_i c_i^\dagger c_i f_i^\dagger f_i, \qquad (17.1)$$

where the creation and annihilation operators satisfy the usual fermionic anticommutation relations. This model is the simplest many-body problem that has a Mott-like metal-insulator transition when the conduction and localized electrons are both half-filled. We work on a hypercubic lattice in infinite-dimensions [12], where the noninteracting density of states is $\rho(\epsilon) = \exp(-\epsilon^2)/\sqrt{\pi}$; we use the hopping energy t^* as the energy unit.

17.2 Formalism

We initially prepare the system in an equilibrium state with a temperature $1/\beta = 0.1$ and turn on a uniform electric field at $t = 0$ (we neglect the transient magnetic field present only at times close to $t = 0$). The uniform electric field is described by a uniform vector potential in the Hamiltonian gauge $[A = -Et\theta(t)]$. We are interested in finding the local many-body density of states (DOS) as a function of time, which is the double-time expectation value $\langle\{c_i^\dagger(t)c_i(t') + c_i(t')c_i^\dagger(t)\}\rangle$ and in finding the distribution of electrons in momentum space as a function of time, which is the equal time expectation value $\langle c_\mathbf{k}^\dagger(t)c_\mathbf{k}(t)\rangle$ that measures how the electrons are distributed over the Brillouin zone. The DOS is normally described as a function of average time $T = (t + t')/2$ and of frequency ω, after Fourier transforming the time-dependent expectation value over the relative time $t_{rel} = t - t'$.

Both the DOS and the distribution functions can be found from the so-called contour-ordered Green's function $G_{ij}(t,t')$

$$G_{ij}(t,t') = -i\mathrm{Tr}e^{-\beta\mathcal{H}(t=-5)}\mathcal{T}_c c_i(t)c_j^\dagger(t')/\mathcal{Z}, \qquad (17.2)$$

where each time argument t and t' lies on the Kadanoff-Baym-Keldysh contour, depicted in Fig. 17.1 for the problems we will be analyzing (the field is described by the spatially uniform vector potential A). The time-dependence of the operators is in the Heisenberg representation, the time-ordering operator

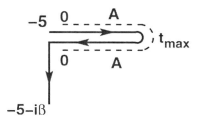

Fig. 17.1. Kadanoff-Baym-Keldysh contour for calculating nonequilibrium Green's functions. The initial five units of time are needed to ensure the system properly represents the equilibrium solution, while the field is turned on at $t = 0$ as represented by the nonzero vector potential A running out to t_{max}. The contour then continues to run back to $t = -5$ and then down the imaginary axis a distance β.

$\mathcal{T}_c$ orders times along the contour, and $\mathcal{Z} = \mathrm{Tr}e^{-\beta\mathcal{H}(t=-5)}$ is the equilibrium partition function (which could be evaluated at *any* time prior to the time the field turns on because the Hamiltonian is time-independent in equilibrium). The Green's functions defined on this contour have a 3×3 matrix structure to them as described by Wagner [13]; this is needed to determine the transient response effects we are interested in.

The DOS is found from the local limit $(i = j)$ of the real-space retarded Green's function

$$G_{ij}^R(t,t') = -i\mathrm{Tr}e^{-\beta\mathcal{H}(t=-5)}\{c_i(t)c_j^\dagger(t') + c_j^\dagger(t')c_i(t)\}/\mathcal{Z}, \qquad (17.3)$$

while the distribution function is found from the equal-time limit of the so-called momentum-dependent lesser Green's function

$$G_{\mathbf{k}}^<(t,t') = i\mathrm{Tr}e^{-\beta\mathcal{H}}c_{\mathbf{k}}^\dagger(t')c_{\mathbf{k}}(t)/\mathcal{Z}, \qquad (17.4)$$

expressed in terms of momentum-dependent creation and annihilation operators. Both of these quantities can be directly extracted from the contour-ordered Green's function.

The vector potential is introduced into the Hamiltonian via the Peierls' substitution [14], and it modifies the kinetic energy operator

$$-\frac{t^*}{2\sqrt{d}}\sum_{\langle ij\rangle}(c_i^\dagger c_j + c_j^\dagger c_i) = \sum_{\mathbf{k}}\epsilon_{\mathbf{k}}c_{\mathbf{k}}^\dagger c_{\mathbf{k}} \to \sum_{\mathbf{k}}\epsilon_{\mathbf{k}-\mathbf{A}(t)}c_{\mathbf{k}}^\dagger c_{\mathbf{k}}, \qquad (17.5)$$

where we have made a Fourier transformation, and $\epsilon_{\mathbf{k}} = -t^*\sum_i\cos k_i/\sqrt{d}$ is the bandstructure for nearest-neighbor hopping on a hypercubic lattice [12].

Because the noninteracting Hamiltonian commutes with itself at different times, the exact noninteracting Green's functions in a field are easy to determine, and form the starting point for solving the nonequilibrium problem [15]. The noninteracting contour-ordered Green's function in a field satisfies

$$G_{\mathbf{k}}^{\text{non}}(t,t') = i\theta_c(t,t') \exp\left[-i \int_{t'}^{t} d\bar{t}\{\epsilon_{\mathbf{k}-\mathbf{A}(\bar{t})} - \mu\}\right] [1 - f(\epsilon_{\mathbf{k}} - \mu)]$$
$$- i\theta_c(t',t) \exp\left[-i \int_{t'}^{t} d\bar{t}\{\epsilon_{\mathbf{k}-\mathbf{A}(\bar{t})} - \mu\}\right] f(\epsilon_{\mathbf{k}} - \mu), \qquad (17.6)$$

where the the theta function $\theta_c(t,t')$ is equal to one if t is farther along the contour than t' (and is equal to zero otherwise), the integral in the exponential function runs along the contour from t' to t (note that $\mathbf{A}(t) = 0$ on the imaginary axis and for real times less than zero), and $f(x) = 1/[1 + \exp(\beta x)]$ is the Fermi-Dirac distribution function.

It turns out that the momentum-dependent Green's function depends on only two scalar quantities when the electric field lies in the diagonal $(1, 1, 1, \ldots)$ direction: the band structure $\epsilon_{\mathbf{k}}$ and a second bandstructure $\bar{\epsilon}_{\mathbf{k}} = -t^* \sum_i \sin \mathbf{k}_i/\sqrt{d}$. The local Green's function can be found by summing the momentum-dependent Green's function over all momentum, which can be replaced by a two-dimensional integral over ϵ and $\bar{\epsilon}$ weighted by the joint density of states which is equal to $\rho(\epsilon)\rho(\bar{\epsilon})$ [15]. All of the Green's functions and self-energies are described by two-time continuous matrix operators, which are discretized into general complex matrices when we perform numerical calculations (because it is a nonequilibrium problem with both a transient and steady-state response, we work in a time formalism). We vary the discretization size along the real time axis, but keep the discretization along the imaginary axis fixed at a size of $\Delta\tau = 0.1$ (100 points for $1/\beta = 0.1$). The initial five units of time are sufficient to allow the system to accurately display its equilibrium properties prior to the field being turned on at $t = 0$. For most calculations presented here, we use $t_{\max} = 35$. The matrices have sizes ranging from 900×900 ($\Delta t = 0.1$ when $t_{\max} = 35$) up to $4{,}100 \times 4{,}100$ ($\Delta t = 0.02$ when $t_{\max} = 35$). The local Green's function is then calculated by evaluating a two-dimensional Gaussian integral of a matrix valued integrand that requires one matrix inversion and two matrix multiplications to calculate. This is the most time-consuming part of the computation.

The nonequilibrium dynamical mean-field theory algorithm is then essentially the same as the equilibrium one [6–8, 10] (all Green's functions and self-energies are dense general complex matrices): (i) begin with a guess for the self-energy (we use the equilibrium self-energy expressed in a time representation); (ii) calculate the local Green's function from the self-energy by using a two-dimensional matrix-valued quadrature {the integrand is $[1 - G^{non}(\epsilon_{\mathbf{k}}, \bar{\epsilon}_{\mathbf{k}})\Sigma]^{-1}G^{non}(\epsilon_{\mathbf{k}}, \bar{\epsilon}_{\mathbf{k}})$, where G^{non} is the exact noninteracting Green's function on the lattice in the electric field (see Eq. (17.6)); (iii) extract the dynamical mean field λ by removing the self-energy from the local Green's function ($\lambda = -G^{-1} - \Sigma + G_{imp}^{non}(\mu)$, with $G_{imp}^{non}(\mu)$ the *free impurity* Green's function with a chemical potential μ and given by Eq. (17.6) with both $\epsilon_{\mathbf{k}} = 0$ and $\mathbf{A} = 0$); (iv) solve the impurity problem in the time-dependent dynamical mean field for the new Green's function $\{G = (1 - w_1)[G_{imp}^{non}(\mu)^{-1} - \lambda]^{-1} + w_1[G_{imp}^{non}(\mu - U)^{-1} - \lambda]^{-1}$, where $w_1 = 1/2$ is the density of the localized electrons$\}$; (v) extract the new self-energy from Dyson's equation $(\Sigma = G_{imp}^{non}(\mu)^{-1} - \lambda - G^{-1})$ and proceed to step (ii) until fully converged.

The nonequilibrium algorithm parallelizes [6, 7, 9] in the master-slave approach the master sends the self-energies to each slave node, which calculates and accumulates the integrands for each quadrature point and then forwards the accumulated results to the master. The master then calculates the impurity part of the algorithm (steps (iii)–(v)), checks the convergence and repeats if necessary. Using BLAS and LAPACK routines makes for a highly efficient algorithm which has achieved over 65% of peak speed on 2032 cores of a SGI Altix supercomputer [9].

In addition to the transient nonequilibrium algorithm described above, one can also examine steady-state properties. At the moment, the formalism has only been developed for the retarded Green's function, and hence for the local DOS. Future work will develop techniques for the lesser Green's function. In a steady-state formalism, we imagine the system has been prepared at a time $-t_1$ in an equilibrium state, and then at time $-t_2$ (with $-t_1 < -t_2$), an electric field is turned on. We then take the limit where $-t_1 \to -\infty$ and $-t_2 \to -\infty$ while maintaining $-t_1 < -t_2$, and examine the Green's functions a long time after the field was turned on. In this case, because all memory of the equilibrium state is lost as we moved into the steady-state response, we can examine the retarded Green's function directly in real time, and do not need to consider the three-part Kadanoff-Baym-Keldysh contour (this statement may not seem to obviously hold, but we will see how the steady-state DOS emerging from this result does appear to be the limit of the transient DOS for large average times).

The starting point is the Dyson equation for the retarded Green's function in real time, which satisfies

$$G_{\mathbf{k}}^{R}(t, t') = G_{\mathbf{k}}^{R,non}(t, t') + \int d\bar{t} \int d\bar{t}' G_{\mathbf{k}}^{R,non}(t, \bar{t}) \Sigma(\bar{t}, \bar{t}') G_{\mathbf{k}}^{R}(\bar{t}', t'), \quad (17.7)$$

with the noninteracting retarded Green's function given by [10]

$$G_{\mathbf{k}}^{R,non}(t, t') = -i\theta(t - t') \exp\left[-i \int_{t'}^{t} d\bar{t}(\epsilon_{\mathbf{k}+E\bar{t}} - \mu)\right] \quad (17.8)$$

$$= -i\theta(t_{rel}) \exp\left[-i \frac{2(\epsilon_{\mathbf{k}} \cos ET - \bar{\epsilon}_{\mathbf{k}} \sin ET)}{E} \sin \frac{Et_{rel}}{2} + i\mu t_{rel}\right],$$

where we used $\mathbf{A}(t) = -Et$ because the field was turned on in the infinite past. The noninteracting Green's function satisfies two important identities. The first we call the *gauge property*, and is

$$G_{\mathbf{k}+E\bar{t}}^{R,non}(t, t') = G_{\mathbf{k}}^{R,non}(t + \bar{t}, t' + \bar{t}) = G_{\mathbf{k}}^{R,non}(T + \bar{t}, t_{rel}), \quad (17.9)$$

where the second equality expresses the Green's function in the Wigner coordinates of average and relative time. The second is the *Bloch periodicity property*

$$G_{\mathbf{k}}^{R,non}(t + t_{Bloch}, t' + t_{Bloch}) = G_{\mathbf{k}}^{R,non}(t, t')$$

$$G_{\mathbf{k}}^{R,non}(T + t_{Bloch}, t_{rel}) = G_{\mathbf{k}}^{R,non}(T, t_{rel}), \qquad (17.10)$$

where $t_{Bloch} = 2\pi/E$ is the Bloch period.

The next step is to make an assumption that the local retarded self-energy is independent of average time. This can be seen to hold from an iterative argument as follows—start with the self-energy equal to zero (which obviously is independent of average time). Then the local Green's function is independent of average time because it is equal to the noninteracting Green's function and the average time dependence for the local noninteracting Green's function vanishes due to the gauge property. Hence the dynamical mean-field will be average-time independent and so will the impurity Green's function and self-energy. Continuing the DMFT iterations will not introduce any average time dependence, so the final converged self-energy will be independent of average time. One can ask whether there could also be a solution where the self-energy depends on average time; we checked this with the transient formalism, and find that as we approach the steady state the self-energy changes more slowly for long times indicating it is becoming average-time independent. While not a proof, this is a strong argument in favor of the initial assumption that we make. If the retarded self-energy is average-time independent, then one can show from the Dyson equation that the momentum-dependent Green's function satisfies both the Bloch periodicity property and the gauge property and hence that the local retarded Green's function is independent of average time. Thus, we have motivated the assumption that the local retarded self-energy and the local interacting retarded Green's function are independent of average time and the momentum-dependent interacting retarded Green's function satisfies both the gauge property and the Bloch periodicity property. This implies that we can perform a Fourier transformation with respect to the relative time, and a discrete Fourier series expansion with respect to the average time, with the only frequencies appearing being the Bloch frequencies $\nu_n = nE$

$$G_{\mathbf{k}}^{R}(T, t_{rel}) = \frac{1}{2\pi} \sum_n \int d\omega G_{\mathbf{k}}^{R}(\nu_n, \omega) e^{-i\nu_n T - i\omega t_{rel}}, \qquad (17.11)$$

and similar expansions for the noninteracting Green's function and the self-energy (note that most researchers assume a much stronger result, that the steady state is independent of average time *by definition*, but there does not appear to be any proof of this on a lattice, and indeed, the lesser Green's functions for the noninteracting system and the momentum-dependent Green's functions for the interacting system *do not* satisfy such an assumption, but they do satisfy the Bloch periodicity property). Using this representation, the Dyson equation becomes

$$G_{\mathbf{k}}^{R}(\nu_n, \omega) = G_{\mathbf{k}}^{R,non}(\nu_n, \omega) + \sum_{m} G_{\mathbf{k}}^{R,non}(\nu_m, \omega + \tfrac{1}{2}\nu_n - \tfrac{1}{2}\nu_m)$$
$$\times \, \Sigma(\omega + \tfrac{1}{2}\nu_n - \nu_m) G_{\mathbf{k}}^{R}(\nu_n - \nu_m, \omega - \tfrac{1}{2}\nu_m), \qquad (17.12)$$

which has an underlying matrix structure to it that allows us to solve for the $G_{\mathbf{k}}^{R}(\nu_n, \omega + \tfrac{1}{2}\nu_m)$ for all m and n (with ω fixed); note that the self-energy is independent of the Bloch frequency because it has no average time dependence. When we solve for the DOS, we need only solve in a frequency range from $0 \leq \omega < E$, because all frequencies outside of that range are coupled together and automatically determined when we solve the Dyson equation. The local Green's function (and hence the local DOS) is found by summing the momentum-dependent Green's function over all momentum, which requires a two-dimensional Gaussian integration (but of a scalar quantity now, rather than a matrix). Because we sum over all momentum, the gauge property tells us the local DOS is independent of average time, and hence we need to evaluate the $\nu_n = 0$ component only

$$\rho^{DOS}(\omega) = -\frac{1}{\pi}\mathrm{Im}\sum_{\mathbf{k}} G_{\mathbf{k}}^{R}(\nu_n = 0, \omega). \qquad (17.13)$$

Note that the steady-state DOS for the Falicov-Kimball model turns out to have no temperature dependence, just like the equilibrium DOS [16]. In our case, the proof is direct, because the temperature never enters the equations that are employed to solve for the retarded Green's function.

17.3 Results

We start by calculating the local DOS for the nonequilibrium system. We will be examining systems described by the Falicov-Kimball model at half filling for the conduction electrons and the localized electrons. The system starts in equilibrium with $1/\beta = 0.1$ and has the field turned on at $t = 0$. The field is directed along the diagonal of the infinite-dimensional hypercube and the magnitude of each Cartesian component is equal to $E = 0.5$. We will plot the DOS as a function of frequency for different average times. But first, we want to verify the accuracy of the calculations, since we have discretized the Kadanoff-Baym-Keldysh contour. This is accomplished by examining sum rules for the DOS [17,18]. The sum rules examine moments of the local DOS which can be related to the equal time Green's function and its derivatives with respect to the relative time. Since the DOS is determined by the retarded Green's function, we examine the sum rules for the retarded Green's function which turn out to be independent of the electric field. In particular, the zeroth

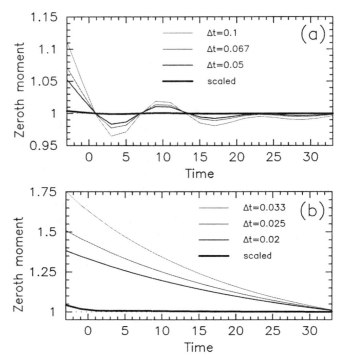

Fig. 17.2. Zeroth moment of the local DOS as a function of average time for the case (a) $U = 0.5$ (strongly scattering metal) and (b) $U = 2$ (moderate gap Mott insulator). We plot results for specific discretizations of the Kadanoff-Baym-Keldysh contour, and for results extrapolated to zero discretization. The extrapolated results are quite accurate, but the accuracy worsens as U increases, and in the equilibrium phase before the field is turned on (the dotted line is the exact result).

moment sum rule is equal to 1 and the second moment sum rule is equal to $0.5 + U^2/4$. Each of these moments can be directly calculated numerically from the contour-ordered Green's functions. In general, we find the results for a given discretization have errors, which can become quite large, but when we extrapolate the results to zero discretization size (usually with a quadratic Lagrange interpolation formula), we find that agreement is better than 1% for $U = 0.5$ and better than 5% for $U = 2$.

The accuracy for the second moment is not as precise, as one might expect because it is more difficult to accurately determine (see Fig. 17.3). The metallic case still has high accuracy (errors less than 1% for the extrapolated result), but the Mott insulator has much reduced accuracy (less than 30% error for the extrapolated result). But if we examine these results more carefully, we see that the largest deviation occurs at early times, in the equilibrium state. Indeed, the system actually improves the accuracy rather dramatically when it is in nonequilibrium, and the overall error is only a few percent for times

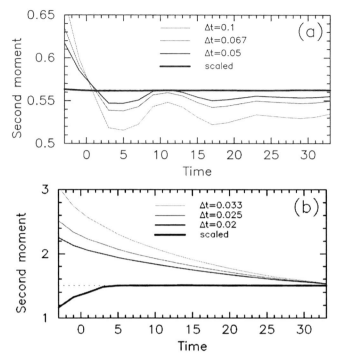

Fig. 17.3. Second moment of the local DOS as a function of average time for the case (a) $U = 0.5$ (strongly scattering metal) and (b) $U = 2$ (moderate gap Mott insulator). We plot results for specific discretizations of the Kadanoff-Baym-Keldysh contour, and for results extrapolated to zero discretization. The dotted line is the exact result.

larger than five units after the field is turned on. This is typically what we see with much of our data, where the real-time formalism is less accurate for equilibrium results than it is for nonequilibrium results.

Having established that we can achieve highly accurate solutions by scaling our data, we next move to examining the local DOS as a function of average time for the metal and the Mott insulator. We continue to scale our data to the zero discretization limit so we can achieve high accuracy (we use $\Delta t = 0.1$, 0.067, and 0.05 for $U = 0.5$ and $\Delta t = 0.033$, 0.025, and 0.02 for $U = 2$). In Fig. 17.4, we plot the DOS for a few different values of average time when $U = 0.5$. Panel (a) is a near equilibrium result (the field is turned on at $T = 0$). As the average time increases, the system first appears to develop broadened peaks at the Bloch frequencies $n/2$; those peaks then evolve into a series of minibands with more complex structure by the steady-state limit. Note that we used the transient nonequilibrium DMFT algorithm for all finite times, and the completely different steady-state nonequilibrium DMFT algorithm for the last panel. The similarity of this data clearly indicates that the transient

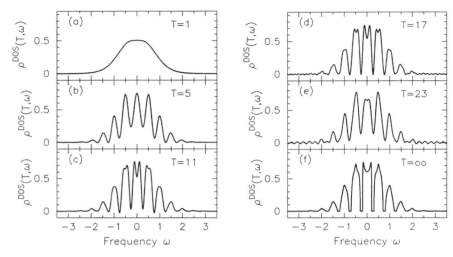

Fig. 17.4. Local DOS as a function of average time for $U = 0.5$ and $E = 0.5$. Note how the system evolves from a near equilibrium DOS to the nonequilibrium steady state.

results are approaching the steady-state results. Note further that a close examination of the $T = 23$ data shows that it appears to deviate more from the steady state than the $T = 17$ data. This is most likely an artefact of the truncation of the range of t_{rel} values that are calculated at that average time. We lose precision for the Fourier transform as we approach the maximal time on the Kadanoff-Baym-Keldysh contour because the range of t_{rel} for which we have data shrinks to zero at the maximal T value. Hence the detailed structure in the DOS becomes hard to represent with the range of relative time values that we have (especially the fine structure present in the peaks).

In Fig. 17.5, we show a similar plot, but now for the Mott insulator with $U = 2$ (by this we mean the system has undergone the Mott transition in equilibrium; on a hypercubic lattice no true gap develops, because the DOS vanishes only at one point in frequency which might more correctly be called a pseudogap, but there is still a wide region, reminiscent of a Mott gap, where the DOS is exponentially small around this point where the DOS vanishes). The field creates a number of additional peaks in the DOS, but does not have actual minibands form, as the minibands now all overlap with one another. The peak structures are all concentrated near the Bloch frequencies, and there is a small peak that develops at $\omega = 0$ in the steady state. This implies that the driving of the system by the electric field creates "subgap states" that have a metallic character to the DOS! This is because the energy pumped into the system by the field is sufficient to overcome the Mott gap formation.

In Fig. 17.6, we plot the steady-state DOS in a field (of strength $E = 0.5$) for different values of U. When $U = 0$, the steady state DOS is the Wannier-

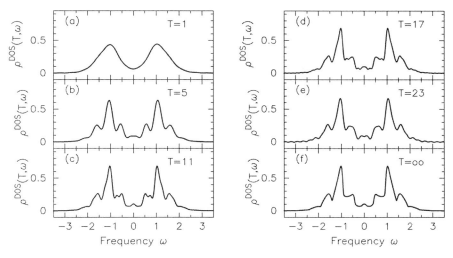

Fig. 17.5. Local DOS as a function of average time for $U = 2$ and $E = 0.5$. Note how the system evolves from a near equilibrium DOS to the nonequilibrium steady state.

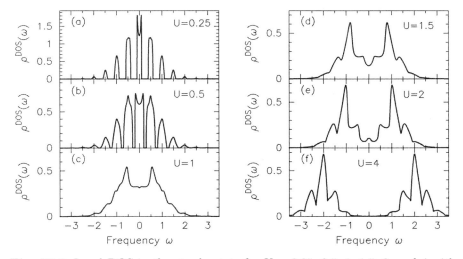

Fig. 17.6. Local DOS in the steady state for $U = 0.25$, 0.5, 1, 1.5, 2, and 4 with $E = 0.5$. Note how the system evolves from a broadened Wannier-Stark ladder to a "perturbed" Mott insulator with upper and lower Hubbard bands.

Stark ladder of delta functions located at the Bloch frequencies, and weighted by a DOS factor. As scattering is turned on, we expect the delta functions to broaden. Indeed, we see this in panel (a), for $U = 0.25$, except the delta function at $\omega = 0$ is broadened with a double-peak structure, separated by U. As U increases further in the metal (panels (b) and (c)), the splitting of

230   J.K. Freericks and A.V. Joura

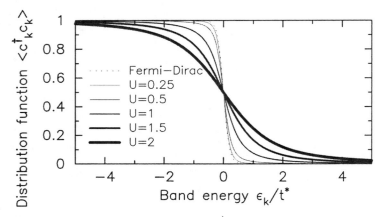

Fig. 17.7. Equilibrium distribution function $\langle c_{\mathbf{k}}^\dagger c_{\mathbf{k}} \rangle = -iG^<(\epsilon_{\mathbf{k}}, t, t)$ for the Falicov-Kimball model at half filling and different U values (the metal-insulator transition takes place at $U = \sqrt{2}$). This data is for $1/\beta = 0.1$.

the "zeroth" miniband continues to increase, but all of the bands broaden and then eventually merge into a complicated DOS by the time $U = 1$. As U increases further into the Mott insulator, the main peak continues to split, and the DOS develops the upper and lower Hubbard bands, with an additional "corrugation" induced by the field. Surprisingly, for $U = 2$, the DOS develops a small metal-like peak around $\omega = 0$.

Next, we focus on calculating the distribution functions in momentum space as they evolve from the equilibrium distribution to the nonequilibrium steady state. But to start, we first want to show the equilibrium distribution functions for different interaction strengths at $1/\beta = 0.1$. This is plotted in Fig. 17.7. The equilibrium distribution function depends only on $\epsilon_{\mathbf{k}}$, and becomes the Fermi-Dirac distribution $1/[1 + \exp(\beta \epsilon_{\mathbf{k}})]$ as $U \to 0$ (dashed line). As the scattering increases, the distribution function deviates more and more from the noninteracting result (the Mott insulator transition occurs at $U = \sqrt{2}$ in this model).

In nonequilibrium, the distribution function depends on two band energies $\epsilon_{\mathbf{k}}$ and $\bar{\epsilon}_{\mathbf{k}}$, so it is more complicated to present the results (for this work we show results for the Green's functions in the Hamiltonian gauge rather than making the transformation to gauge-invariant Green's functions—for this field, the transformation [10,19] involves just a rotation in the $\epsilon - \bar{\epsilon}$ plane). We will show the behavior for some specific values of $\epsilon_{\mathbf{k}}$ and $\bar{\epsilon}_{\mathbf{k}}$. There is one technical detail we use for calculating the distribution functions. In this formalism, we can find both the retarded and the lesser Green's functions at equal times. The retarded Green's functions should have a value of 1 everywhere because they are equal to the equal time anticommutator of the creation and annihilation operators. But for a given discretization, we often find that the retarded Green's function does not equal one at equal times (see Fig. 17.2). Hence, we

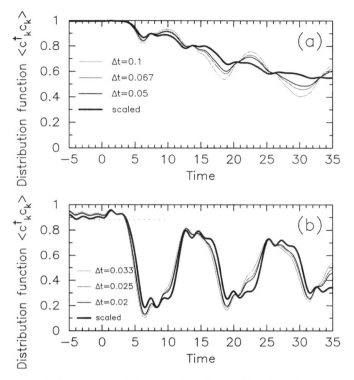

Fig. 17.8. Raw data and scaled data for the distribution function at $\epsilon_{\mathbf{k}} = -2$ and $\bar{\epsilon}_{\mathbf{k}} = -2$ for (a) $U = 0.5$ and (b) $U = 2$. Both data sets are scaled with a quadratic Lagrange extrapolation. In panel (b), the dotted line is the equilibrium distribution function (the initial temperature is $1/\beta = 0.1$). The electric field has a magnitude of 0.5, and is turned on at time $t = 0$.

calculate the distribution function by examining the *ratio* of the lesser Green's function to the retarded Green's function, which turns out to be much more accurate than examining the lesser Green's function directly. When we scale our results, it is the ratio that is scaled to the zero discretization size limit.

In Fig. 17.8, we plot the distribution function at $\epsilon = \bar{\epsilon} = -2$ for different values of the discretization size of the Kadanoff-Baym-Keldysh contour (panel a is for a metal with $U = 0.5$ and panel b is for an insulator with $U = 2$). Also included in that figure are scaled results, which are extrapolated using a quadratic Lagrange interpolation formula. One can see that in the metal, the extrapolation tends to reduce the oscillations, while in the insulator, the extrapolation is less severe of a change from the raw data with a finite discretization. In both cases we start at $1/\beta = 0.1$ and turn on a field in the diagonal direction with a magnitude of $E = 0.5$ for each Cartesian component at $t = 0$. Hence the curves should be completely flat, and agree with the results in Fig. 17.7. This is true for the metal, and can be seen to work fairly well

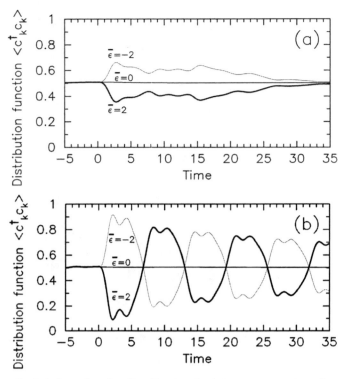

Fig. 17.9. Scaled data for the distribution function at $\epsilon_k = 0$ and $\bar{\epsilon}_k = -2$, 0, and 2 for (a) $U = 0.5$ and (b) $U = 2$. Both data sets are scaled with a quadratic Lagrange extrapolation. The initial temperature is $1/\beta = 0.1$; the electric field has a magnitude of 0.5, and is turned on at time $t = 0$. Note the larger amplitude oscillations for the Mott insulator (panel b) than the metal (panel a).

for the insulator (there is an oscillation of magnitude a few percent for the insulator). One surprising result is that the distribution function has much larger amplitude oscillations in the insulator than it has in the metal (when the field is on).

Having shown that we can achieve good accuracy by scaling our results, we next focus on the $\bar{\epsilon}$ dependence of the distribution function. In Fig. 17.9, we show results for $\epsilon = 0$ and various $\bar{\epsilon}$ in the (a) metal and (b) insulator. Once again one can see much larger amplitude oscillations in the insulator, and one can see the effect of particle-hole symmetry where results for negative $\bar{\epsilon}$ values are mirror reflected from the positive ones. Note further that the oscillations seen in the equilibrium region ($t < 0$) in the previous figure are much reduced here. The nonequilibrium oscillations are expected on general grounds as the system generically has an oscillating current develop for short times after the field is turned on. Since the oscillating current arises from (quasi)periodic changes of the electron distribution through the Brillouin zone,

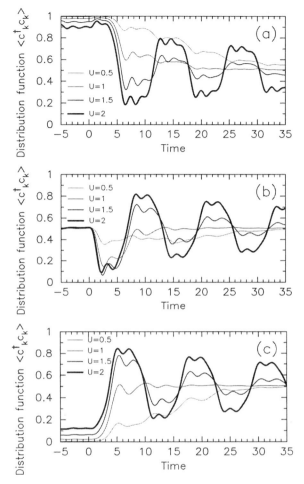

Fig. 17.10. Scaled data for the distribution functions for different U values. Panel (a) is $\epsilon_{\mathbf{k}} = -2$ and $\bar{\epsilon}_{\mathbf{k}} = -2$, panel (b) is $\epsilon_{\mathbf{k}} = 0$ and $\bar{\epsilon}_{\mathbf{k}} = 2$, and panel (c) is $\epsilon_{\mathbf{k}} = 2$ and $\bar{\epsilon}_{\mathbf{k}} = 0$.

such oscillations in the distribution functions are consistent with an oscillating current.

Next, we show scaled results for specific points in the Brillouin zone withdifferent values of U to see how the system evolves from a metal to an insulator. The most striking behavior is that the amplitude of the oscillations grow rather dramatically as the scattering increases. They are not simple sinusoidal oscillations, having complex structure to them. One can also see, by comparing with the equilibrium results for negative times, that the accuracy is quite good too (the curves are essentially flat for negative times).

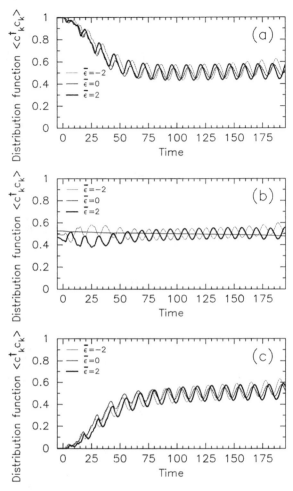

Fig. 17.11. $\Delta t = 0.1$ (unscaled) data for the distribution functions for $U = 0.25$ extended to much longer times. Panel (a) is $\epsilon_{\mathbf{k}} = -2$ and $\bar{\epsilon}_{\mathbf{k}} = -2$, 0, and 3; panel (b) is $\epsilon_{\mathbf{k}} = 0$ and $\bar{\epsilon}_{\mathbf{k}} = -2$, 0, and 2; and panel (c) is $\epsilon_{\mathbf{k}} = 2$ and $\bar{\epsilon}_{\mathbf{k}} = -2$, 0, and 2.

Finally, we show plots for the different distribution functions we have been examining for the case of weak coupling $U = 0.25$ and large times in Fig. 17.11. The data is not scaled, but rather is calculated with a fixed discretization $\Delta t = 0.1$. This could have the tendency to slightly overestimate the magnitude of the oscillations, as we have seen with our previous data. Note how in all cases except for $\epsilon = \bar{\epsilon} = 0$ we have long-lived oscillations that develop with what appears to be a well defined phase difference for different $\bar{\epsilon}$ values. The period of the oscillations is essentially the Bloch period (equal to 4π in this case).

We now consider the lesser Green's function sum rules for the Falicov-Kimball model in a field [17, 18]. The distribution functions are always equal to the occupation of the conduction electrons over the Brillouin zone. Unfortunately, unlike the retarded Green's functions, where the equal time limit is equal to one, here we do not know *a priori* what the occupations are, so we cannot use the sum rules to verify the accuracy of the calculations. The only test we can perform is to see that the results are constant for negative times and that the constant value is equal to the value calculated in an independent equilibrium formalism. Here we find the accuracy of the scaled results is pretty good for these values, with errors on the order of a few percent. Since we generically find the time formalism is less accurate for the equilibrium results than the nonequilibrium ones, we anticipate that the scaled results for the nonequilibrium distribution functions have errors that are also in the few percent range.

17.4 Conclusions

In this work, we have illustrated how dynamical mean-field theory can be generalized to nonequilibrium and applied the formalism to the spinless Falicov-Kimball model in a large electric field. We showed results for the local DOS, which plot the distribution of quantum-mechanical states, and for the distribution functions, which plot how the electrons occupy those quantum states in the Brillouin zone. Our main focus was on showing how one can scale the results to the zero discretization size limit, and how the scaled results have high accuracy. For the DOS, we compared a transient formalism to a steady-state formalism and found good agreement at long times. The distribution functions also show interesting behavior—they develop oscillations which tend to grow in amplitude as the system becomes more strongly correlated and passes from a metal to an insulator across the Mott transition. In the future, we will further investigate these distribution functions by examining their behavior through the entire Brillouin zone, rather than at just a few selected points.

Acknowledgements

This work was supported by the National Science Foundation under grant number DMR-0705266 and by DARPA under grant number W911NF-07-1-0576. Supercomputer time was provided by the NASA NLCS program and the HPCMP of the DOD including two CAP phase II projects (one at ERDC and one at ARSC). The author also acknowledges the hospitality of NIST's Gaithersburg campus, where this work was completed. Finally, useful discussions and collaborations with A. Hewson, V. Turkowski and V. Zlatić are acknowledged.

References

1. L.P. Kadanoff, G. Baym, *Quantum Statistical Mechanics* (W. A. Benjamin, New York, 1962)
2. L.V. Keldysh, J. Exptl. Theoret. Phys. **47**, 1515 (1964)
3. A.P. Jauho, J.W. Wilkins, Phys. Rev. B **29**, 1919 (1984)
4. U. Brandt, C. Mielsch, Z. Phys. B–Condens. Mat. **75**, 365 (1989)
5. A. Georges, G. Kotliar, W. Krauth, M.J. Rozenberg, Rev. Mod. Phys. **68**, 13 (1996)
6. J.K. Freericks, V.M. Turkowski, V. Zlatić, in *Proceedings of the HPCMP Users Group Conference 2006, Denver, CO, June 26–29, 2006*, ed. by D.E. Post (IEEE Computer Society, Los Alamitos, CA, 2006), pp. 218–226
7. J.K. Freericks, in *Proceedings of the HPCMP Users Group Conference 2007, Pittsburgh, PA, June 18–21, 2007*, ed. by D.E. Post (IEEE Computer Society, Los Alamitos, CA, 2007), p. *to appear*
8. J.K. Freericks, V.M. Turkowski, V. Zlatić, Phys. Rev. Lett. **97**, 266408 (2006)
9. J.K. Freericks, Y.T. Chang, J. Chang, Int. J. High Perf. Comp. Appl. Submitted (2007)
10. V.M. Turkowski, J.K. Freericks, in *Strongly Correlated Systems: Coherence and Entanglement*, ed. by J.M.P. Carmelo, J.M.B.L. dos Santos, V.R. Vieira, P.D. Sacramento (World Scientific, Singapore, 2007), pp. 187–210
11. L.M. Falicov, J.C. Kimball, Phys. Rev. Lett. **22**, 997 (1969)
12. W. Metzner, D. Vollhardt, Phys. Rev. Lett. **62**, 324 (1989)
13. M. Wagner, Phys. Rev. B **44**, 6104 (1991)
14. R.E. Peierls, Z. Phys. **80**, 763 (1933)
15. V.M. Turkowski, J.K. Freericks, Phys. Rev. B **71**, 085104 (2005)
16. P.G. van Dongen, Phys. Rev. B **45**, 2267 (1992)
17. S.R. White, Phys. Rev. B **44**, 4670 (1991)
18. V.M. Turkowski, J.K. Freericks, Phys. Rev. B **73**, 075108 (2006)
19. R. Bertoncini, A.P. Jauho, Phys. Rev. B **44**, 3655 (1991)

NON-EQUILIBRIUM PHYSICS IN SOLIDS: HOT-ELECTRON RELAXATION

K.H. BENNEMANN

Institut für Theoretische Physik, Freie Universität Berlin, Germany.
`khb@physik.fu-berlin.de`

Abstract. We discuss the time scales for relaxation of excited electrons in transition–metals, ferromagnets like Ni, Co, and semiconductors like diamond, graphite. Ultrafast relaxation faster than ps may occur as a result of strong electron-electron interactions, while magnetoelastic forces controlling magnetic reorientation, domain dynamics, for example, involve relaxation times of the order of 100 ps or more.

Keywords: non equlibrium process,ultrafast relaxation, domain dynamics

18.1 Introduction

Due to recent advances in time resolved optical studies, laser physics, one may observe the non-equilibrium behaviour of solids, in particular metals and semiconductors, in which many electrons have been excited by strong laser irradiation, for example. Then, using pump-probe spectroscopy one observes the time-resolved response. Thus, two-photon photoemission (2PPE) studied the lifetimes of excited electrons in Cu [1]; in the ferromagnets Ni, Co [2], the relaxation of the magnetization in Ni with many hot electrons [3], the relaxation of excitations in high T_c–superconductors [4] and laser induced phase transitions in carbon, graphitization of diamond by electron-hole pairs [5]. Thus, ultrafast fs-responses were observed, while earlier experiments studying magnetic reorientation dynamics and spin lattice relaxation at surfaces and in thin films observed a slower response of the order of 100ps [6]. In the following we discuss the general situation and the time scales for the various relaxation phenomena. For simplicity, we use "golden-rule" type arguments.

The physical situation is illustrated in Fig. 18.1. Upon laser irradiation many excited electrons are present and the solid is in a non-equilibrium state. After a short time the hot electrons thermalize again due to electron-electron interactions. This thermalization may involve up to 100 fs or more. d-Electrons

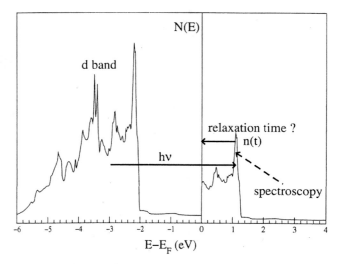

Fig. 18.1. Illustration of the scenario for the non-equilibrium state produced by exciting optically electrons out of the ground state (the Fermi sea). The hot electrons relax first due to electron-electron interactions. The electron thermalization is characterized by the electron temperature $T_{el}(t)$. Energy transfer from the electrons to the lattice occurs via the electron-lattice coupling and which controls $T_{el}(t)$ for longer times.

will thermalize faster than s-electrons, for example. As a result the electrons will aquire an electronic temperature $T_{el}(t)$ which will change with time and become different from the lattice temperature $T_{latt}(t)$, $T_{el} > T_{latt}$ As time progresses energy is transferred from the hot electron system to the lattice via the electron lattice coupling. This occurs during a time of $t \gtrsim 200 \div 300$ fs, roughly, and continues for several ps and more, and finally $T_{el}(t) \rightarrow T_{latt}(t)$ again after many ps. Hence, after laser irradiation T_{el} increases first, reaches a maximum at a time which is set by the intensity of the laser light and the duration of the laser pulse, the electron-electron interaction strength and the one of the electron-lattice coupling [3, 7].

Due to the strong electron-electron interaction in transition-metals and noble metals one expects for the excited electrons relaxation times $\tau(E) \sim (E-E_F)^{-2} \sim N^{-1}(0)$, hence of the order of $100 \div 10$ fs depending on $(E-E_F)$ of course, the matrix-elements and the density of states at the Fermi-energy E_F, N(o). Thus, τ in Cu, Ag is expected to be larger than in Ni, Co, Fe [8].

Due to the hot electrons the magnetization in itinerant ferromagnets will change and becomes time dependent, M(t). On general grounds, one expects

$$M(t) \longrightarrow M(T_{el}(t))$$

as soon as the excited electrons have thermalized again. Consequently response times of the electron relaxation and of the (itinerant) magnetization should be of the same order, possibly faster than 100 fs in view of the strength of the electron-electron interaction and the exchange interaction. Note, changes of the magnetization must obey angular momentum conservation and thus possibly may involve atomic spin-orbit coupling (which is of the order of 50 ÷ 70 meV in Ni, Fe, for example). We expect the magnetization to decrease upon laser irradiation up to a minimum, when $T_{el}(t)\to$max. and then to relax again when $T_{el}(t)\to T_{latt}(t)$. Note, hot electrons affect also the exchange coupling J and then the coupling between magnetic films, for example.

In contrast, magnetoelastic responses controlled by spin-lattice coupling and magnetic anisotropy energies (of the order of 10^{-5}eV in transition-metals) will occur during times of the order of 100 ps or longer, for example. This will be the case for magnetic reorientation at surfaces and in thin films spin-relaxations in hot lattices, and magnetic domain dynamics.

In semiconductors like graphite, diamond electron-hole excitations cause a change of bonding, for example of $sp^3 \to s^2p^2$ bonding in graphitization of diamond upon laser irradiation? Again, such structural changes involve relatively strong electron-electron interactions and, in terms of phonons, Brillouin-zone boundary phonons. Thus response times of the order of several 100 fs and faster than ps-times are expected. The optical control of $sp^3 \rightleftarrows s2p^2$ bonding is of course a fascinating problem and may lead to optical switching of amorphous$\leftrightarrows$crystalline transitions and diamond$\leftrightarrows$graphite under special conditions [10].

For high T_c–superconductors one expects an interesting dynamical behaviour, since Cooper-pairing and antiferromagnetic interactions are involved [11]. Thus, upon laser irradiation destroying superfluid density n_s, Cooper-pair density, and also causing antiferromagnetic excitations one expects corresponding relaxations as where recently observed by Kaindl et al. [4]. On general grounds one expects $\tau_1 \sim n_s^{-1} \sim T_c^{-1} \sim$ several ps for the superfluid relaxation, and $\tau_3 \Delta E_{mag}^{-1} \sim (T^*)^{-1}$ for the antiferromagnetic relaxation time, $\tau_3 \simeq (T_c/T^*)\tau_1$. Here, ΔE_{mag} is the characteristic energy for a magnetic excitation also characterized by the temperature T^*. Since $(T_c/T^*) \ll 1$ one expects $\tau_3 < \tau_1$, as is observed [4].

In summary, we have outlined the time-scales for important relaxation processes in solids occuring upon laser irradiation. This is illustrated in Fig. 18.2. In the following we present results for the dynamics in non-equilibrium solids, supposedly illustrating the general situation. For calculating the non-equilibrium distribution of electrons we use the Boltzman equation [2]. Similarly, $T_{el}(t)$ is calculated using two coupled equations for $T_{el}(t)$ and $T_{latt}(t)$ [3].

First, in Fig. 18.3 we show for Cu the relaxation time τ referring to the dynamics of the electron occupation $n(E,t)$ as deduced from 2PPE [2]. Note, τ includes contributions from secondary electrons, in particular Auger-electrons. Of course, one expects τ to depend on the lifetime τ_h of the holes near E_F. From experiment one estimates $\tau_h \simeq 15$ fs. The structure on $\tau(E)$ results from the contribution of the Cu–d band which begins approximately 2 eV below

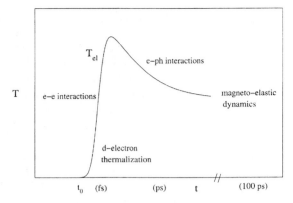

Fig. 18.2. Time scales for the dynamics of the non-equilibrium state. Electron thermalization occurs during $10 \div 100$ fs. Thus, $T_{el}(T) \to T_{latt}$ after some ps. After a longer time, $t \gtrsim 100$ ps, magnetoelastic dynamics involving domains, magnetic reorientation and spin-relaxation in a warm lattice occurs. Note, for itinerant metals changes of the magnetization happen on the same time scale as those of the electronic distribution.

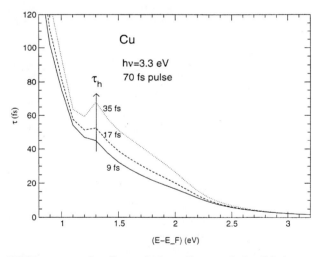

Fig. 18.3. 2PPE spectra for Cu and dependence of the lifetime τ referring to the level occupation dynamics on the lifetime τ_h of holes. The latter controls the contribution of Auger electrons to τ. The structure may not appear for Au, for example, where τ_h is smaller.

E_F. The exciting photons have the energy $\hbar\omega \simeq 3.6$ eV. The position of the side peak is hence as expected. Height and position of the structure is in fair agreement with experiment. The results indicate the role played by Auger electrons [12].

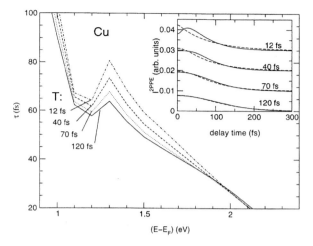

Fig. 18.4. Dependence of relaxation of hot electrons in Cu on duration T of the exciting laser pulse.

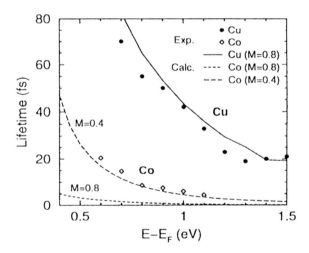

Fig. 18.5. Lifetime τ of excited electrons in Cu and Co. Note, the electron–electron interaction matrix–element M is taken to be smaller for Co then for Cu due to d–electron screening.

In Fig. 18.4 we show how 2PPE results depend on the duration T of the exciting laser pulse. Experiments should verify this. We conclude that Auger electrons are not negligible for the 2PPE spectrum of Cu, smaller fingerprint of this is also expected for Au, but not for Ni, Co, etc. [12].

In Fig. 18.5 we show results for the lifetime τ of hot electrons in Cu and Co, for example [2]. The smaller values of τ in Co result from the larger density of states $N(E)$ near E_F. However, note agreement with experimental results by

Aeschlimann et al. [2] is only obtained if the characteristic matrix-element M for electronic collisions is ~ 0.5 in Co rather than 0.9 as in Cu. This may result from the fact that the d-electrons in Co contribute strongly to the screening of the Coulomb interactions between the electrons [2]. Note, for $(E - E_F) \sim 1$ eV one finds already fast relaxations of the order of 40 fs or smaller. This sheds also light on the time required for thermalization of hot electrons [2].

The magnetic response upon laser irradiation is illustrated in the following. Generally we expect in itinerant ferromagnets

$$M(T_{latt}) \to M(T) \to M(T_{el}(t)) \to M(T_{latt}) \qquad (18.1)$$

for non-equilibrium. Here, the electronic temperature is determined from the coupled master-type equations $T_{el}(t)$ and $T_{latt}(t)$. To compare with experiments we analyze the dynamics expected for nonlinear magneto-optics, magnetic dichroism, Kerr-rotation, etc. [13]. The second harmonic signal (SHG) $I(2\omega)$ is analyzed in the form [3, 13]

$$\delta^- = I(M) - I(-M), \qquad (18.2)$$

where $I \propto |E_i(2\omega)|^2$ and $E_l(2\omega) \propto F_{ijl}\chi_{ijl}E_jE_l$. Here, F_{ijl} refer to Fresnel factors and χ_{ijl} is the nonlinear susceptibility. Note, in general different sensor elements χ_{ijl} having generally different phases may contribute to $E_i(2\omega)$ and to the non-linear light intensity $I(2\omega)$. This involves phases of χ_{ijl} which may depend on the excitations and on time [3, 4]. Also, in principle χ_{ijl} should be calculated for the non-equilibrium distribution of electrons. In order to simplify, quasi as a test of the situation, we assume [7, 13]

$$\chi_{ijl}(M) = \chi_{ijl}^1 + \chi_{ijl}^2 M + \ldots, \qquad (18.3)$$

splitting χ into an even and odd contribution in the magnetization M. Then, approximately,

$$\delta^-(t) \approx \frac{D(t)}{D(t_0)} - 1 \underset{t}{\to} \frac{M(t)}{M(t_0)} - 1, \qquad (18.4)$$

where the probe laser heating up the electrons starts at time t_0 and where $D(t) = I(M) - I(-M)$. The analysis assumes no time dependent phase effects, and essentially a few dominating χ_{ijl} determining $E_i(2\omega)$ [3, 7].

In Fig. 18.6 we show results for $T_{el}(t)$ and the SHG-signal δ^- [7]. One concludes that for somewhat longer delay times in the pump probe analysis, $\Delta(t) \gtrsim 100 \div 200$ fs, $\delta^-(t) \sim \frac{M(T_{el})}{M(t_0)} - 1$ holds approximately.

The magnetic response is very fast. Note, $M(t) \to 0$ upon laser irradiation for times $t \gtrsim 50$ fs and $M(t) \to$ min. when $T_{el} \to$ max [3, 7].

In Fig. 18.7 we show that $M(t) \to 0$ for $T_{el} \gtrsim T_c$, T_c is the Curie-temperature. This is in agreement with recent experiments [15].

High-T_c superconductors like $La_{i_x}Sr_xCuO_4$, YBCO, etc. are of particular interest. Antiferromagnetic excitations are present. Fig. 18.8 characterizes the phase diagram with T_c, being the superconducting transition temperature,

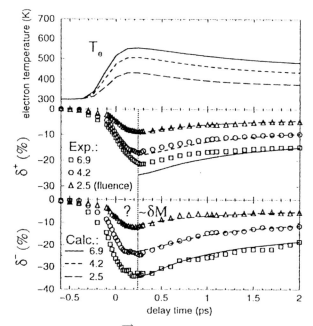

Fig. 18.6. Results for $\delta^-(t) \sim I(\overrightarrow{M}) - I(-M)$ using SHG. $T_e(t)$ is the electronic temperature, $\delta^+(t) \sim I(\overrightarrow{M}) + I(-\overrightarrow{M})$. Note, T_e is max. when $\delta^-(t)$ is minimal. Theory and experiment compare well for $t \gtrsim 200$ fs, but less for small t.

$T_c \propto n_s$ for underdoped superconductors, and T^* being a critical temperature characteristic for a.f. excitations [16]. Hence, in such systems one expects dynamics upon laser irradiation due to breaking up Cooper-pairs and due to antiferromagnetic excitations.

In Fig. 18.8 we illustrate the possible dynamics. Regarding relaxation towards superfluidity, for the underdoped superconductors with doping $x < 0.15$, we have

$$\tau_1 \sim (\Delta \overline{e^{i\varphi}})^{-1} \sim T_c^{-1}, \tag{18.5}$$

where φ refers to the phase of the Cooper pairs and approximately $\overline{e^{i\varphi}} = 1$ for $T < T_c$ and $\overline{e^{i\varphi}} = 0$ for $T > T_c$. In accordance with experiments we estimate $\tau_1 \gtrsim 1 \div 3$ ps. Furthermore, one expects for the magnetic excitations (ΔE_{af})

$$\tau_3 \sim \frac{1}{\Delta E_{af}} \sim \frac{1}{T^*}, \tag{18.6}$$

which may be rewritten as

$$\tau_3 \simeq \frac{T_c}{T^*}\tau_1. \tag{18.7}$$

Hence, $\tau_3 < \tau_1$ and one gets a fast ($\tau_3 \lesssim 1$ps) and a slower relaxation τ_1 as is observed [4]. If phase coherent Cooper pairs continue to exist above T_c,

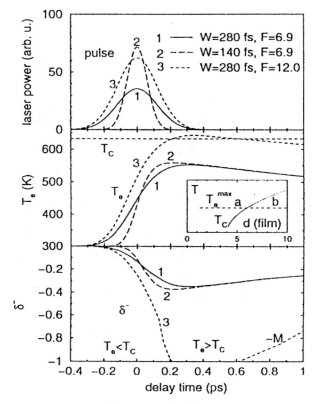

Fig. 18.7. SHG results showing the dependence of the magnetic response $\delta^-(t)$ on the duration T of the exciting laser pulse and its intensity. Behaviour of the magnetic response $\delta^-(t)$ for $T > T_c$ being the Curie-temperature, is also shown.

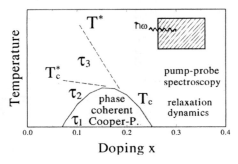

Fig. 18.8. Dynamics in underdoped high-T_c–superconductors with transition temperature $T_c \sim n_s$ and temperature T^* characterizing a.f. excitations. τ_1 refers to relaxation of phase-coherent Cooper pairs and τ_3 to relaxation of a.f. excitations.

(a) t = 50 fs   (b) t = 70 fs   (c) t = 80 fs

(d) t = 90 fs   (e) t = 100 fs   (f) t = 100 fs (rotated)

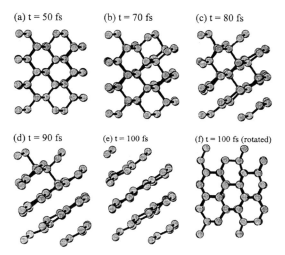

Fig. 18.9. Graphitization of diamond upon laser irradiation producing electron-hole excitations. Note, the diamond structure has changed significantly for $t \gtrsim 80$ fs.

one would expect further relaxations characterized by τ_2. This interpretation needs be checked by further experiments. We note again, that this dynamics sheds light on the important physics behind high T_c–superconductivity.

Finally, we discuss bond-breaking dynamics, in particular the one related to $sp^3 \rightleftarrows s^2p^2$ bond formation and controlled optically. Obviously, $sp^3 \rightleftarrows s^2p^2$ involves metallization and structural change like graphitization of diamond [5,9].

In Fig. 18.9 we show by using a tight-binding approximation for the s, p electrons and molecular-dynamics for the atomic structure, the ultrafast structural changes in diamond upon laser irradiation creating many electron-hole pairs and thus $sp^3 \rightleftarrows s^2p^2$ bond changes. After times of the order of 80 fs or more the structure changes.

In Fig. 18.10 we show how the light absorption in diamond and graphite depends on the duration T of the exciting laser pulse [17]. This sheds light on the role played by T for the non-equilibrium behaviour.

Structural changes which occur after some time explain the different behaviour observed for diamond and graphite. In the case of graphite more energy is absorbed from short pulses than from long ones, for diamond the opposite is true. Atomic disorder becoming effective for longer pulses causes in graphite less absorption due to bond changes, while in diamond disorder consisting of partial graphitization and creation of states in the diamond gap between valence and conduction band causes more absorption. This shows again how time-dependent behaviour on a fs-time scale may reveal interesting physics.

In summary, the discussion of the ultrafast response of metals and semi-conductors shows that pump and probe spectroscopy and relaxation of

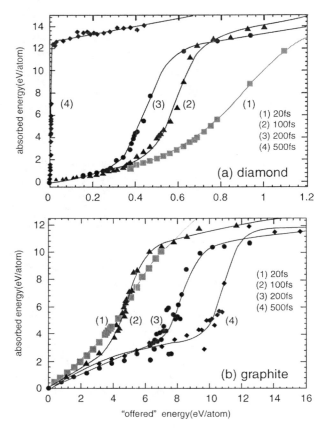

Fig. 18.10. Absorbed energy from a laser pulse in diamond (a) and in graphite (b).

non-equilibrium states produced by laser irradiation may help to understand better important non-equilibrium physical processes. For magnetic recording in particular magnetic reorientation transitions due to $T_{el}(t)$ and domain dynamics is important [18].

I have to thank many people for help, in particular P. Stampfli, R. Knorren, H. Jeschke, Prof M.Garcia, Prof. E. Matthias, Dr. D. Manske and C. Bennemann, in particular for technical assistance.

References

1. M. Aeschlimann, M. Bauer, S. Pawlick, W. Weber, R. Burgemeister, D. Oberli and H.C. Siegmann, Phys. Rev. Lett. **79**, 5158 (1997); E. Knoesel, A. Hotzel and M. Wolf, Phys. Rev. B **57**, 12812 (1998).
2. R. Knorren, K.H. Bennemann, R. Burgemeister, M. Aeschlimann, Phys. Rev. B, April 2000.

3. J. Hohlfeld, E. Matthias, R. Knorren, K.H. Bennemann, Phys. Rev. Lett. **78**, 4861 (1997); R. Knorren, K.H. Bennemann, Appl. Phys. B **68**, 501 (1999).
4. R.A. Kaindl, M. Woerner, T. Elsaesser et al., Science, **287**, 470 (2000).
5. H. Jeschke, FU-Berlin thesis (2000); J. Jeschke, M.E. Garcia, K.H. Bennemann, Appl. Phys. A **69**, 49 (1999).
6. W. Hiibner and K.H. Bennemann, Phys. Rev. B **53**, 1 (1996); H.J. Siegmann et al., private communication; see also D.R. Penn, S.P. Apell, and S.M. Girvin, Phys. Rev. B **32**, 7753 (1985); M. Aeschlimann, M. Bauer, S. Pawlik, W. Weber, R. Burgermeister, D. Oberli, and H.C. Siegmann, Phys. Rev. Lett. **79**, 5158 (1997).
7. R. Knorren and K.H. Bennemann, Appl. Phys. B **68**, 401 (1999); Proc. Stockholm, Conf. Magnetism, August (1999), (ed. K.V. Rao); R. Knorren, FU-Berlin thesis (2000).
8. H.C. Siegmann, M. Aeschlimann, private communication.
9. H. Jeschke, M. Garcia, K.H. Bennemann, see thesis H. Jeschke (FU-Berlin) (2000).
10. During crystal growth electron-hole excitations may play an important role, see also H. Jeschke, FU-Berlin thesis (2000).
11. D. Manske, K.H. Bennemann, Low. Temp. Conference, February (2000), Houston
12. R. Knorren, G. Bouzerar, K.H. Bennemann, Phys. Rev. B. **63**, 125122 (2001).
13. T.A. Luce and K.H. Bennemann, in Nonlinear Optics in Metals (Clarendon Press, Oxford 1998), p. 437.
14. J. Hohlfeld, R. Knorren, J. Güdde, E. Matthias, private communication (1999).
15. U.Konrad , J. Güdde, V. Janke, E. Matthias, Appl. Phys. B **68**, 51 (1999).
16. D. Manske, T. Dahm, K.H. Bennemann, cond-mat/9912062.
17. H. Jeschke, M. Garcia, K.H. Bennemann, see thesis H. Jeschke (FU-Berlin)(2000).
18. Clearly, SHG and pump probe spectroscopy can be used to study such dynamics, one estimates times of the order of 100 ps.

19

FUNCTIONAL RENORMALIZATION GROUP APPROACH TO NON-EQUILIBRIUM PROPERTIES OF MESOSCOPIC SYSTEMS

T. PRUSCHKE, R. GEZZI, AND A. DIRKS

Institute for Theoretical Physics, University of Göttingen,
Friedrich-Hund-Platz 1, D-37077 Göttingen, Germany.
pruschke@theorie.physik.uni-goettingen.de

Abstract. We present an extension of the concepts of the functional renormalization group approach to quantum many-body problems in non-equilibrium situations. The approach is completely general and allows calculations for both stationary and time-dependent situations. As a specific example we study the stationary state transport through a quantum dot with local Coulomb correlations. We discuss the influence of finite bias voltage as well as magnetic field and temperature on the current and conductance. For finite bias and magnetic fields we compare our results to recent experimental observations on a quantum dot in an external magnetic field.

Keywords: non-equilibrium properties, mesoscopic systems, quantum dot, local Coulomb correlations

19.1 Introduction

The investigation of transport through mesoscopic systems has developed into a very active research field in condensed matter during the past decade due to their possible relevance for next-generation electronic devices and quantum computing [1,2]. The advance in preparation and nano-structuring of layered semiconductors [3] or the handling of molecules respectively nano-tubes has led to an increasing amount of experimental knowledge about such systems [4].

The simplest realization of a mesoscopic system is the quantum dot [3,5,6]. It can be viewed as artificial atom coupled to an external bath, whose properties can be precisely manipulated over a wide range [3]. The transport properties of quantum dots in the linear response regime are very well understood from the experimental as well as theoretical point of view [3]. On the other hand, a reliable theoretical description of even the stationary transport in non-equilibrium is still a considerable challenge.

J. Bonča, S. Kruchinin (eds.), *Electron Transport in Nanosystems.*
© Springer Science + Business Media B.V. 2008

The reason is that a reliable calculation of physical properties of interacting quantum mechanical many-particle systems presents a formidable task. Typically, one has to cope with the interplay of different energy-scales possibly covering several orders of magnitude even for simple situations. Approximate tools like perturbation theory, but even numerically exact techniques can usually handle only a restricted window of energy scales and are furthermore limited in their applicability by the approximations involved or the computational resources available. In addition, due to the divergence of certain classes of Feynman diagrams, some of the interesting many-particle problems cannot be tackled by straight forward perturbation theory.

The situation becomes even more involved if one is interested in properties off equilibrium, in particular time-dependent situations. A standard approach for such cases is based on the Keldysh formalism [7–9] for the time evolution of Green functions, resulting in a matrix structure of propagators and self-energies. This structure is a direct consequence of the fact that in non-equilibrium we have to calculate averages of operators taken not with respect to the ground state but with respect to an arbitrary state. Therefore, the Gell-Mann and Low theorem [10] is not valid any more. Other approaches attempt to treat the time evolution of the non-equilibrium system numerically, for example using the density matrix renormalization group [11], the numerical renormalization group (NRG) [12,13] or the Hamiltonian based flow-equation method [14].

The Keldysh technique shows a big flexibility and one therefore can find a wide range of applications such as transport through atomic, molecular and nano devices under various conditions [15], systems of atoms interacting with a radiation field in contact with a bath [16,17], or electron-electron interaction in a weakly ionized plasma [18].

One powerful concept to study interacting many-particle systems is the rather general idea of the renormalization group [19] (RG), which has already been applied to time-dependent and stationary non-equilibrium situations recently [20–24]. In the RG approach one usually starts from high energy scales, leaving out possible infrared divergences and works ones way down to the desired low-energy region in a systematic way. However, the precise definition of "systematic way" does in general depend on the problem studied.

In order to resolve this ambiguity for interacting quantum mechanical many-particle systems in equilibrium, two different schemes attempting a unique, problem independent prescription have emerged during the past decade. One is Wegner's Hamiltonian based flow-equation technique [25, 26], the second a field theoretical approach, which we want to focus on in the following. This approach is based on a functional representation of the partition function of the system and has become known as functional renormalization group (fRG) [27–30].

A detailed description of the various possible implementations of the fRG and its previous applications in equilibrium can be found e.g. in Refs. [31, 32]. In the present contribution we introduce an extension of the fRG to

non-equilibrium [33]. Within a diagrammatic approach a similar set of fRG
flow equations has already been derived by Jakobs and Schoeller and applied
to study nonlinear transport through one-dimensional correlated electron
systems [34, 35]. We believe that this method will enable us to treat a variety of non-equilibrium problems within a scheme which is well established
in equilibrium and in contrast to other approaches is comparatively modest
with respect to the computer resources required. Our framework for non-
equilibrium will turn out to be sufficiently general to allow for a treatment of
systems disturbed by arbitrary external fields (bias voltage, laser field, etc.),
which can be constant or time-dependent.

We apply the formalism to the single impurity Anderson model (SIAM)
[36]. This model represents the paradigm for correlation effects in condensed
matter physics and is at the heart of a large range of experimental [3,5,37–41]
and theoretical investigations [42,43]. It is furthermore the standard model for
the description of the transport properties of interacting single-level quantum
dots. Therefore, we study stationary transport for this model in the presence
of a finite bias voltage V_B and under the influence of an external magnetic
field B at both $T = 0$ and finite T.

Quantum dots in external magnetic field have been the subject of interest
for some time and experimental studies of these systems have been performed
by several groups [44–47]. From a theoretical point of view, non-equilibrium
properties in magnetic field were investigated by Meir and Wingreen [15]
combining different methods such as non-crossing approximation (NCA),
equations of motion (EOM) and variational wavelength approach in the limit
in which the Coulomb repulsion U is very large. Rosch et al. [24] used a perturbative renormalization group, which permits the description of the transport
properties in single-level quantum dots for large bias and large magnetic fields.
König et al. [48] studied tunneling through a single-level quantum dot in
the presence of strong Coulomb repulsion beyond the perturbative regime
by means of a real-time diagrammatic formulation. However, a theory that
allows to access intermediate coupling, bias voltage, small to intermediate
magnetic field strengths and finite temperatures on a unique footing is missing
so far.

The paper is organized as follows: In the next section we introduce the
single impurity Anderson model which we use to study transport through a
single-level quantum dot. Based on the derivation in Ref. [33] we will provide
an expression for the non-equilibrium fRG equations used to calculate the
Keldysh components of the self-energy and the transport parameters following from them. In Sect. 19.3 we present our results for the stationary transport
properties. The limit $V_B \to 0$ is investigated first to make contact to previous
work [49] and find the regime, where our method is applicable. We then discuss how the transport parameters current J and conductance G behave as
functions of temperature and applied bias-voltage with and without magnetic
field. A summary and conclusions will finish the paper.

19.2 Method and model

19.2.1 KELDYSH APPROACH TO NON-EQUILIBRIUM

The standard many-body theory for situations away from thermodynamic equilibrium has been introduced by Keldysh [7–9]. One starts from the usual definition of the two-time Green function [50]

$$G(\xi', \xi) = -i \langle S^{-1} \psi_I(\xi') \psi_I^\dagger(\xi) S \rangle \ , \tag{19.1}$$

in the interaction picture, where ξ comprises a set of single-particle quantum numbers and time t. The time evolution operators are given as

$$S = \mathrm{T} \exp \left\{ -i \int\limits_{-\infty}^{\infty} V_I(t) dt \right\} \tag{19.2a}$$

$$S^{-1} = \tilde{\mathrm{T}} \exp \left\{ i \int\limits_{-\infty}^{\infty} V_I(t) dt \right\} \ , \tag{19.2b}$$

with T the usual time ordering and $\tilde{\mathrm{T}}$ the anti time-ordering operator. The interaction term $V_I(t)$ is arbitrary, including possible explicit time dependence. In a non-equilibrium situation, the propagation of the system from $-\infty \to \infty$ is not any more equivalent to the propagation from $\infty \to -\infty$, i.e. one has to distinguish whether the time arguments in Eq. (19.1) belong to the former or latter [50–52]. This scheme is usually depicted by the Keldysh double-time contour shown in Fig. 19.1, where C_{K_-} represents the propagation $-\infty \to \infty$ (upper branch of the Keldysh contour) and C_{K_+} the propagation $\infty \to -\infty$ (lower branch of the Keldysh contour). Consequently, one has to introduce four distinct propagators, namely the time-ordered, anti time-ordered and mixed Green function

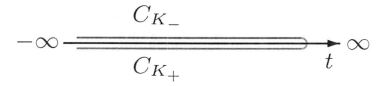

Fig. 19.1. (color online) Keldysh contour.

$$G^{--}(\xi',\xi) = -i\langle T\psi(\xi')\psi^{\dagger}(\xi)\rangle$$
$$= -i\theta(t'-t)\langle\psi(\xi')\psi^{\dagger}(\xi)\rangle - \zeta i\theta(t-t')\langle\psi^{\dagger}(\xi)\psi(\xi')\rangle\,,\text{(19.3a)}$$
$$t\,,\,t' \in C_{K_-}$$

$$G^{++}(\xi',\xi) = -i\langle\tilde{T}\psi(\xi')\psi^{\dagger}(\xi)\rangle$$
$$= -i\theta(t-t')\langle\psi(\xi')\psi^{\dagger}(\xi)\rangle - \zeta i\theta(t'-t)\langle\psi^{\dagger}(\xi)\psi(\xi')\rangle\,,\text{(19.3b)}$$
$$t\,,\,t' \in C_{K_+}$$

$$G^{+-}(\xi',\xi) = -i\langle\psi(\xi')\psi^{\dagger}(\xi)\rangle\,,\,t' \in C_{K_+}, t \in C_{K_-} \tag{19.3c}$$

$$G^{-+}(\xi',\xi) = -\zeta i\langle\psi^{\dagger}(\xi)\psi(\xi')\rangle\,,\,t' \in C_{K_-}, t \in C_{K_+}\,. \tag{19.3d}$$

The "statistical indicator" $\zeta = +1$ for bosons, while $\zeta = -1$ for fermions. Concerning the Keldysh indexes we here follow the notation of Ref. [50]. $G^{--}(\xi',\xi)$ and $G^{++}(\xi',\xi)$ take into account the excitation spectrum while $G^{+-}(\xi',\xi)$ and $G^{-+}(\xi',\xi)$ describe the thermodynamic state of the system. Only three of the Green functions are independent and one commonly introduces the linear combinations

$$G^{R}(\xi',\xi) = \theta(t'-t)\left[G^{+-}(\xi',\xi) + \zeta G^{-+}(\xi',\xi)\right] \tag{19.4a}$$

$$G^{A}(\xi',\xi) = \theta(t-t')\left[G^{-+}(\xi',\xi) + \zeta G^{+-}(\xi',\xi)\right] \tag{19.4b}$$

$$G^{K}(\xi',\xi) = G^{-+}(\xi',\xi) + G^{+-}(\xi',\xi)\,, \tag{19.4c}$$

named retarded, advanced and Keldysh component, respectively.

In a general disequilibrium situation the Green functions will depend on the two times t and t' separately, and not on their difference only as in equilibrium. Hence the structure of a perturbation expansion becomes quite tedious. The notable exception is the case when one is not interested in the transient behavior but only on the long-time limit. Commonly one assumes that this long time limit is described by a time-independent situation again, which one calls stationary state. In such a stationary state the above Green functions again only depend on $t - t'$. In this case, one can employ a standard Fourier transform and work in frequency instead in time space. Note that there does not exist a general proof that for a particular model such a (equilibrium) stationary state must at all exists.

Without going into detail here, we note that, in contrast to the imaginary-time Matsubara approach, in the Keldysh theory dynamics and thermodynamics enter the formalism differently [50, 53]. While at first sight this may rather be viewed as an additional complication, the results presented in Sect. 19.3.2 show that within the fRG discussed next this feature turns out to be extremely helpful for calculating transport coefficients at finite temperatures.

19.2.2 FUNCTIONAL RENORMALIZATION GROUP EQUATIONS

The detailed derivation of the fRG flow equations in a general non-equilibrium situation has been extensively discussed in Ref. [33, 54]. One starts from a functional integral representation of a generalized partition function [55], where all quantities are suitable tensor objects in the indexes defined by the Keldysh contour. Introducing source fields into the action, one can go through the standard procedures of many-body theory [56] to obtain quantities like m-particle Green functions $G_m(\ldots)$ or m-particle scattering vertexes $\gamma_m(\ldots)$ via functional derivatives of the generalized partition function [33,55]. By introducing an additional "cutoff" parameter Λ into the free part of the action, all quantities formally become Λ-dependent, i.e. $G_m(\ldots) \rightarrow G_m^\Lambda(\ldots)$ and $\gamma_m(\ldots) \rightarrow \gamma_m^\Lambda(\ldots)$. Differentiating the expressions with respect to Λ and collecting equal orders in an expansion in the source fields one obtains an infinite hierarchy of differential equations for the irreducible m-particle vertexes [31–33]. Note that no reference has been made up to now on how the parameter Λ is actually introduced. Furthermore, no properties of the Green function enter at any point, i.e. the scheme is completely general and encompasses fermionic, bosonic and mixed systems as well as time dependent or stationary situations.

Obviously, an infinite set of differential equations cannot be solved in general, thus a truncation is necessary for actual calculations. Typically, this truncation is realized by setting the three-particle vertex $\gamma_3^\Lambda \equiv 0$ [30,32].

19.2.3 SINGLE IMPURITY ANDERSON MODEL

We consider in the following a quantum dot consisting of a single level. Such a system can be described schematically as in Fig. 19.2. The quantum dot is depicted by the central region of the figure and the left (L) and right (R) leads are modeled by a continuum of noninteracting fermionic states described by a dispersion $\varepsilon_{k\alpha}$, where k denotes the wave vector and $\alpha = L, R$. In equilibrium, both leads have a common chemical potential which we denote by μ. The energy of the quantum dot level relative to the Fermi energy μ of the leads is controlled by a gate voltage V_G and denoted as $\varepsilon_\sigma = -eV_G - g\mu_B B\sigma/2$ with spin quantum number $\sigma = \pm 1$. The quantity B is the external magnetic field and e the elementary charge. Since the magnetic fields applied are much smaller than the Fermi energy of the leads, its effect on the electrons in the leads can be ignored for the present purpose.

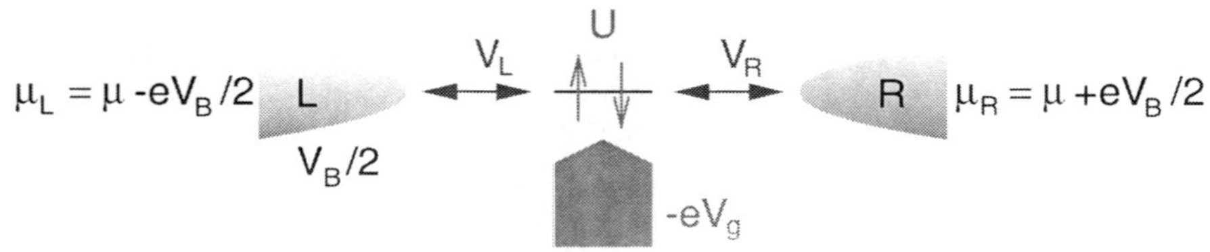

Fig. 19.2. Sketch of a single level quantum dot. The electron filling in the left and right leads can be controlled through the chemical potential. If $\mu_L \neq \mu_R$ a current can flow and the system is in a non-equilibrium situation.

The dot and the leads are coupled through an energy and spin independent hybridization V_α. Finally, if two electrons occupy the dot, they experience a Coulomb repulsion U. This situation is described by the long-known single impurity Anderson model, given by the Hamiltonian [36]

$$H = \sum_{k\sigma\alpha} \varepsilon_{k\sigma\alpha} c^\dagger_{k\sigma\alpha} c_{k\sigma\alpha} + \sum_\sigma \varepsilon_\sigma d^\dagger_\sigma d_\sigma + U \left(n_\uparrow - \frac{1}{2} \right) \left(n_\downarrow - \frac{1}{2} \right)$$
$$+ \frac{1}{\sqrt{N}} \sum_{k\sigma\alpha} \left[V_\alpha c^\dagger_{k\sigma\alpha} d_\sigma + h.c. \right] \tag{19.5}$$

in standard notation.

To study transport properties, the system can be subject to a finite bias voltage V_B, which enters through a modification of the chemical potential of the leads, resulting in an unequal occupancy and thus a current flowing. We model such a situation by introducing a symmetric shift in the chemical potentials of the left and right lead, μ_L and μ_R, such that $\mu_L - \mu_R = -eV_B$ (see Fig. 19.2).

19.2.4 EXAMPLE: STATIONARY STATE GREEN FUNCTION OF THE SIAM FOR $U = 0$

Let us give an instructive example, which will also be useful later on, namely the Keldysh components for the single-particle Green function for the model (19.5) with interaction $U = 0$. We can concentrate in particular on the Green functions for the local or d-states, as the others follow from them via a simple sequence of equations-of-motion. Furthermore, we will present results for the stationary state only, because we will focus on that situation later anyway. The full time dependence can be calculated, too, but the resulting expressions are pretty lengthy [57]. A further standard assumption we will make is that the electronic properties of left and right lead are identical and that the hybridization obeys $V_L = V_R = V/\sqrt{2}$.

In that case the coupling between dot and leads is characterized by the quantities

$$\Gamma_\alpha = \frac{\pi|V|^2 N_F}{2} \equiv \frac{\Gamma}{2} \ , \tag{19.6}$$

with N_F the local density of states of the leads at the dot site. The result for the Keldysh components of the Green function then reads [33]

$$G^{--}_{d\sigma,0}(\omega) = \frac{\omega - \varepsilon_\sigma - i\Gamma\left[1 - f_L(\omega) - f_R(\omega)\right]}{(\omega - \varepsilon_\sigma)^2 + \Gamma^2} \ , \tag{19.7a}$$

$$G^{++}_{d\sigma,0}(\omega) = -[G^{--}_{d\sigma,0}(\omega)]^* \ , \tag{19.7b}$$

$$G^{-+}_{d\sigma,0}(\omega) = i\frac{\Gamma\left[f_L(\omega) + f_R(\omega)\right]}{(\omega - \varepsilon_\sigma)^2 + \Gamma^2} \ , \tag{19.7c}$$

$$G^{+-}_{d\sigma,0}(\omega) = -i\frac{\Gamma\left[f_L(-\omega) + f_R(-\omega)\right]}{(\omega - \varepsilon_\sigma)^2 + \Gamma^2} \ , \tag{19.7d}$$

where $f_\alpha(\pm\omega) := f(\pm(\omega - \mu_\alpha))$ are the Fermi functions of the leads, and we realize the stationary non-equilibrium situation by introducing different chemical potentials μ_α for the left and right reservoirs through an applied bias voltage $V_B = (\mu_L - \mu_R)/(-e)$.

An important detail can be read off Eqs. (19.7a–d). The temperature enters only as a parameter via the Fermi function of the leads. As we will see later, this feature will enable us to calculate transport coefficients at finite temperature straightforwardly with the same computational effort.

19.2.5 DIFFERENTIAL EQUATIONS FOR THE SIAM

In the following we will restrict the discussion to the SIAM (19.5) in a steady-state equilibrium situation. In this case translational invariance in time is restored and one can make use of energy conservation and Fourier transform to frequency or energy space. As a result, the two-particle scattering vertex γ_2^Λ has four indexes for each spin quantum number and Keldysh index and depends on three frequency arguments. The hierarchy of differential equations is truncated at the three-particle level, i.e. we set $\gamma_3^\Lambda \equiv 0$. In addition to this truncation, we further neglect the energy dependence of the two-particle scattering vertex $\gamma_2^\Lambda \equiv \gamma^\Lambda$. The latter approximation is, strictly speaking, not mandatory for the SIAM (see e.g. the discussion in Ref. [32]). However, as we will see further below, the complexity of the remaining set of coupled differential equations is rather high and the neglect of the energy dependence reduces the computational effort necessary to integrate the system quite dramatically. Moreover, this approximation has turned out to be rather successful in its application to the calculation of transport coefficients in the linear response regime using the imaginary-time fRG [49].

Last but not least, we have to specify how the cutoff Λ is introduced. As usual [32] we choose a Θ-cutoff on the level of the non-interacting dot here, i.e.

$$G_{d\sigma,0}^{\alpha\beta,\Lambda}(\omega) := \Theta(\Lambda - |\omega|)G_{d\sigma,0}^{\alpha\beta}$$

with $G_{d\sigma,0}^{\alpha\beta}(\omega)$ the Green function matrix of the dot for $U = 0$ given by Eqs. (19.7a–d).

With these definitions and approximations, the resulting system of differential equations for the spin-dependent single particle self-energy ($\hat{=}$ one-particle scattering vertex) and two-particle scattering vertex becomes (for details see e.g. Appendix B in Ref. [33])

$$\frac{d}{d\Lambda}\Sigma_\sigma^{\alpha\beta,\Lambda} = -\frac{1}{2\pi}\sum_{\sigma'}\sum_{\omega=\pm\Lambda}\sum_{\mu\nu}\tilde{G}_{\sigma'}^{\mu\nu,\Lambda}(\omega)\gamma_{\sigma,\sigma',\sigma,\sigma'}^{\alpha\nu\beta\mu,\Lambda}\,,\tag{19.8a}$$

$$\frac{d}{d\Lambda}\gamma_{\sigma_1',\sigma_2';\sigma_1,\sigma_2}^{\alpha\beta\gamma\delta,\Lambda} = \frac{1}{4\pi}\sum_{\omega=\pm\Lambda}\sum_{\sigma_3,\sigma_4}\sum_{\mu,\nu,\rho,\eta}\Bigg(\tag{19.8b}$$

$$\tilde{G}_{\sigma_3}^{\rho\eta,\Lambda}(-\omega)\tilde{G}_{\sigma_4}^{\nu\mu,\Lambda}(\omega)\gamma_{\sigma_1',\sigma_2';\sigma_3,\sigma_4}^{\alpha\beta\rho\nu,\Lambda}\gamma_{\sigma_3,\sigma_4;\sigma_1,\sigma_2}^{\eta\mu\gamma\delta,\Lambda}-$$

$$\tilde{G}_{\sigma_3}^{\eta\rho,\Lambda}(\omega)\tilde{G}_{\sigma_4}^{\nu\mu,\Lambda}(\omega)\Big[\gamma_{\sigma_1',\sigma_4;\sigma_1,\sigma_3}^{\alpha\mu\gamma\eta,\Lambda}\gamma_{\sigma_3,\sigma_2';\sigma_4,\sigma_2}^{\rho\beta\nu\delta,\Lambda}+$$

$$\gamma_{\sigma_1',\sigma_3;\sigma_1,\sigma_4}^{\alpha\rho\gamma\nu,\Lambda}\gamma_{\sigma_4,\sigma_2';\sigma_3,\sigma_2}^{\mu\beta\eta\delta,\Lambda}-$$

$$\gamma_{\sigma_2',\sigma_4;\sigma_1,\sigma_3}^{\beta\mu\gamma\eta,\Lambda}\gamma_{\sigma_3,\sigma_1';\sigma_4,\sigma_2}^{\rho\alpha\nu\delta,\Lambda}-$$

$$\gamma_{\sigma_2',\sigma_3;\sigma_1,\sigma_4}^{\beta\rho\gamma\nu,\Lambda}\gamma_{\sigma_4,\sigma_1';\sigma_3,\sigma_2}^{\mu\alpha\eta\delta,\Lambda}\Big]\Bigg)\,.$$

In expressions (19.8)

$$\tilde{G}_\sigma^{\mu\nu,\Lambda}(\omega) = \left[\frac{1}{\hat{G}_{d\sigma,0}(\omega)^{-1}-\hat{\Sigma}_\sigma^\Lambda}\right]^{\mu\nu}\,,$$

and the upper small greek indexes refer to the branches of the Keldysh contour. The quantities $\hat{G}_{d\sigma,0}$ and $\hat{\Sigma}_\sigma^\lambda$ denote the 2×2 matrices built from $G_{d\sigma,0}^{\alpha\beta}$ and $\Sigma_\sigma^{\alpha\beta}$, respectively.

The initial conditions at $\Lambda=\infty$ are are given by

$$\Sigma_\sigma^{\Lambda=\infty} = 0$$

and

$$\gamma_{\sigma_1',\sigma_2';\sigma_1,\sigma_2}^{\alpha\alpha\alpha\alpha,\Lambda=\infty} = i\alpha U(\delta_{\sigma_1,\sigma_1'}\delta_{\sigma_2,\sigma_2'}-\delta_{\sigma_1,\sigma_2'}\delta_{\sigma_2,\sigma_1'})\,.$$

All other components of $\gamma^{\Lambda=\infty}=0$.

The integration of Eqs. (19.8) can be done with a standard Runge-Kutta solver, starting at sufficiently large Λ and integrating down to $\Lambda=0$. We must note that there exist certain combination of parameters, where during the integration of the differential equations a branching of solutions may occur around $\Lambda\approx 1$ and the Runge-Kutta solver will pick the wrong branch. This is a purely numerical artifact and not a problem of the fRG itself, as it can be shown for analytically solvable cases that the exact solution at $\Lambda=0$ is the correct one [33].

Some more thorough remarks are in order concerning the neglect of the energy dependence in the vertex γ_2^Λ and hence the self-energy. As has been discussed extensively in Ref. [33], this approximation leads to violations of causality and charge conservation on the dot for finite bias voltage V_B. However, these artifacts of the approximation are controlled in the sense that they evolve smoothly with increasing Coulomb parameter U and bias voltage V_B

proportional to U^2 and V_B^2 and remain small well into the intermediate coupling regime $U \gtrsim \pi\Gamma$ [33]. Moreover, the violation of charge conservation is only present for $V_G \neq 0$.

19.3 Results

19.3.1 AN EXPRESSION FOR THE CURRENT

In this section we intend to present results for current and differential conductance in the weak and intermediate coupling regime of the SIAM. The explicit formula

$$J = \frac{ie\Gamma}{4\pi\hbar} \sum_\sigma \int d\epsilon \, [f_L(\epsilon) - f_R(\epsilon)] \left(\tilde{G}_{d\sigma}^{+-}(\epsilon) - \tilde{G}_{d\sigma}^{-+}(\epsilon) \right) \tag{19.9}$$

for the total current of the model (19.5) can be found in Refs. [15,58]. For an energy-independent self-energy and $T = 0$ it can be evaluated further with the result

$$J = \frac{e\Gamma}{2\pi\hbar} \sum_\sigma \frac{\tilde{\Gamma}_\sigma}{\Gamma_\sigma^*} \sum_{s=\pm 1} s \arctan \left(\frac{eV_{G,\sigma}^* + s\frac{eV_B}{2}}{\Gamma_\sigma^*} \right) \tag{19.10}$$

where we introduced the abbreviations

$$eV_{G,\sigma}^* = eV_G + \Re e\Sigma_\sigma^{--} \,, \tag{19.11a}$$

$$\tilde{\Gamma}_\sigma = \Gamma - \Im m\Sigma_\sigma^{-+} > \Gamma \tag{19.11b}$$

$$\Gamma_\sigma^* = \sqrt{\tilde{\Gamma}^2 + (\Im m\Sigma_\sigma^{--})^2} \,. \tag{19.11c}$$

This expression nicely demonstrates how the different Keldysh components of the self-energy modify the physical properties: The diagonal components lead to a renormalization of the local energy level position and the level-width, while the off-diagonal components modify the tunneling rate given by Γ in the bare system.

In the following we use the quantity $J_0 = e\Gamma/(2\pi\hbar)$ as unit for the current and $G_0 = 2e^2/h$ as unit for the differential conductance

$$G = \frac{dJ}{dV_B} \,.$$

The latter has to be determined numerically, as the effective gate voltage and scattering rate defined in (19.11) both depend on V_B. An example, which we will discuss in detail in Sect. 19.3.3 for the current calculated with the fRG using expression (19.9) is given in Fig. 19.3. The calculations were done for three different values of $U/\Gamma = 1$, $U/\Gamma = 6$ and $U/\Gamma = 15$, i.e. the weak, intermediate and strong coupling regime of the SIAM, as function of V_B at $T = 0$. The gate voltage was chosen as $V_G = 0$.

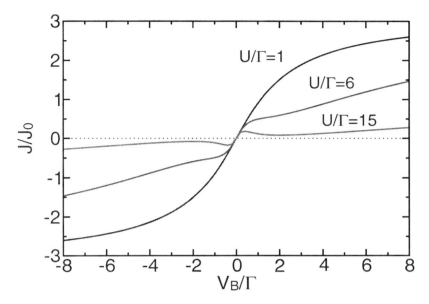

Fig. 19.3. Current calculated with fRG for $T = 0$ and $V_G = 0$ versus bias voltage V_B for different values of the Coulomb repulsion U.

19.3.2 CONDUCTANCE IN THE LINEAR RESPONSE REGIME

Let us start to discuss our results by presenting a comparison of the conductance in the linear response regime $V_B \to 0$ calculated with fRG for $U/\Gamma = 5$ for various temperatures with and without magnetic field to data obtained from numerical renormalization group (NRG) [19,59,60] calculations. In both cases, viz for zero magnetic field (Fig. 19.4) and an intermediate magnetic field $B/\Gamma = 0.58$ (Fig. 19.5) we find rather good agreement for the conductance as function of V_G between both approaches. Note that at high temperatures the NRG tends to wash out structures, so that the increasing discrepancy is hard to pinpoint to the fRG.

This good agreement between fRG and other, more precise methods, has been noted before by Karrasch et al. [49] using the imaginary-time formulation of the fRG. It must be emphasized, though, that the agreement in their case is restricted to $T = 0$. At finite T, the calculations within imaginary-time fRG are not able to reproduce the temperature dependence of the structures in the conductance. The reason why this is apparently possible by the Keldysh-fRG lies in the fact that dynamics and thermodynamics are decoupled. Note, however, that dynamical quantities like the density of states are *not at all* reproduced by either fRG version. To achieve this goal one has to also take into account the full energy dependence of the two-particle vertex [32]. Note also that for finite V_G the results agree with NRG as well. This shows that the fRG,

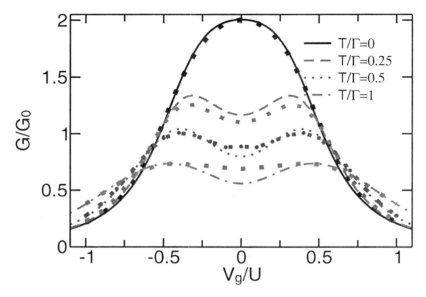

Fig. 19.4. Conductance versus V_G for different temperatures T and $U/\Gamma = 5$. Dots are results computed with NRG.

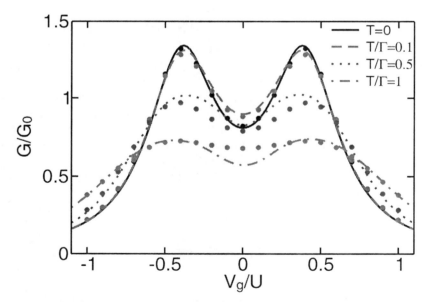

Fig. 19.5. Conductance versus V_G for different temperatures T and $U/\Gamma = 5$ at finite magnetic field $g\mu_B B/\Gamma = 0.58$. Dots are results computed with NRG.

even with the additional approximations introduced here,is a much better approximation than for example straight low-order perturbation theory, as the latter notoriously fails for situation away from particle-hole symmetry [59].

19.3.3 FINITE BIAS FOR $T = 0$ AND $B = 0$

Keeping $V_G = 0$ fixed and varying V_B for $T = 0$ results in the current profiles already depicted in Fig. 19.3. As expected one finds a linear regime as $V_B \rightarrow 0$ with a common slope due to the fact that at $T = 0$ the conductance at $V_G = 0$ is always G_0. As the Coulomb interaction U increases, the extent of this linear regime decreases and a shoulder appears, which turns into an extremum for very large U. The occurrence of such a shoulder in an intermediate bis regime has been observed before [61–63]. It will lead to a strong suppression of the differential conductance as one can see from Fig. 19.6, which for large U turns into a regime o negative differential conductance. The energy scale for this suppression is strongly decreasing with increasing U, and one would expect it to roughly follow the Kondo scale [63]. It is not clear whether the presence of negative differential conductance is a real feature of the model or an artifact of the approximations used here. It is quite interesting to mention that similar features were observed by other authors, too [63, 64]. On the other hand, preliminary fRG calculations using the full energy-dependence of the vertex lead to a significant softening of the shoulder. Further work along that line is in progress.

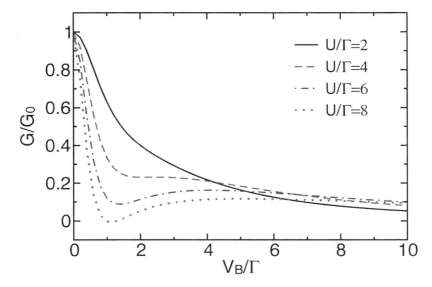

Fig. 19.6. Differential conductance versus bias voltage V_B for $T = 0$, $V_G = 0$ and several values of U. Due to the symmetry $V_B \rightarrow -V_B$ only the part with $V_B \geq 0$ is shown.

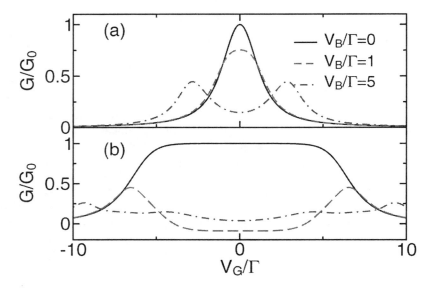

Fig. 19.7. Conductance versus V_G for different V_B and (a) $U/\Gamma = 1$ and (b) $U/\Gamma = 15$.

The general features observed here prevail also for finite gate voltage V_G. The conductance as function of V_G for two values of U and different values of V_B is collected in Fig. 19.7. As $V_B \rightarrow 0$, one observes for large U (Fig. 19.7b) the well-known Kondo plateau, which is due to the pinning of the scattering resonance to the Fermi energy by the Kondo effect [59]. Increasing V_B, small and large U behave distinctively different, too. While in the former case the conductance smoothly decreases and peaks at $\approx \pm V_B$ start to form, the conductance in the latter case becomes strongly suppressed for V_G within the Kondo plateau even for small bias and peaks form at $\pm U/2$ rather than V_B. Only with strongly increased V_B additional structures at $\approx \pm V_B$ form, but the conductance already has been washed out quite dramatically in that regime.

Again we would like to note that, although the occurrence of a negative differential conductance is possibly an artifact of the method, the strong suppression around $V_G = 0$ is consistent with experiment and thus should be a real feature. The mere possibility to calculate $G(V_G, V_B, \ldots)$ with the fRG in the intermediate coupling regime of the SIAM already must be viewed as substantial progress, because the conventionally used perturbational treatments of the model (19.5) typically fail for $V_G \neq 0$ [59].

19.3.4 FINITE MAGNETIC FIELD AT $T = 0$

In a recent publication, Quay et al. investigated experimentally the behavior of the nonlinear conductance of a spin 1/2 quantum dot [65]. They observe, for a gate voltage chosen to be in the middle of the Kondo mesa, a splitting of the central resonance as function of V_B. Analyzing this splitting they observe that the splitting δ of the peak positions at large magnetic field follows the naive expectation $\delta = 2\mu_B B$, but drops significantly below this for small magnetic fields, the low-field limit seemingly consistent with predictions by Logan and Dickens [66].

As the fRG to our knowledge currently is the only theory that allows to calculate transport coefficient for finite V_B and small to intermediate magnetic fields in the intermediate coupling regime of the SIAM, we present in Fig. 19.8 results for the development of the conductance at $V_G = 0$ as function of the bias voltage V_B and magnetic field B. The value $U/\Gamma = 5$ for the Coulomb interaction was chosen because for larger U the numerical instability briefly discussed in the end of Sect. 19.2.5 does only permit calculations for small respectively large magnetic field here. Nevertheless, the structures found are qualitatively very similar to those seen by Quay et al. (see Fig. 19.2d in Ref. [65]).

A more direct comparison can be done by looking at the peak positions in Fig. 19.8 as function of magnetic field. While it is straightforward to define the peak position for large field, a certain ambiguity remains for small fields. Here, we used the superposition of two Lorentians to fit the data at low bias. Several initial guesses were used to obtain an estimate of the error bars. The results of this procedure are collected in Fig. 19.9. For the lowest field

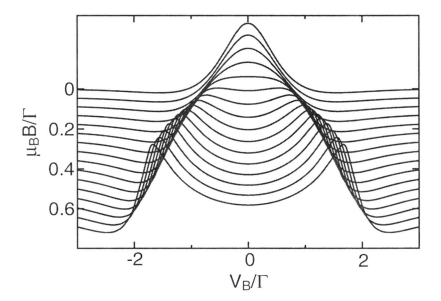

Fig. 19.8. Conductance versus V_B for different magnetic fields B and $U/\Gamma = 5$ at $T = 0$.

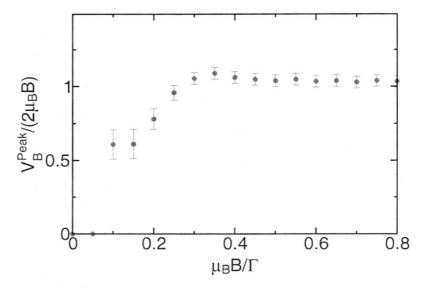

Fig. 19.9. Peak positions as function of B for the data from Fig. 19.8.

values a reasonable fit was not possible. The results show, that indeed for small fields a strong reduction from the Zeeman value $\delta_Z = 2\mu_B B$ to roughly $2/3\delta_Z$ is found, consistent with experiment and predictions from equilibrium theory [65, 66], while for large fields the Zeeman splitting is recovered. Note that the reduction for small fields is by no means trivial and an subtle interplay of the quasi-particle renormalization and spin-susceptibility enhancement [66]. That the fRG can reproduce this reduction shows that, although from its structure it may look like being similar to second order perturbation theory, it includes diagrams that are necessary to produce these type of strong coupling physics [49]. It is furthermore far from obvious that this interplay also holds out of equilibrium. Our results thus are the first to our knowledge that show this behavior for finite bias.

19.3.5 FINITE TEMPERATURE

Last but not least one can quite obviously also study the effects of a bias voltage and magnetic field at finite temperatures. A representative set of calculations for $U/\Gamma = 2\pi$ and some combination of V_B and magnetic field B are collected in Fig. 19.10. For $V_B = B = 0$, one starts to see a logarithmic dependence on T of G in the region above $T/\Gamma \approx 0.1$. Finite V_B of B suppress G at $T \to 0$, as expected. Increasing T slowly increases G again, and when either $k_B T \approx \mu_B B$ or $k_B T \approx eV_B$, the temperature becomes the relevant energy scale and the influence of a magnetic field or non-equilibrium effects disappear, i.e. the conductance behaves like the one for $V_B = B = 0$

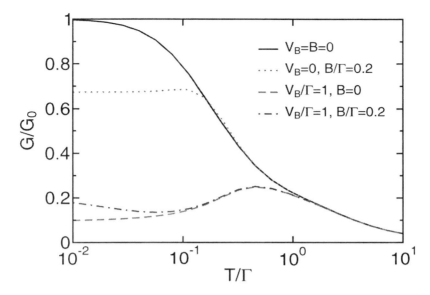

Fig. 19.10. Conductance as function of temperature for $U/\Gamma = 2\pi$ for typical values of V_B and B.

again. Similarly, for the chosen combination of finite magnetic field and bias voltage one initially observes an enhancement G at $T \to 0$ (c.f. Fig 19.8). As $k_B T \approx \mu_B B$, the effects of the magnetic field have been eliminated, the bias voltage remaining the relevant energy scale. Hence the curve follows the one for $B = 0$ and $V_B/\Gamma = 1$ from here, until the temperature takes over as the only left relevant energy scale at $k_B T \approx eV_B$.

19.4 Summary and outlook

We have presented a novel approach to study non-equilibrium properties for models of mesoscopic systems based on a generalization of the functional renormalization group concept. This approach leads to an infinite hierarchy of coupled differential equations for the m-particle scattering vertexes. The necessary truncation of this hierarchy is done at $m = 3$, i.e. we generate something akin to the usual one-loop approximation in conventional scaling approaches. However, due to the fundamentally different structure of the fRG, the resulting approximation goes well beyond the standard one-loop treatment and is in fact able to capture also non-trivial strong-coupling physics [49].

Including non-equilibrium effects in the theory enhances the complexity of the equations quite dramatically. We thus employed as further approximation the neglect of energy dependence in the flow of the two-particle vertex and hence also the one-particle self-energy. We are well aware that such an approximation may be quite severe in connection with non-equilibrium. We indeed observe several artifacts of this approximations, viz violation of causality and charge conservation of the dot. It is thus mandatory to view the results very critically.

We demonstrated the usefulness of the method by first presenting results for the differential conductance as function of temperature and magnetic field in equilibrium. These calculations can be compared to precise methods like the numerical renormalization group and the agreement was found to be very good. Note that these calculation also involved finite gate voltage. The good agreement here also shows once again, that the method is far superior to standard second order perturbation theory as this approach does not allow to do calculation away from particle-hole symmetry [59].

Our results for transport coefficients off equilibrium compare at minimum qualitatively well with known results from e.g. weak-coupling perturbation theory, but go well beyond them in the sense that we can also study finite magnetic field, gate voltage and temperature effects. The comparison of the behavior of the conductance as function of magnetic field and bias voltage allowed to make at least qualitative contact to experiment and also find a behavior in the low-field regime, the reduction of the splitting well below its Zeeman value, that is manifestly a strong coupling feature and cannot be captured to our knowledge by standard perturbational approaches.

Of course the approximations made are quite severe and the question remains how far one can trust the results quantitatively, for example the occurrence of a negative differential conductance for large Coulomb parameter U at small bias voltage. To answer this question unambiguously a reliable many-body tool for calculations off equilibrium in the strong-coupling regime is necessary. Unfortunately, such a tool does not yet exist. We thus have to rely on consistency checks within the method, for example by introducing the full energy-dependence in the vertex. Calculations along this line are currently in progress. They support the general features found by the approximate calculations while curing deficiencies like the violation of charge conservation and, at least partially, the violation of causality relations. We are not yet in a position to make definite statements about for example the existence of negative differential conductance, but do observe a weakening of the shoulders in the current at larger values of U, hinting towards a suppression of this effect.

References

1. D. Goldhaber-Gordon, I. Goldhaber-Gordon, Nature **412**, 594 (2001)
2. I. Zutić, J. Fabian, S.D. Sarma, Rev. Mod. Phys. **76**, 323 (2004)
3. L.P. Kouwenhoven, D.G. Austing, S. Tarucha, Rep. Prog. Phys. **64**(6), 701 (2001)
4. M.E. Gershenson, V. Podzorov, A.F. Morpurgo, Rev. Mod. Phys. **78**, 973 (2006)
5. M.A. Kastner, Rev. Mod. Phys. **64**(3), 849 (1992)

6. L.P. Kouwenhoven, et al., in *Mesoscopic Electron Transport*, ed. by L.L. Sohn, et al. (Dordrecht, Kluwer, 1997), p. 105
7. J. Schwinger, J. Math. Phys. **2**, 407 (1961)
8. L.P. Kadanoff, G. Baym, *Quantum Statistical Mechanics: Green's Function Methods in Equilibrium and Nonequilibrium Problems* (Addison-Wesley, New York 1962)
9. L.P. Keldysh, JETP **20**, 1018 (1965)
10. A. Fetter, J. Walecka, *Quantum Theory of Many-Particle Systems*. International Series in Pure and Applied Physics (McGraw-Hill, New York, 1971)
11. U. Schollwöck, Rev. Mod. Phys. **77**, 259 (2005)
12. T. Costi, Phys. Rev. B **55**, 3003 (1997)
13. F.B. Anders, A. Schiller, Spin precession and real time dynamics in the kondo model: A time-dependent numerical renormalization-group study (2006). Cond-mat/0604517
14. D. Lobaskin, S. Kehrein, Phys. Rev. B **71**, 193303 (2005)
15. Y. Meir, N.S. Wingreen, Phys. Rev. Lett. **68**, 2512 (1992)
16. V. Koremann, Ann. of Phys. **39**, 72 (1966)
17. D. Langreth, in *NATO advanced study institute Series B*, vol. 17, ed. by J. Devreese, E. van Doren (Plenum, New York/London, 1967), vol. 17
18. B. Altshuler, Y. Aharonov, JETP **48**, 812 (1978)
19. K.G. Wilson, Rev. Mod. Phys. **47**, 773 (1975)
20. H. Schoeller, in *Low-Dimensional Systems*, vol. 17, ed. by T. Brandes (Springer Verlag, 1999), vol. 17, p. 137
21. H. Schoeller, J. König, Phys. Rev. Lett. **84**, 3686 (2000)
22. M. Keil, H. Schoeller, Phys. Rev. B **63**, 180302 (2001)
23. A. Rosch, J. Paaske, J. Kroha, P. Wölfle, Phys. Rev. Lett. **90**, 076804 (2003)
24. A. Rosch, J. Paaske, J. Kroha, P. Wölfle, J. Phys. Soc. Jpn. **74**, 118 (2005)
25. F. Wegner, Ann. Physik (Leipzig) **3**, 77 (1994)
26. S.D. Glazek, P.B. Wiegmann, Phys. Rev. D **48**, 5863 (1993)
27. J. Polchinski, Nucl. Phys. B **231**, 269 (1984)
28. C. Wetterich, Phys. Lett. B **301**, 90 (1993)
29. T.R. Morris, Int. J. Mod. Phys. A **9**, 2411 (1994)
30. M. Salmhofer, *Renormalization* (Springer, Berlin, 1998)
31. M. Salmhofer, C. Honerkamp, Prog. Theor. Phys. **105**, 1 (2001)
32. R. Hedden, V. Meden, T. Pruschke, K. Schönhammer, J. Phys.: Condens. Matter **16**, 5279 (2004)
33. R. Gezzi, T. Pruschke, V. Meden, Phys. Rev. B **75**, 045324 (2007)
34. S. Jakobs, Renormierngsgruppen-methoden für nichtlinearen transport. Diploma thesis, RWTH Aachen (2003)
35. S.G. Jakobs, V. Meden, H. Schoeller, Phys. Rev. Lett. **99** (2007)
36. P.W. Anderson, Phys. Rev. **124**, 41 (1961)
37. J. Nygård, D.H. Cobden, P.E. Lindelof, Nature **408**, 342 (2000)
38. V. Madhavan, W. Chen, T. Jamneala, M.F. Crommie, N.S. Wingreen, Science **280**, 567 (1998)
39. J. Li, W.D. Schneider, R. Berndt, B. Delley, Phys. Rev. Lett. **80**, 2893 (1998)
40. G.A. Fiete, E.J. Heller, Rev. Mod. Phys. **75**, 933 (2003)
41. M. Pustilnik, L. Glazman, J. Phys.: Condens. Matter **16**, R513 (2004)
42. A. Georges, G. Kotliar, W. Krauth, M.J. Rozenberg, Rev. Mod. Phys. **68**, 13 (1996)

43. T. Maier, M. Jarrell, T. Pruschke, M.H. Hettler, Rev. Mod. Phys. **77**, 1027 (2005)
44. S. De Franceschi, R. Hanson, W.G. van der Wiel, J.M. Elzerman, J.J. Wijpkema, T. Fujisawa, S. Tarucha, L.P. Kouwenhoven, Phys. Rev. Lett. **89**, 156801 (2002)
45. D. Goldhaber-Gordon, H. Shtrikman, D. Mahalu, D. Abusch-Magder, U. Meirav, M.A. Kastner, Nature **391**, 156 (1998)
46. D.C. Ralph, R.A. Buhrman, Phys. Rev. Lett. **69**, 2118 (1992)
47. J. Schmid, J. Weis, K. Eberl, K. v. Klitzing, Phys. Rev. Lett. **84**, 5824 (2000)
48. J. König, J. Schmid, H. Schoeller, G. Schön, Phys. Rev. B **54**, 16820 (1996)
49. C. Karrasch, T. Enss, V. Meden, Phys. Rev. B **73**, 235337 (2006)
50. L. Landau, E. Lifshitz, *Physical Kinetics* (Akademie-Verlag, Berlin, 1983)
51. J. Rammer, H. Smith, Rev. Mod. Phys. **58**, 323 (1986)
52. H. Haug, A.P. Jauho, *Quantum Kinetics and Optics of Semiconductors* (Springer, Berlin 1996)
53. M. Wagner, Phys. Rev. B **44**(12), 6104 (1991)
54. R. Gezzi, A. Dirks, T. Pruschke, arXiv:cond-mat/0707.0289 (2007)
55. A. Kamenev, in *Les Houches, Volume Session LX*, ed. by H. Bouchiat, Y. Gefen, S. Guéron, G. Montambaux, J. Dalibard (Elsevier, North-Holland, 2004). Cond-mat/0412296
56. J. Negele, H. Orland, *Quantum Many-Particle Physics* (Addison-Wesley, New York, 1988)
57. A.P. Jauho, N.S. Wingreen, Y. Meir, Phys. Rev. B **50**, 5528 (1994)
58. N.S. Wingreen, Y. Meir, Phys. Rev. B **49**, 11040 (1994)
59. A.C. Hewson, *The Kondo Problem to Heavy Fermions*. Cambridge Studies in Magnetism (Cambridge University Press, Cambridge, 1993)
60. R. Bulla, T. Costi, T. Pruschke, The numerical renormalization group method for quantum impurity systems (2007). Rev. Mod. Phys. (in press)
61. S. Hershfield, J.H. Davies, J.W. Wilkins, Phys. Rev. Lett. **67**, 3720 (1991)
62. S. Hershfield, J.H. Davies, J.W. Wilkins, Phys. Rev. B **46**, 7046 (1992)
63. T. Fujii, K. Ueda, Phys. Rev. B **68**, 155310 (2003)
64. J. Takahashi, S. Tasaki, Nonequilibrium steady states and fano-kondo resonances in an ab ring with a quantum dot (2006). Cond-mat/0603337
65. C.H.L. Quay, J.C. an S. J. Gamble, R. de Picciotto, H. Kataura, D. Goldhaber-Gordon, Phys. Rev. B **76**, 73404 (2007)
66. D.E. Logan, N.L. Dickens, J. Phys.: Condens. Matter **13**, 9713 (2001)

20

MICROSCOPIC PROXIMITY EFFECT PARAMETERS IN S/N AND S/F HETEROSTRUCTURES

S.L. PRISCHEPA[1], V.N. KUSHNIR[1], E.A. ILYINA[1],
C. ATTANASIO[2], C. CIRILLO[2], AND J. AARTS[3]

[1] *Belarus State University of Informatics and Radioelectronics, P.Brovka str.6, Minsk 220013, Belarus.* `prischepa@bsuir.by`
[2] *Dipartimento di Fisica "E.R. Caianiello" and Laboratorio Regionale SuperMat CNR/INFM-Salerno, Università degli Studi di Salerno, Baronissi (Sa), I-84081, Italy*
[3] *Kamerling Onnes Laboratory, Leiden University, P.O. Box 9504, 2300 RA Leiden, The Netherlands*

Abstract. Superconducting proximity effect in Cu/Nb/Cu, Nb/Cu/Nb trilayers and in $Nb/Pd_{0.81}Ni_{0.19}$ bilayers has been studied. The dependence of the superconducting transition temperature T_c versus Nb thickness in Cu/Nb/Cu trilayers and versus Cu thickness in Nb/Cu/Nb trilayers is described by different sets of material parameters. We attribute this discrepancy to the influence on the superconducting properties of the external edges of Nb/Cu/Nb hybrids. This conclusion was confirmed from the results obtained for $Nb/Pd_{0.81}Ni_{0.19}$ bilayers, in which the top layer was always $Pd_{0.81}Ni_{0.19}$. In this case, in fact, the T_c dependence as a function of both Nb and $Pd_{0.81}Ni_{0.19}$ thicknesses is described by unique set of microscopic parameters.

Keywords: proximity effect; microscopic parameters; interface transparency; heterostructures

20.1 Introduction

The investigation of the properties of superconducting multilayers is a widely studied subject in the physics of superconductivity [1]. Given the geometrical and the material parameters of layered samples (composition, number of layers and their thickness, interfaces quality) it is possible to obtain superconductors with the wishful thermodynamics properties. The control of the above mentioned parameters is easier in superconductor/normal metal (S/N) hybrids, which are superconducting due to the proximity effect [2]. In general, the

proximity effect consists in the penetration of the Cooper pairs from the superconductor to the adjacent normal metal [2]. As a result, a superconducting
state with a non homogeneous distribution of Cooper pairs appears in the
layered structure.

In recent years the promising field of superconducting spintronics caused a
great interest also in artificially fabricated layered ferromagnet/superconductor
(F/S) systems (for a review see Ref. [3]). In magnetic metals pair breaking
mechanism is mainly due to the exchange energy E_{ex} which acts on the spin of
the Cooper pairs. Due to this the superconductivity in S/F heterostructures
is observed under more rigid requirements and the superconducting correlations decay length in ferromagnet ξ_F does not exceed few nanometers. (In
the dirty limit and for $E_{ex} \gg k_B T$ $\xi_F = (\hbar D_F/E_{ex})^{1/2}$, where D_F is the
diffusion constant of the ferromagnet, k_B is the Boltzmann constant, T is
the temperature.) The presence of the exchange field is responsible of a series
of different phenomena peculiar of S/F structures: the nonmonotonic dependence of the critical temperature T_c on the ferromagnetic layer thickness d_F,
as well as the formation of π junctions in S/F/S structures, the inversion of
the superconducting density of states induced in F [3].

In this work, the problem of the influence of the external boundaries,
namely, the top surface of the sample, on the critical temperature in S/N
and S/F heterostructures is examined. It is usually considered that trilayers,
fabricated in the same conditions, are characterized by the same parameters
of the superconducting state. In particular, it is supposed a priori, that the
critical temperature T_c of N/S/N structure with layer thicknesses d_S and d_N
(the superconducting and the normal layer thicknesses, respectively) is equal
to the critical temperature of S/N/S sample with layer thicknesses $d_S/2$ and
$2d_N^4$. We show that it is not possible to theoretically describe the experimental
$T_c(T_S)$ dependence for N/S/N and the $T_c(d_N)$ dependence for S/N/S samples
using the same numbers. In particular, it is impossible to take into account,
within reasonable calculation accuracy, the influence of the external boundaries in S/N/S structures. On the other hand, the absence of free Nb edges
in Nb/Pd$_{0.81}$Ni$_{0.19}$ bilayers (ferromagnet is always the top layer) allows us to
determine precisely the microscopic parameters of S/F structure.

20.2 Experiment

High quality S/N samples were grown by molecular beam epitaxy (MBE)
on Si(100) substrates kept at room temperature. The initial pressure was
1×10^{-10} mbar while the pressure during the deposition was around 10^{-8} mbar.
Cu/Nb/Cu trilayers with external Cu layers of fixed thickness ($d_N = 150$ nm)
and internal Nb layers with thicknesses in the range $d_S = 20, \ldots, 110$ nm, were
prepared to determine the dependence $T_c(d_S)$. For $T_c(d_N)$ measurements a
set of Nb/Cu/Nb samples with fixed Nb thickness ($d_S = 22$ nm) and Cu layer
thicknesses in the range $d_N = 10, \ldots, 160$ nm was prepared. The details of the
sample preparation and characterization were published elsewhere [5].

Nb/Pd$_{0.81}$Ni$_{0.19}$ bilayers were grown on Si(100) substrates in a UHV dc diode magnetron sputtering system with a base pressure less than 10^{-9} mbar and a sputtering Ar pressure of 4×10^{-3} mbar. The Nb and Pd$_{0.81}$Ni$_{0.19}$ layers were deposited at typical rates of 0.1 and 0.2 nm/s, respectively, measured by a quartz crystal monitor calibrated by low-angle X-ray reflectivity measurements. The Ni content (19%) was obtained by Rutherford-backscattering analysis. The top layer was always PdNi. In order to study the T_c dependence as a function of the ferromagnetic layer thickness samples were deposited with constant Nb thickness ($d_S = 14$ nm) and variable thicknesses of the Pd$_{0.81}$Ni$_{0.19}$ layers ($d_F = 1.1,\ldots,7.1$ nm). The behavior of $T_c(d_S)$ was investigated on another set of bilayers consisting of a Pd$_{0.81}$Ni$_{0.19}$ layer with constant thickness ($d_F = 19$ nm) and a Nb layer with variable thicknesses ($d_S = 10,\ldots,95$ nm). Moreover, one set of Nb single films was deposited ($d_S = 14,\ldots,200$ nm) in order to study the thickness dependence of both critical temperature and the electrical resistance. Samples of Pd$_{0.81}$Ni$_{0.19}$ with different thicknesses were also fabricated to study the electrical properties of the alloy [6].

The superconducting properties, transition temperatures T_c and perpendicular upper critical magnetic fields $H_{c2\perp}$ were resistively measured using a standard dc four-probe technique. T_c was defined as the midpoint of the transition curve.

The superconducting properties, transition temperatures Tc and perpendicular upper critical magnetic fields Hc2 were resistively measured using a standard dc four-probe technique. Tc was defined as the midpoint of the transition curve.

20.3 Results and discussion

20.3.1 N/S/N AND S/N/S TRILAYERS

The experimental $T_c(d_S)$ and $T_c(d_N)$ dependences for Cu/Nb trilayers are shown in Figs. 20.1 and 20.2, respectively, by closed circles. The diffusive limit of the quasiclassical approach to the microscopic theory [4] was used for consideration of the critical state. In this case it is necessary to solve the boundary problem for linearized Usadel equations [7]. The solution of the boundary problem depends on five parameters [8]: T_S, the critical temperature of bulk Nb, the coherence lengths of superconducting $\xi_N = \sqrt{\hbar D_N/2\pi k_B T_S}$ metals, the transparency parameter of S/N interface t_N [9] and the parameter

$$p = \frac{D_N N_N}{D_S N_S} = \frac{\rho_S}{\rho_N} \qquad (20.1)$$

which determines the jump at the S/N interfaces of first derivatives of Gor'kov quasiclassical anomalous Green functions. In Eq. 20.1 $N_{S(N)}$ are the density

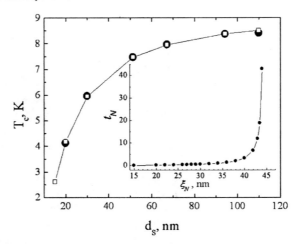

Fig. 20.1. Critical temperature, T_c, versus d_S for Cu/Nb/Cu structure with $d_N = 150$ nm. Closed circles (open squares) correspond to the measured (calculated) values. The values of the used parameters are reported in the text. Inset: dependence of the S/N interface transparency t_N on the normal metal coherence length ξ_N.

of states at the Fermi level in S(N) metal and $\rho_{S(N)}$ are the low temperature ($T = 10$ K) resistivity values of the superconductor (normal) metal, respectively.

The coherence length of the superconducting material, ξ_S, was determined by measuring the $H_{c2\perp}(T)$ dependence. The value obtained for a single Nb MBE film 100 nm thick was $\xi_S = (6.4 \pm 0.2)$ nm.

In the case of Cu/Nb/Cu the value of the parameter p is determined quite unambiguously because the external Cu layers can be considered infinite ($d_N = 150$ nm) and because the resistivity of the internal Nb layer in our thickness range ($d_S > 20$ nm) practically corresponds to the resistivity of a bulk sample [10]. As a result, using $\rho_S = 3.6\ \mu\Omega \times$ cm and $\rho_N = 1.3\ \mu\Omega \times$ cm [5], for the Cu/Nb/Cu samples we get the value $p = 2.8 \pm 0.1$. On the other hand, for the Nb/Cu/Nb trilayers only a rough estimation could be done for the resistivities, and the value of the parameter p can be only roughly estimated. In fact, the low temperature resistivity values increase as the film thickness is reduced, mainly due to the electron scattering at the film surfaces [11]. In multilayers surface effects are more relevant at the external boundaries of the sample, the scattering being weaker at the S/N internal interfaces. As a result we have a range of possible values for the parameter p: that is $p \approx 2.0, \ldots, 8.5$.

The parameter T_S can be accurately determined for the N/S/N system from the asymptotic $T_c(d_S)$ dependence. As a result, for Cu/Nb/Cu we have $T_S = (9.0 \pm 0.2)$ K, which gives us the calculated superconducting coherence

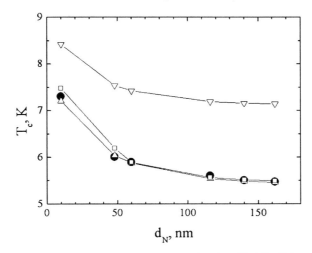

Fig. 20.2. Critical temperature, T_c, versus d_N for Nb/Cu/Nb structure with $d_S = 22$ nm. Closed circles correspond to the measured values. Open symbols correspond to the results of calculations, as described in the text.

length $\xi_S = 6.4$ nm [8], in accordance with the value estimated from $H_{c2\perp}(T)$ measurements. On the other hand, for the Nb/Cu/Nb system we do not directly measure the value of T_S. In this case we can only establish a lower limit for T_S, which corresponds to the measured T_c for the trilayers in the limit $d_N \to 0$. On the basis of Ref. [5] we get $7.5\,\text{K} < T_S < 9.2\,\text{K}$.

So while for the Cu/Nb/Cu system only two fitting parameters have to be estimated, t_N and ξ_N, for the Nb/Cu/Nb structure four parameters remain undetermined: t_N, ξ_N, p and T_S.

In Fig. 20.1 the $T_c(d_S)$ dependence calculated in the one mode approximation is shown by open squares using the fitting parameter $t_N = 0.98$ and the above mentioned parameters, $T_S = 9.0$ K, $\xi_S = 6.4$ nm, $p = 2.8$. The good agreement with the experimental data is evident. As it was shown in Ref. [8], the parameters t_N and ξ_N in N/S/N structure are functionally linked. The $t_N(\xi_N)$ curve, all points of which give the same $T_c(d_S)$ dependence, is shown in the inset of Fig. 20.1. From this curve it can be seen that the value of $t_N = 0.98$ corresponds to $\xi_N = 34$ nm.

At this point we tried to reproduce the experimental dependence $T_c(d_N)$ for Nb/Cu/Nb by solving the boundary problem for Usadel equations using the same set of fitting parameters, which have been already determined from the $T_c(d_S)$ curve. We observe that, for $p = 2.77$ and $T_S = 9$ K, it does not exist the pair of (t_N, ξ_N) values from the $t_N(\xi_N)$ dependence (see inset to Fig. 20.1), which reproduces the experimental $T_c(d_N)$ curve. In Fig. 20.2 the theoretical $T_c(d_N)$ curve calculated with the parameters $t_N = 0.98$ (consequently, $\xi_N = 34$ nm), $p = 2.77$ and $T_S = 9$ K, obtained from the $T_c(d_S)$ fitting procedure, is shown by down open triangles. The complete disagreement with the experimental data for the Nb/Cu/Nb structure is evident, the measured $T_c(d_N)$ dependence for S/N/S structure lying well below the theoretical prediction. As discussed above we believe that one of the main reasons of such discrepancy is related to the presence of thin external Nb layers with suppressed superconducting properties in Nb/Cu/Nb system.

It is possible to reproduce the experimental $T_c(d_N)$ dependence only changing the values of p and T_S. As a first attempt we considered T_S as a free parameter. In Fig. 20.2 the result for the Nb/Cu/Nb structure calculated for $t_N = 0.98$ ($\xi_N = 34$ nm), $p = 2.77$ and $T_S = 8$ K is shown by up triangles. In the same figure by open squares we show the calculated $T_S(d_N)$ dependence for $\xi_S = 6.4$ nm, $t_N = 0.98$ ($\xi_N = 34$ nm), $T_S = 9$ K, but this time the p value was considered as a free parameter, $p = 9.8$, which is out of the range of the allowable values. In Fig. 20.3 we show the $T_S(p)$ curve, calculated using $\xi_S = 6.4$ nm and $t_N = 0.98$ ($\xi_N = 34$ nm). This curve determines the range of (p, T_S), which accurately reproduces the experimental data for the structure Nb/Cu/Nb. The result has been obtained with the accuracy of 0.15 K. Analogous curves can be also obtained for other ranges of fitting parameters.

What it is worth to note is that if we try to reproduce the $T_c(d_N)$ curve using the bulk $T_S = 9$ K we get an unphysical value for the parameter p. Finally, it is interesting to point out that even if we force T_S and p to change about 10% of their values in the Nb/Cu/Nb system we cannot reproduce

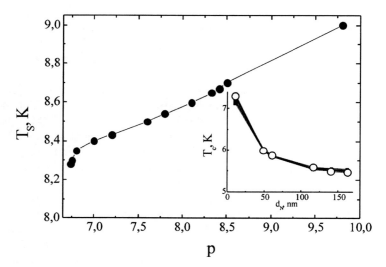

Fig. 20.3. Critical temperature of bulk Nb, T_S, versus p for Nb/Cu/Nb system. Inset: T_c versus d_N dependences for Nb/Cu/Nb trilayers. Open circles correspond to the measured values while solid lines correspond to the theoretical calculations from which the corresponding T_S and p values on the main plot were obtained.

both $T_c(d_S)$ and $T_c(d_N)$ curves fixing the values of the other microscopical parameters.

20.3.2 S/F HYBRIDS

Superconducting transition temperatures T_c were measured for all Nb/Pd$_{0.81}$Ni$_{0.19}$ samples. Fig. 20.4 shows the data for $T_c(d_S)$, which demonstrate the standard behavior for S/F bilayers. The data for the $T_c(d_F)$ of set with variable PdNi thickness ($d_S = 14$ nm) are given in Fig. 20.5. The curve exhibits a rapid drop as the ferromagnetic thickness is increased until a shallow minimum is reached for $d_F \approx 3.0, \ldots, 3.5$ nm. A thickness-independent saturation value is observed for d_F above 5 nm.

The behavior of both $T_c(d_S)$ and $T_c(d_F)$ dependences was analyzed in the framework of the proximity effect model based on the linearized Usadel equations [12]. For S/F case the boundary problem for T_c determination is similar to boundary task for S/N case, but the solution for S/F bilayers depends on six parameters: T_S, ξ_S, ξ_F, t_F, p, and E_{ex}.

From the analysis of the result of Fig. 20.4 we obtain $T_S = 8.8$ K. As we noticed above, the value of p can be only roughly estimated from resistivity measurements, since the low temperature resistivity values increase as the film thickness is reduced. In Fig. 20.6 we show both the $\rho_S(d_S)$ and $\rho_F(d_F)$ dependences measured on bare Nb and Pd$_{0.81}$Ni$_{0.19}$ films of different thicknesses. From Fig. 20.6 a rough estimation of thin layers resistivities gives $p \approx 0.1, \ldots,$

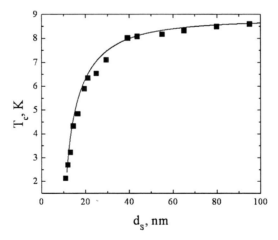

Fig. 20.4. Critical temperature, T$_c$, versus d$_S$ for Nb/Pd$_{0.81}$Ni$_{0.19}$ hybrids with $d_F = 19$ nm. Squares correspond to the measured values. The solid line is the result of two-mode calculations. The values of the fitting parameters are reported in the text.

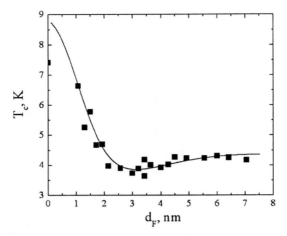

Fig. 20.5. Critical temperature, T_c, versus d_F for Nb/Pd$_{0.81}$Ni$_{0.19}$ hybrids with $d_S = 14$ nm. Squares correspond to the measured values. The solid line is the result of two-mode calculations. The values of the fitting parameters are reported in the text.

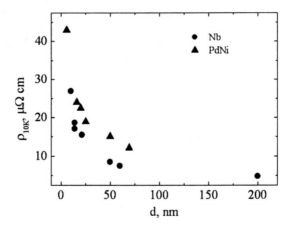

Fig. 20.6. Low temperature resistivity versus thickness for Nb (circles) and Pd$_{0.81}$Ni$_{0.19}$ (triangles) thin films.

0.6. The calculated Nb coherence length is $\xi_S = 5.8$ nm [10]. Finally, the value of E_{ex} for Pd$_{0.81}$Ni$_{0.19}$ samples was obtained in Ref. [6] as $E_{ex} = 230$ K.

In such a way we leave as a completely free parameters only ξ_F and t_F. For simplicity we apply the two mode approximation for fitting the experimental data. We obtain that parameters t_F and ξ_F in S/F hybrids are functionally linked. In Fig. 20.4 by solid line we show the result of the theoretical simulations for $T_c(d_S)$ data obtained for $E_{ex} = 230$ K, $T_S = 8.8$ K, $\xi_S = 5.8$ nm, $p = 0.3$, and using the fitting parameters $t_F = 3.2$, $\xi_F = 3.2$ nm. Similarly to the S/N case, fixing the first four parameters it is possible to obtain the set of (t_F, ξ_F) couples, which reproduces the same curve shown in Fig. 20.4 as a solid line. This condition is fulfilled in the ranges: $0.5 \text{ nm} \leqslant \xi_F \leqslant 4.3 \text{ nm}$ and $0.028 \leqslant t_F \leqslant 220$.

The same procedure was followed for the $T_c(d_F)$ data, using the same fixed parameter reported above. The result is shown in Fig. 20.5 by solid line. It is worth to note that in this case the set of (t_F, ξ_F) couples which reproduces this curve lies in a much smaller interval: $3.0 \text{ nm} \leq \xi_F \leq 3.4 \text{ nm}$, $2.7 \leq t_F \leq 6.4$. Moreover these values fall completely in the interval of the fitting parameters which restore the $T_c(d_S)$ dependence. The error is within the 5 %. A further check of the ξ_F estimation can be obtained from the position of the T_c minimum in $T_c(d_F)$ dependence, d_{min}. According to Ref. [13] $d_{min} = 0.7\pi\xi_F/2$ and with $d_{min} = 3.4$ nm, we find in fact $\xi_F = 3.1$ nm, in very good agreement with the value obtained from the fit.

At this point it is worth to introduce the quantum mechanical transparency coefficient T. The relation between T and t_F is [9]

$$T = \frac{t_F}{1 + t_F}$$

From the last expression it is evident that as the transparency parameter t_F changes from zero (completely reflecting S/F interface) to infinity (completely transparent S/F interface), T varies between 0 (negligible transparency) and 1 (perfect interface).

In Fig. 20.7 we show the calculated $T(\xi_F)$ dependence for $Nb/Pd_{0.81}Ni_{0.19}$ system. Again we remind that all the points of this curve give the same $T_c(d_S)$ dependence. As we mentioned above, the analysis of $T_c(d_F)$ curve allows us to reduce significantly the number of allowed couples (t_F, ξ_F) (open symbols in Fig. 20.7). Such complete correspondence of microscopic parameters obtained from $T_c(d_S)$ and $T_c(d_F)$ curves is related to the absence of free surfaces in the studied $Nb/Pd_{0.81}Ni_{0.19}$ system.

20.4 Conclusion

The main purpose of this work was to examine the values of the microscopic parameters which describe the proximity effect both in S/N and S/F heterostructures. The data for S/N system were derived by analyzing the experimental data of Cu/Nb/Cu and Nb/Cu/Nb trilayers, while for S/F system we used $Nb/Pd_{0.81}Ni_{0.19}$ bilayer structures. The $T_c(d_S)$ dependence in Cu/Nb/Cu samples was theoretically reproduced and a set of numbers was extracted, which however, did not describe the $T_c(d_N)$ behavior of Nb/Cu/Nb trilayers. This result is mostly due to the different properties of the internal S layers in Cu/Nb/Cu with respect to the external S layers in Nb/Cu/Nb samples. For $Nb/Pd_{0.81}Ni_{0.19}$ bilayers the S/F interface is always the same for both the $T_c(d_S)$ and $T_c(d_F)$ experiments. This leads to the same set of

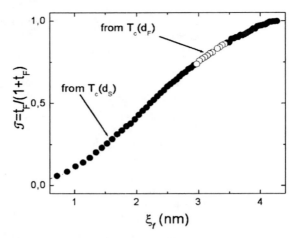

Fig. 20.7. The quantum mechanical transparency coefficient T versus ξ_F for Nb/Pd$_{0.81}$Ni$_{0.19}$ bilayers as obtained from the T$_c$(d$_S$) dependence (closed symbols). Open symbols corresponds to the T(ξ_F) values followed from the T$_c$(d$_F$) dependence.

fitting parameters, which can be determined in this case with higher accuracy with respect to the S/N case with free Nb surface.

References

1. See, e.g. R. Gross, A. Sidorenko, L. Tagirov (Eds.), Nanoscale devices - fundamentals and applications, in: NATO Science Series II. Mathematics, Physics and Chemistry, vol. 233, Springer, 2006, p. 399.
2. G. Deutscher and P.G. de Gennes, Proximity effects, in Superconductivity, edited by R.D. Parks (Marcel, Dekker, New York), 1005–1034 (1969).
3. A.I. Buzdin, Proximity effects in superconductor-ferromagnet heterostructures, Rev. Mod. Phys. 77(3), 935–976 (2005).
4. J.B. Ketterson and S.N. Song, Superconductivity, Cambridge University Press, 1999, p. 497.
5. A. Tesauro, A.Aurigemma, C. Cirillo, S.L. Prischepa, M. Salvato, and C. Attanasio, Interface transparency and proximity effect in Nb/Cu triple layers realized by sputtering and molecular beam epitaxy, Supercond. Sci. Technol. 18(1), 1–8 (2005).
6. C. Cirillo, Superconducting proximity effect in Nb/PdNi hybrids: probing the role of the ferromagnet, PhD thesis, University of Salerno, 2005, p.105.
7. L. Usadel, Generalized diffusion equation for superconducting alloys, Phys. Rev. Lett. 25(8), 507–509 (1970).
8. V.N. Kushnir, S.L. Prischepa, C. Cirillo, and C. Attanasio, Critical temperature and interface transparency of N/S/N triple layers:theory and experiment, Eur. Phys. J. B 52(1), 9–14 (2006).

9. L.R. Tagirov, Proximity effect and superconducting transition temperature in superconductor/ferromagnet sandwiches, Physica C 307, 145–163 (1998).
10. C. Cirillo, A. Rusanov, C. Bell, and J. Aarts, Depairing current behavior in superconducting Nb/Pd81Ni19 bilayers, Phys. Rev. B 75(17), 174510(7) (2007).
11. E.H. Sondheimer, The influence of a transverse magnetic field on the conductivity of thin metallic films, Phys. Rev. 80(3), 401–406 (1950).
12. A.I. Buzdin and M.Y. Kupriyanov, Transition temperature of a superconductor-ferromagnet superlattices, Pis'ma Zh. Eksp. Teor Fiz. 52(9), 1089–1091 (1990) [JETP Lett. 52(9), 487–491 (1990)].
13. Ya.V. Fominov, N.M. Chtchelkatchev, and A.A. Golubov, Nonmonotonic critical temperature in superconductor/ferromagnet bilayers, Phys. Rev. B 66(1), 014507(13) (2002).

21

CONDUCTANCE OSCILLATIONS WITH MAGNETIC FIELD OF A TWO-DIMENSIONAL ELECTRON GAS-SUPERCONDUCTOR JUNCTION

N.M. CHTCHELKATCHEV[1] AND I.S. BURMISTROV[2]

[1] *L.D. Landau Institute for Theoretical Physics, Russian Academy of Sciences, 117940 Moscow, Russia.* `nms@itp.ac.ru`

[2] *L.D. Landau Institute for Theoretical Physics, Russian Academy of Sciences, 117940 Moscow, Russia.* `burmi@itp.ac.ru`

Abstract. We develop the theory for the current voltage characteristics of a two-dimensional electron gas - superconductor interface in magnetic field at arbitrary temperatures and in the presence of the surface roughness. Our theory predicts that in the case of disordered interface the higher harmonics of the conductance oscillations with the filling factor are strongly suppressed as compared with the first one; it should be contrasted with the case of the ideal interface for which amplitudes of all harmonics involved are of the same order. Our findings are in qualitative agreement with recent experimental data.

Keywords: conductance, junction, magnetic field

21.1 Introduction

The study of hybrid systems consisting of superconductors (S) in contact with high mobility two-dimensional electron gas (2DEG) in magnetic field has attracted considerable interest in recent years [1–3]. The quantum transport in this type of structures can be investigated in the framework of Andreev refection theory [4]. When an electron quasiparticle in a normal metal (N) reflects from the interface of the superconductor (S) into a hole, Cooper pair transfers into the superconductor. A number of very interesting phenomena based on Andreev reflection had been studied in the past [5]. For example, if the normal metal is surrounded by superconductors, so we have a SNS junction, a number of Andreev reflections appear at the NS interfaces. In the equilibrium, it leads to Andreev quasiparticle levels in the normal metal that carry considerable part of the Josephson current; out of the equilibrium,

J. Bonča, S. Kruchinin (eds.), *Electron Transport in Nanosystems.*
© Springer Science + Business Media B.V. 2008

when superconductors are voltage biased, quasiparticles Andreev reflect about $2\Delta/eV$ times transferring large quanta of charge from one superconductor to the other. This effect is called Multiple Andreev Reflection (MAR) [5].

Effect similar to MAR appears at a long enough N-S interface in the presence of magnetic field; it bends quasiparticle trajectories and forces quasiparticles to reflect many times from the superconductor. If phase coherence is maintained interference between electrons and holes can result in periodic, Aharonov–Bohm-like oscillations in the magnetoresistance. The conductance g of a S–2DEG interface in the presence of the magnetic field has been measured in recent experiments [6–10]. At large filling factors it demonstrated highly non-monotonic dependence with the magnetic field B. The most interesting effect was the oscillations of g with the filling factor ν in a somewhat similar manner as in Shubnikov-de Haas effect.

Recently, a phenomenological analytical theory of these phenomena based on an "athese ph with the Aharonov-Bohm effect was suggested in Ref. [11]. Numerical simulation was performed in Ref. [12]. It was theoretically shown that the transport along the infinitely long 2DEG-S interface can be described in the framework of electron and hole edge states [13]. The 2DEG-S interfaces investigated in the experiments were not infinitely long. Their length, L, was typically of the order of few cyclotron orbits, R_c, of an electron in 2DEG at the Fermi energy E_F. Quasiclassical theory of the charge transport through 2DEG-S interface of arbitrary length was developed in Ref. [14] for the case of large filling factors and vanishing temperature. In most theoretical papers mentioned above the 2DEG-S interface with no roughness (ideal interface) was considered. As it was shown [14], when $L \sim 2R_c$, $g(\nu)$ oscillates nearly harmonically with ν, as $\cos(2\pi\nu)$. For $L \sim 4R_c$ the next harmonics with $\cos(4\pi\nu)$ appears and so on. In experiments $L > 6R_c$, therefore, one would expect to observe $\cos(2\pi n\nu)$-harmonics in the conductance with $n = 1, 2, \ldots$. However, if we try to compare theoretically predicted $g(\nu)$ with the experimentally measured one then we find: (i) although $L > 6R_c$, the lowest harmonic $\cos(2\pi\nu)$ in the conductance survives whereas higher harmonics are absent; (ii) the amplitude of the conductance oscillations is much smaller than it is predicted by the theories. The reason of this disagreement is probably the roughness of the 2DEG-S interface in experiments and the perfect flatness of this interface in theory.

The main objective of the present paper is to develop the theory for the current voltage characteristics of the 2DEG-S interface in magnetic field at finite temperature which takes into account the surface roughness. Our approach with the surface roughness possibly helps to make a step towards explanation of the experimental results.

In this paper we find the current voltage characteristics of a 2DEG-S interface in the magnetic field in the presence of the surface roughness. Theories based on the assumption of the interface perfectness predict the conductance $g = g_0 + g_1 \cos(2\pi\nu + \delta_1) + g_2 \cos(4\pi\nu + \delta_2) + \ldots$; the amplitudes of the harmonics are of the same order: $g_1 \sim g_2 \sim \ldots$. This result cannot qualitatively describe the visibility of the conductance oscillations in experiments. As we have mentioned above, experimentally, the conductance behaves as $g = g_0 + g_1 \cos(2\pi\nu + \delta_1)$; higher harmonics: $g_2 \cos(4\pi\nu + \delta_2), \ldots$, are hardly observed. Our approach that takes into account the surface roughness allows to eliminate discrepancy between experiments and the theory. Due to the presence of a disorder at a 2DEG-S interface, the higher harmonics in the conductance oscillations with ν are suppressed in qualitatively agreement with experiments.

21.2 Formalism

We consider a junction consisting of a superconductor, 2DEG and a normal conductor segments (see Fig. 21.1). Magnetic field B is applied along z direction, perpendicular to the plain of 2DEG. It is supposed that quasiparticle transport is ballistic; the mean free path of an electron $l_{tr} \gg L$. The current I flows between normal (N) and superconducting (S) terminals provided the voltage V is applied between them.

Following Ref. [15–18], we shall describe the transport properties of the junction in terms of electron and hole quasiparticle scattering states, which satisfy Bogoliubov-de Gennes (BdG) equations. Then the current through the 2DEG-S surface is given as

$$I(V) = \frac{e}{h} \int_0^\infty dE \left\{ f_e \mathrm{Tr}\left[\hat{1} - R_{ee} + R_{he}\right] - f_h \mathrm{Tr}\left[\hat{1} - R_{hh} + R_{eh}\right] \right\}, \quad (21.1)$$

where $f_{e(h)}$ denotes the Fermi-Dirac distribution function:

$$f_{e(h)} = \frac{1}{e^{(E \mp eV)/T} + 1}.$$

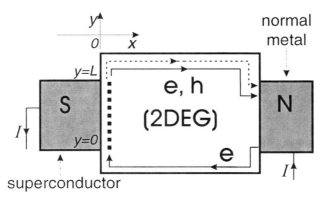

Fig. 21.1. The device which we investigate consists of a superconductor, 2DEG and a normal conductor. An electron injected from the normal conductor to 2DEG in the integer quantum Hall regime transfers through an edge state to the superconductor. It reflects into a hole and an electron which return to the normal contact through the other edge states.

The energy E is measured from the chemical potential μ of the superconductor, $R_{ee}(E, n_o, n_i)$ is the probability of the (normal) reflection of an electron with the energy E incident on the superconductor in the edge channel with quantum number n_i to an electron going from the superconductor in the channel n_o. The trace in Eq. (21.1) is taken over the channel space provided that spin degrees of freedom are included into the channel definition.

According to Eq. (21.1) if we find the probabilities R_{ab}, $a, b = e, h$, then we shall be able to evaluate the current, the conductance, the current noise and so on [18]. We shall focus on the case $R_c < L$. Then, quasiparticles reflected from the superconductor due to normal and Andreev reflection return to the superconductor again due to bending of their trajectories by the magnetic field.

At large filling factors the quasiclassical approximation is applicable. An electron (hole) quasiparticle in 2DEG can be viewed in the quasiclassics as a beam of rays [19, 20]. In a similar way, propagation of the light is described in optics within the eikonal approximation in terms of the ray beams [21]. Trajectories of the quasiparticle rays can be found from the equations of the classical mechanics. In terms of the wave functions, this description means that we somehow make wave-packets from wave functions of the edge states. Reflection of an electron from the superconductor is schematically shown in Figs. 21.2–21.5. In what follows we demonstrate that within the quasiclassical approximation the matching problem can be solved even in the presence of an interface disorder and the probabilities R_{ab} can be explicitly evaluated.

21.3 Ideally flat 2DEG-S interface

We start from the transport properties of the ideally flat 2DEG-S interface which can be most simply described provided that the edge channels do not mix at such interface, $R_{he}(E; n_o, n_i) \propto \delta_{n_o, n_i}$. In the quasiclassical language it means that electrons (holes) skip along the 2DEG edge along the same arc-trajectories, like it is shown in Fig. 21.2.

The probability of Andreev reflection can be found as follows for the trajectories shown in Figs. 21.2–21.3:

$$R_{he}(y_0; n_o, n_i) =$$
$$\delta_{n_o, n_i} \left| e^{i(S_e - \pi/2)} \left\{ r_{he} r_{ee} r_{ee} r_{ee} e^{3iS_e - i3\pi/2 - i\phi(y_3)} + \right.\right.$$
$$\left.\left. r_{hh} r_{he} r_{ee} r_{ee} e^{iS_h + 2iS_e - i\pi/2 - i\phi(y_2)} + \ldots \right\} \right|^2, \quad (21.2)$$

where r_{ba} is the amplitude of reflection of a quasiparticle (ray) a into a quasiparticle (ray) b from the superconductor. The quantity $S_{e(h)}$ is the quasiclassical action of an electron (hole) taken along the part of the trajectory connecting the adjacent points of reflection. The additional phase $\pm\pi/2$ is the Maslov index [22] of the electron trajectory. The phase $\phi(y)$ arises due to

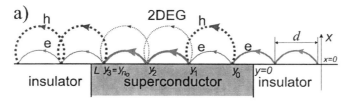

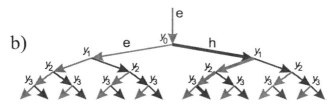

Fig. 21.2. (Color online) The propagation of quasiparticles in the quasiclassical approximation can be described in terms of rays. The state of a ray can be found from the equations of the classical mechanics. The figures (a) and (b) show what happens if an electron ray from an edge state of 2DEG comes to the superconductor. The electron ray reflects at the point $y = y_0$ from the superconductor into electron and hole rays (normal and Andreev reflections). They reflect in turn at the position y_1 from the superconductor generating two other electron (red lines) and two other hole rays (blue lines), and so on. To find the probabilities, e.g., $R_{he}(E, n_o, n_i)$, it is necessary to know the sum of the amplitudes of the eight holes that appear after the last beam-reflection at the point y_3. Drawing Fig. 21.2a we assume that Andreev approximation [4] is applicable and the 2DEG-S interface is ideally flat.

the screening supercurrents. We assume that the superconductor satisfies the description within the London theory [it is usually the case in experiments] then $\phi(y) = \phi(0) + (2m/\hbar) \int_0^y d\tilde{y} v_s(\tilde{y})$ [23], where v_s is the superfluid velocity evaluated at $x = 0$ and m is electron mass in the superconductor. Here, we used the following property of London superconductors: the spatial dependence of the vector potential and v_s are small in the direction perpendicular to the superconductor edge on the length scale ξ.

Summation of the amplitudes on the tree shown in Fig. 21.2 can be performed analytically using the matrix approach; the result can be expressed through Chebyshev polynomials, U_n [24]. If the amplitudes of local reflection r_{ab} are given then we can find the probabilities R_{ab} for $n + 1$ reflections:

$$R_{ee} = |r_{ee}r_{hh} - r_{eh}r_{he}|^{n+1} \left| \frac{r_{ee}e^{i\Omega_0}U_n(a)}{\sqrt{r_{ee}r_{hh} - r_{eh}r_{he}}} - U_{n-1}(a) \right|^2$$

$$R_{hh} = |r_{ee}r_{hh} - r_{eh}r_{he}|^{n+1} \left| \frac{r_{hh}e^{-i\Omega_0}U_n(a)}{\sqrt{r_{ee}r_{hh} - r_{eh}r_{he}}} - U_{n-1}(a) \right|^2$$

$$R_{eh} = |r_{ee}r_{hh} - r_{eh}r_{he}|^n |r_{eh}|^2 |U_n(a)|^2$$

$$R_{he} = |r_{ee}r_{hh} - r_{eh}r_{he}|^n |r_{he}|^2 |U_n(a)|^2. \tag{21.3}$$

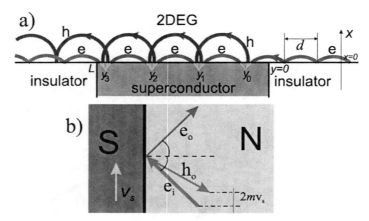

Fig. 21.3. (Color online) When the conditions of the Andreev approximation are violated we can not use the assumption that an Andreev-reflected hole velocity is exactly opposite to the velocity of the incident at the superconductor electron. Then, the quasiparticle rays propagate along the 2DEG-S interface as sketched in the figure. The orbits are organized as if the scattering occurs not at the 2DEG-S interface but from the interface in the superconductor lying at some distance from the 2DEG-S interface. According to Ref. [13], Andreev reflection couples electron and hole orbits with the guiding center x-coordinates $-\delta \pm X$ where $\delta \simeq l_B^2 m v_s / \hbar$, $l_B = \sqrt{\hbar c / eB}$ and the value of v_s should be taken at the 2DEG-S interface. For superconductors wider than the London penetration length λ_M, $\delta = \lambda_M$. Fig. 21.3b illustrates schematically how Andreev and normal quasiparticle reflections occur at the superconducting interface along which a supercurrent flows. Indices i and o label the incident and reflected quasiparticle respectively.

Here, $U_n(a) = \sin[(n-1)\arccos a]/\sqrt{1-a^2}$ denotes the Chebyshev polynomial of the second kind and

$$a = \frac{r_{ee}e^{i\Omega_0 - i\delta\phi/2} + r_{hh}e^{-i\Omega_0 + i\delta\phi/2}}{2\sqrt{r_{ee}r_{hh} - r_{eh}r_{he}}}, \qquad (21.4)$$

with $\Omega_0 = (S_e - S_h - \pi)/2$.

The probabilities R_{ab} depend on the position of the first reflection from the 2DEG-S interface, y_0, that varies in the range $(0, d)$, see Fig. 21.2a. The number of reflections n depends on the choice of y_0 as $n = 1 + [(L - y_0)/d]$ where $[x]$ denotes the integer part. So, the solution strategy is to calculate the current using Eq. (21.1) with the probabilities R_{ab} defined in Eq. (21.3) and, then, average the result over y_0. The natural choice for the distribution of y_0 is the uniform distribution: $P_{n_i}[y_0] = \theta[d(n_i) - y_0]/d(n_i)$. Hence, we find

$$I(V) = \frac{e}{h} \int_0^\infty dE \sum_{n_i} \int dy_0 P_{n_i}(y_0)\{[1 - R_{ee} + R_{he}]f_e -$$

$$[1 - R_{hh} + R_{eh}]\}f_h. \quad (21.5)$$

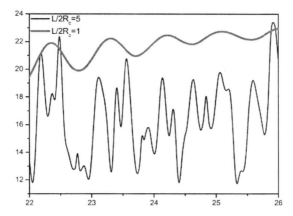

Fig. 21.4. The oscillations with ν of the dimensionless zero-bias conductance at $T = 0$. The parameters are as follows. Fermi wave vector $k_F^{(2\mathrm{DEG})} = 2 \cdot 10^6\ \mathrm{cm}^{-1}$ and $\delta\phi/4 = 20$ that corresponds to NbN film with the width of the order of 100 nm. The $L = 3\,\mu\mathrm{m}$ $[L/2R_c \simeq 5$ at $\nu = 25]$ for the lower (black thin) curve and $L = 0.6\,\mu\mathrm{m}$ $[L/2R_c \simeq 1$ at $\nu = 25]$ for the upper (red thick) curve. We neglect the Zeeman splitting which is typically small. The 2DEG-S interface scattering amplitudes r_{ab} were taken according to the BTK model [15] with $Z = 0.6$. The curves were produced using Eq. (21.6).

Eqs. (21.3) and (21.5) constitute one of the main results of the paper. They allow to find the current through the 2DEG-S surface at arbitrary temperature provided the amplitudes of local reflection r_{ab} are given.

At $T = 0$ the amplitudes of local reflection r_{ab} are related with each other as $r_{hh} = r_{ee}^\star$, $r_{eh} = -r_{he}^\star$ and $|r_{ee}|^2 + |r_{he}|^2 = 1$. Then, using Eq. (21.5), we find the conductance at zero temperature and voltage:

$$g = \frac{2e^2}{h} \sum_{n_i} \sum_s P_s \frac{|r_{eh}|^2 \sin^2[s \arccos(\sqrt{|r_{ee}|^2} \cos(\Omega)]}{1 - |r_{ee}|^2 \cos^2(\Omega)}, \qquad (21.6)$$

where spin is combined with the channel index, $\Omega = \pi\nu + \theta - \delta\phi/2$, $\theta = \arg(r_{ee})$ is the phase of the amplitude of electron – electron reflection from the superconductor. If the superconductor characteristic dimensions in x-direction are larger than the Meissner penetration length λ_M then $\delta\phi/2 = 2\lambda_M k_\perp$, where $k_\perp = k_\perp(n_i)$ is the perpendicular component of the quasiparticle momentum when it reflects from the superconductor. The function P_s is the probability that the orbit experiences s reflections from the surface of the superconductor. The function P_s originates from the integration over y_0 in Eq. (21.5). It can be expressed through the maximum number of jumps, $[L/d]$, over the S-2DEG surface as

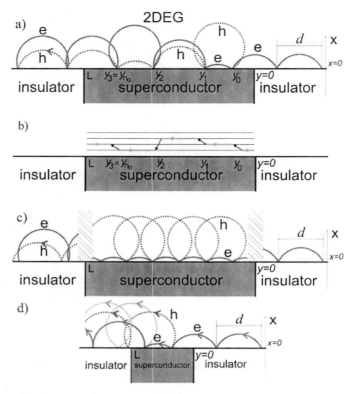

Fig. 21.5. (Color online) The propagation of qusiparticles in the quasiclassical approximation along the 2DEG-S interface in the presence of disorder, see Fig. 21.5a. It corresponds to the transitions between the Andreev edge states, as shown in Fig. 21.5b, which are induced by relatively weak short-range disorder. Disorder at the edges of the 2DEG-S interface leads to the orbits depicted in Fig. 21.5c. Strong disorder destroys Andreev edge states. There are no oscillations in the conductance because quasiparticles reflected from the strongly disordered 2DEG-S surface are incoherent. Their orbits are shown in Fig. 21.5d.

$$P_s = \begin{cases} \frac{L}{d} - \left[\frac{L}{d}\right] & \text{if } s = 1 + [L/d], \\ 1 - \frac{L}{d} + \left[\frac{L}{d}\right] & \text{if } s = [L/d], \\ 0 & \text{otherwise.} \end{cases} \tag{21.7}$$

The conductance, Eq. (21.6), as well as the current, Eq. (21.5) is an oscillating function of ν:

$$g(\nu) = \sum_{n=0}^{\infty} g_n \cos(2\pi\nu n + \delta_n), \tag{21.8}$$

where g_n are the Fourier coefficients and δ_n the "phase shifts". For the length of the interface $L \lesssim 2R_c$ the leading contribution to the conductance (current) is given by the zero harmonic g_0. In the case $2R_c \lesssim L \lesssim 4R_c$ the conductance is determined by the zero and first harmonics, $g \approx g_0 + g_1 \cos(2\pi\nu + \delta_1)$. If $4R_c \lesssim L \lesssim 6R_c$ the second harmonics $g_2 \cos(4\pi\nu + \delta_2)$ becomes relevant and so on.

How the conductance changes at $T = 0$ with ν is illustrated in Fig. 21.4a. Thin black curve in Fig. 21.4 corresponds to $L/2R_c \gtrsim 3$ that is a typical value in experiments. Applying BTK model for extracting $r_{ee}, \ldots$, one can plot figures of the zero bias conductance $g(\nu)$ at finite temperatures using such parameters that many harmonics are presented as in the case $T = 0$ shown in Fig. 21.4. It is worthwhile mentioning that the amplitudes $g_1, g_2, g_3, \ldots$ of all visible harmonics of the conductance oscillations decrease more or less equally when temperature grows from zero to T_c; finally, they vanish at $T = T_c$. However, contrary to these predictions where a large number of harmonics are well seen, experimentally, the zeroth and first harmonics: g_0 and g_1, survive only. The reason for the discrepancy between our theory and the experiment is the assumption that the 2DEG-S interface is ideally flat. Indeed, as we shall demonstrate in next section, disorder at the 2DEG-S interface makes $g_0 > g_1 > g_2, \ldots$.

21.4 Disordered 2DEG-S interface

Usually 2DEG-S interface is not ideally flat. The disorder at the interface can be divided at two classes: long range and short range with the respect to the characteristic wavelength, λ_F, in 2DEG. Presence of the long range disorder implies that the 2DEG-S interface position fluctuates around the line $x = 0$ at length scales much larger than the $\lambda_F^{(2DEG)} \sim 10^{-6}\mu m$. Photographs of the experimental setups do not allow to think that 2DEG-S interface bends strongly from the line $x = 0$. Therefore, in the experiments of Refs. [6–10] this kind of the disorder is likely not very important.

The short-range disorder includes the fluctuations of the surface at length scales smaller than $\lambda_F^{(2DEG)}$. Usually, this type of disorder is provided by impurities, clusters of atoms at the surface due to defects of the lithography and etc. When, for example, an electron ray falls on the disordered 2DEG-S surface the reflected electron rays go off the surface not at a fixed angle but they may go at any angle with certain disorder-induced probability distribution (diffusive reflection). The phases that carry the reflected electron rays going off the surface at different angles may be considered random, so the reflected electron rays can be considered as incoherent [25]. But to any reflected electron ray an Andreev reflected hole ray is attached that is coherent with the electron. So the interference of rays that produces the conductance oscillations may be not suppressed completely by the short range disorder.

"Weak" short range disorder at 2DEG-S interface does not destroy the Andreev edge states but it induces transitions between the edge states, see Fig. 21.5b. Andreev edge states in quasiclassics fix electron-hole orbit-arcs with the same beginning and end. The quasiclassical picture of the disorder-induced transitions is shown in Fig. 21.5a.

After some diagrammatic technique calculations [see Ref. [24]], we obtain the desired result for the zero-bias conductance at $T = 0$:

$$g = \frac{4e^2}{h} \nu \sum_s P_s \left\{ 1 - (|r_{ee}|^2 - |r_{eh}|^2)^s \right.$$

$$\left. \times \left[1 + e^{-\eta} \frac{4(s-1)|r_{ee}|^2 |r_{eh}|^2}{(|r_{ee}|^2 - |r_{eh}|^2)^2} \cos(2\Omega) \right] \right\}, \tag{21.9}$$

where P_s is defined in Eq. (21.7), but with d being substituted by $2R_c$ and $\exp(-\eta)$ is the small parameter describing the disorder suppression of the quasiparticle interference [see Ref. [24]]. As one can see, in the limit $e^{-\eta} \ll 1$ the conductance oscillations are described by the first harmonics $g_1 \cos(2\pi\nu + \delta_1)$ in Eq. (21.8); higher harmonics are exponentially suppressed by disorder as $g_{n+1}/g_n \propto e^{-\eta} \ll 1$.

The conductance oscillations, Eq. (21.9), are illustrated in Fig. 21.6. The conductance behavior agrees qualitatively with the experimental data (see for detailed discussion Ref. [8]).

Finally, we mention that the result similar to (21.9) could be obtained by calculation of the influence of the disorder at the edges of 2DEG-S interface on the magnetoconductance (see Fig. 21.5c).

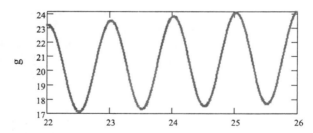

Fig. 21.6. (Color online) The figure shows the dependence of the conductance on ν at $T = 0$ for the case of the disordered 2DEG-S interface. The parameters used for this graph: length L of the S-2DEG interface and the normal conductance of the interface, are the same as for the black thin curve in Fig. 21.4. The comparison with the data of Refs. [6, 8] suggests $e^{-\eta} \simeq 0.1$ that we choose for this figure. The harmonics $g_2, g_3, \ldots$ do not contribute to the conductance in the case of the disordered interface in contrast to the clean interface case of Fig. 21.4. Remarkably, such features of the conductance oscillations as suppression of the harmonics higher than g_1, the order of the oscillation amplitude and etc. agree qualitatively with the experimental data.

21.5 Conclusions

To summarize, in this paper we develop the theory for the current voltage characteristics of a 2DEG-S interface in the magnetic field at finite temperature and in the presence of the surface roughness (diffusive reflection). We predict that the surface roughness at the 2DEG-S interface suppresses higher harmonics in the conductance oscillations with the filling factor of 2DEG. We believe that it removes the contradiction between the theory and the experiment that existed so far. The magnetoconductance of the 2DEG-S boundary seems to be sensitive to the degree of the boundary roughness, which then offer an independent way of probing the interface quality.

Acknowledgments

We would like to thank I. E. Batov for the detailed discussions of the experimental data. The research was funded in part by RFBR, grant #07-02-00998, the Russian Ministry of Education and Science, Council for Grants of the President of Russian Federation[grant MK-4401.2007.2], Russian Science Support Foundation, Dynasty Foundation and the Program of RAS "Quantum Macrophysics".

References

1. B. J. van Wees and H. Takayanagi, in *Mesoscopic electron transport*, ed. by L. L. Son et al. (Kluwer, The Netherlands, 1997), pp. 469–501.
2. H. Kroemer and E. Hu, in *Nanotechnology*, ed. by G. L. Timp (Springer, Berlin, 1999).
3. T. Schäpers, *Superconductor/semiconductor junctions*, vol. 174 of Springer Tracts on Modern Physics (Springer, Berlin/Heidelberg, 2001).
4. A. F. Andreev, Zh. Eksp. Teor. Fiz. **46**, 1823 (1964) [Sov. Phys. JETP **19**, 1228 (1964)].
5. M.Tinkham, *Introduction to superconductivity*, (Mc.Graw-Hill, 1996).
6. H. Takayanagi and T. Akazaki, Physica B **249–251**, 462 (1998).
7. D. Uhlisch et.al., Phys. Rev. B **61**, 12463 (2000).
8. I. E. Batov, Th. Schapers, N.M. Chtchelkatchev, A.V. Ustinov, and H. Hardtdegen, Andreev reflection and strongly enhanced magnetoresistance oscillations in $Ga_x In_{1-x} As/In P$ heterostructures with superconducting contacts, Phys. Rev. B **76**, 115313 (2007).
9. T. D. Moore and D. A. Williams, Phys. Rev. B **59**, 7308 (1999).
10. J. Eroms et.al., Phys. Rev. Lett. **95**, 107001 (2005).
11. Y. Asano, Phys. Rev. B **61**, 1732 (2000); Y. Asano, T. Yuito, Phys. Rev. B **62**, 7477 (2000); Y. Asano and T. Kato, J. Phys. Soc. Jap. **629**, 1125 (2000).
12. Y. Takagaki, Phys. Rev. B **57**, 4009 (1998).
13. H. Hoppe, U. Zülicke, and G. Schön, Phys. Rev. Lett. **84**, 1804 (2000); F. Giazotto, M. Governale, U. Zülicke, and F. Beltram, Phys. Rev. B **72**, 54518 (2005).

14. N. M. Chtchelkatchev, JETP Lett. **73**, 94 (2001) [Pis. Zh. Eksp. Teo Fiz. Vol. **73**, 100 (2001)];
15. G. E. Blonder, M. Tinkham, and T. M. Klapwijk, Phys. Rev. B **25**, 4515 (1982).
16. C. J. Lambert, J. Phys.: Condens. Matter **3**, 6579(1991); Y. Takane and H. Ebisawa, J. Phys. Soc. Jpn. **61**, 1685 (1992).
17. S. Datta, P. F. Bagwell, M. P. Anatram, Phys. Low-Dim. Struct., **3**, 1 (1996).
18. Ya. M. Blanter and M. Büttiker, Phys. Rep. **336**, 1 (2000).
19. H. U. Baranger, D. P. DiVincenzo, R. A. Jalabert, A. D. Stone, Phys. Rev. B **44**, 10637 (1991).
20. K. Richter, *Semiclassical theory of mesoscopic quantum systems*, Springer tracts in modern physics, vol. 161 (Springer, Berlin/Heidelberg, 2000), pp. 63–68.
21. M. Born, E. Wolf, *Principles of optics*, Pergamon Press, 1986, p. 341.
22. V. P. Maslov, M. V. Fedoruk, *Quasiclassical approximation for equations of quantum mechanics*, (Nauka publishing, Moscow, 1976).
23. A. V. Svidzinsky, *Space nongomogenius problems of supercondcutivity*, (Nauka publishing, Moscow 1982).
24. N. M. Chtchelkatchev, I. S. Burmistrov, Conductance oscillations with magnetic field of a two-dimensional electron gas-superconductor junction, Phys. Rev. B 75, 214510 (2007).
25. This is usual assumptions that the reflected electron (hole) rays can be considered as incoherent with the incident ray on a (sligtly) disorderd interface. This assumption is widely used in nanophysics. Well-known Zaitsev boundary conditions for quasiclassical Green functions were derived with this assumption. With the help of this assumption the so-called "nondiagonal", quckly oscillating components of the Green functions that carry information about the coherence were regarded as quickly fading [if we go off the surface] [?].
26. A. V. Zaitsev, Sov. Phys. JETP **59**, 1163 (1984).

22

UNIVERSALITIES IN SCATTERING ON MESOSCOPIC OBSTACLES IN TWO-DIMENSIONAL ELECTRON GASES

V. YUDSON[1] AND D. MASLOV[2]
[1] *Institute of Spectroscopy, Russian Academy of Sciences, Troitsk, Moscow region, 142190, Russia.* `yudson@isan.troitsk.ru`
[2] *Department of Physics, University of Florida, P. O. Box 118440, Gainesville, FL 32611-8440 USA.* `maslov@phys.ufl.edu`

Abstract. Scattering of electrons propagating through a disordered conductor is studied for the case when the scatterers are large compared the de Broglie wavelength of electrons but small compared the electron mean free path. If such randomly located and oriented mesoscopic scatterers are of an arbitrary convex shape, electron scattering has a universal feature: the ratio of the "quantum" (total) to transport scattering rates $\eta = \tau_{\mathrm{tr}}/\tau_{\mathrm{q}}$ does not depend on the shape of the scatterer but only on the nature of scattering (specular vs. diffuse) and on the spatial dimensionality. In particular, for specular scattering in two-dimensional (2D) electron gas $\eta = 3/2$. Possible experimental manifestations of this universality are discussed.

Keywords: scattering, two-dimensional gas, mesoscopic

22.1 Introduction

We study electron transport in very high-mobility heterostructures with large-scale potential fluctuations assuming they provide the dominant scattering mechanism for electrons. Such fluctuations may be either due to intrinsic imperfections, or they may be introduced intentionally, as antidot arrays [2]. Randomness in positions and/or shapes of antidots leads to scattering of electrons. The semiclassical motion of electrons in the presence of large-scale inhomogeneities has attracted a significant interest due to non-Boltzmann effects in magneto- and ac transport [1]. Here, we report novel basic universalities in the effective scattering cross-sections by large-scale disorder.

We assume that disorder is represented by an ensemble of mesoscopic scatterers (of a typical size a being much larger than the electron de Broglie

Fig. 22.1. Model of disorder.

wavelength λ but smaller than the electron mean free path l). Scatterers are randomly placed and oriented along the conducting plane and are of irregular but smooth shape (see Fig. 22.1). Important parameters of electron transport in a disordered media are the transport and the total ("quantum") scattering rates, $\tau_{\mathrm{tr}}^{-1} \sim \sigma_{\mathrm{tr}}$ and $\tau_{\mathrm{q}}^{-1} \sim \sigma_{\mathrm{tot}}$, associated with the transport and the total scattering cross-sections, σ_{tr} and σ_{tot} determined by

$$\sigma_{\mathrm{tr}} = \int d\Omega \frac{d\sigma}{d\Omega} (1 - \cos\theta) \qquad (22.1)$$

and

$$\sigma_{\mathrm{tot}} = \int d\Omega \frac{d\sigma}{d\Omega}, \qquad (22.2)$$

correspondingly. Here, $d\sigma/d\Omega$ is the differential scattering cross-section. Experimentally, τ_{tr} is extracted from the conductivity while τ_{q} is obtained from the damping of the de Haas-van Alfen or Shubnikov-de Haas oscillations.

The spatial structure of disorder can be characterized by a "figure of merit", that is, the ratio of transport and "quantum" mean free times,

$$\eta = \frac{\tau_{\mathrm{tr}}}{\tau_{\mathrm{q}}} = \frac{\sigma_{\mathrm{tot}}}{\sigma_{\mathrm{tr}}}. \qquad (22.3)$$

For long-range disorder, like in a GaAs heterostructure with modulation doping [3], $\eta \gg 1$. For isotropic impurities, $\eta = 1$. The case of $\eta < 1$ corresponds to enhanced backscattering. A minimum value of $\eta = 1/2$ is achieved for the limiting case of strict backscattering, when $d\sigma/d\Omega \propto \delta(\theta - \pi)$.

Here we show that for a wide class of randomly placed and oriented plane scatterers of mesoscopic size ($a \gg \lambda$), there is surprising universality in the value of the parameter η. Namely, $\eta = 3/2$ and $\eta = 4/3$ for specular and diffuse scattering, respectively.

Universality of η manifests itself both at the classical and true quantum levels of consideration. In Section 9.2, we consider classical scattering of electrons on smooth mesoscopic obstacles. In Section 9.3, we discuss the important modifications caused by the quantum nature of scattering particles.

22.2 Classical scattering

For the 2D case of interest, the differential cross-section for specular scattering by a disk of radius $a \gg \lambda$ is given by:

$$\frac{d\sigma^{\mathrm{cl}}}{d\Omega} = \frac{d\sigma^{\mathrm{cl}}(\theta)}{d\theta} = \frac{a}{2}\sin\frac{\theta}{2}, \quad \theta \in (0, 2\pi), \tag{22.4}$$

so that

$$\sigma^{\mathrm{cl}}_{\mathrm{tot}} = 2a, \quad \sigma^{\mathrm{cl}}_{\mathrm{tr}} = 8a/3, \quad \text{and} \quad \eta^{\mathrm{cl}} = 3/4. \tag{22.5}$$

Somewhat unexpectedly, a 2D array of randomly oriented ellipses, squares and even rectangles is characterized by the same value of η^{cl}. Moreover, such universality turns out to be a general feature for scattering on a wide class of (randomly oriented) plane objects. Namely, this is a property of all convex objects. To prove this statement, consider scattering of a particle beam (parallel to the x-axis) by a small element (of length dl) of the boundary (see Fig. 22.2).

For a general case of partially diffusive scattering, the outgoing angle β differs from the angle of incidence α, the latter is determined as the angle between the external normal $\mathbf{n}$ to the element dl and the x-axis. The distribution probability of angles $\beta \in (-\pi/2, \pi/2)$ is characterized by a function $P(\beta; \alpha)$. In particular cases of specular and absolutely diffuse scattering reflection $P(\beta; \alpha) = \delta(\beta - \alpha)$ and $P(\beta; \alpha) = 1/\pi$, correspondingly. The scattering angle θ is given by $\theta = \pi - \alpha - \beta$ and the flux of particles colliding with the element dl is proportional to $\cos\alpha$. Assuming random orientations of scatterers in the conduction plane, we average the cross-section over all possible orientations of $\mathbf{n}$, i.e. over the values of $\alpha \in (-\pi, \pi)$. Taking into account

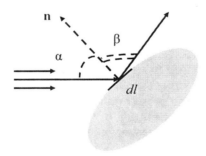

Fig. 22.2. Scattering by a convex obstacle.

that scattering by element dl occurs only for $\alpha \in (-\pi/2, \pi/2)$, we obtain its contribution to the averaged differential cross-section as

$$\frac{d\sigma^{\mathrm{cl}}(\theta)}{d\theta} = dl \int_{-\pi/2}^{\pi/2} \frac{d\alpha}{2\pi} \cos \alpha$$

$$\times \int_{-\pi/2}^{\pi/2} d\beta \, P(\beta; \alpha) \delta(\theta - \pi + \alpha + \beta) \,. \tag{22.6}$$

Assuming the distribution function $P(\beta; \alpha)$ to be uniform along the scatterer boundary and integrating over dl, we replace dl in Eq. (22.6) by the perimeter of the object, P. Therefore, the total and transport cross-sections are given by

$$\sigma_{\mathrm{tot}}^{\mathrm{cl}} = P \int_{-\pi/2}^{\pi/2} \frac{d\alpha d\beta}{2\pi} P(\beta; \alpha) \cos \alpha \tag{22.7}$$

and

$$\sigma_{\mathrm{tr}}^{\mathrm{cl}} = P \int_{-\pi/2}^{\pi/2} \frac{d\alpha d\beta}{2\pi} P(\beta; \alpha)[1 + \cos (\alpha + \beta)] \cos \alpha \,. \tag{22.8}$$

The ratio of the two cross-sections, η^{cl}, is determined entirely by the function $P(\beta; \alpha)$ and is independent of a particular geometry of the scattering object.

Consider two particular situations of special interest. For specular scattering, Eqs. (22.6–22.8) reduce to

$$\frac{d\sigma^{\mathrm{cl}}(\theta)}{d\theta} = \frac{P}{4\pi} \sin \frac{\theta}{2} \,, \tag{22.9}$$

and

$$\sigma_{\mathrm{tot}}^{\mathrm{cl}} = P/\pi \,, \quad \sigma_{\mathrm{tr}}^{\mathrm{cl}} = 4P/3\pi \,, \quad \text{and} \quad \eta = 3/4 \,. \tag{22.10}$$

For a disk, these expressions coincide with Eqs.(22.4) and (22.5), respectively. For absolutely diffuse scattering, we arrive at

$$\frac{d\sigma^{\mathrm{cl}}(\theta)}{d\theta} = \frac{P}{\pi^2} \sin^2 \frac{\theta}{2} \,, \tag{22.11}$$

so that

$$\sigma_{\mathrm{tot}}^{\mathrm{cl}} = P/\pi \,, \quad \sigma_{\mathrm{tr}}^{\mathrm{cl}} = 3P/2\pi \,, \quad \text{and} \quad \eta^{\mathrm{cl}} = 2/3 \,. \tag{22.12}$$

Universality results from the averaging over random orientation of scatterers under the condition of convexity of their shapes. The convexity condition prevents repeated scattering. As an illustration of the importance of this requirement, consider specular scattering by randomly oriented "right angles" of size a. The repeated scattering is not excluded in this geometry and we have for averaged total and transport cross-sections: $\sigma_{\mathrm{tot}}^{\mathrm{cl}} = a(2 + \sqrt{2})/\pi$ and $\sigma_{\mathrm{tr}}^{\mathrm{cl}} = a(4 + \sqrt{2})/\pi$, so that $\eta^{\mathrm{cl}} = (3 + \sqrt{2})/7 \approx 0.63$ differs from the universal value or scatterers of a convex form.

The reasoning that has led us to universality is not restricted to a specific dimensionality. In 3D, averaging over random orientations of a convex scatterer is equivalent to averaging of a contribution of a particular surface patch which is rotated over the whole solid angle. For instance, for specular scattering in 3D we arrive at the universal expressions: $\sigma_{\text{tot}}^{\text{cl}} = \sigma_{\text{tr}}^{\text{cl}} = S/4$ (S is a surface area), and $\eta^{\text{cl}} = 1$. In fact, the universal connection $\sigma_{\text{geom}} = S/4$ between the geometrical cross-section and the surface area of randomly oriented convex scatterers is well known in the field of light scattering by dust particles (see e.g., Ref. [4]).

22.3 Quantum scattering

Here we discuss modifications of the classical results of Section 9.2 caused by the wave nature of scattering particles (in optical language, we are going beyond the geometrical optics). As long as λ/a is much smaller than one, these modifications are negligible for the transport cross-section, but are very substantial for the total cross-section. For instance, the total quantum-mechanical cross-section of a sphere (of radius $a \gg \lambda$) is *twice larger than the classical value* $\sigma_{\text{tot}}^{\text{cl}} = \pi a^2$ determined simply by the geometrical cross-section (see, e.g., [4,5]). Such a dramatic discrepancy with the classical result (the so-called Extinction Paradox) stems from a sharp peak in the differential cross-section for forward scattering ($\theta \to 0$) which cannot be described semiclassically [5]. On the other hand, this peak does not contribute to the transport cross-section due to the factor $1 - \cos\theta$ ($\to 0$, at $\theta \to 0$) and $\sigma_{\text{tr}} \approx \sigma_{\text{tr}}^{\text{cl}}$.

The relation between the classical and quantum cross-sections, described in the previous paragraph, is not specific either to spherically symmetric scatterers or to a particular spatial dimensionality but is a feature of scattering by any opaque object of size $a \gg \lambda$.

Directly behind this object there is a shadow region with a vanishing amplitude of the wave field, $A = 0$. According to the superposition principle, $A = A_i + A_s$, where A_i and A_s are the amplitudes of the incident and scattered waves, correspondingly. Therefore, $A_s = -A_i$ within the shadow region (Babinet principle, 1837). This means, that in addition to the flux scattered in the directions outside the shadow region, an opaque object also scatters the incoming radiation in the forward direction, within a very narrow diffraction cone of angle $\theta \sim \lambda/a$. Obviously, the flux of the forward-scattered wave equals to the flux of the incident wave through the geometric cross-section σ_{geom} of the scatterer. This leads to an additional contribution $\delta\sigma_{\text{tot}} = \sigma_{\text{geom}}$ to the total scattering cross-section. As a result, the true total scattering cross-section σ_{tot} is a sum of its semi-classical value $\sigma_{\text{tot}}^{\text{cl}} = \sigma_{\text{geom}}$ and the forward-scattering part $\delta\sigma_{\text{tot}} = \sigma_{\text{geom}}$. Thus, we arrive at the universal relationships

$$\sigma_{\text{tot}} = 2\sigma_{\text{tot}}^{\text{cl}} , \tag{22.13}$$

$$\sigma_{\text{tr}} = \sigma_{\text{tr}}^{\text{cl}} \tag{22.14}$$

valid to the leading order in the small parameter λ/a for an arbitrary opaque scatterer with a well-defined boundary (we will discuss the latter condition below). These relations refer to an arbitrary orientation of the scatterer, hence they remain valid after averaging over orientations.

Finally, combining Eqs. (22.13) and (22.14) with Eqs. (22.10) and (22.12), we obtain the announced results

$$\eta = \frac{3}{2} \tag{22.15}$$

for the case of specular scattering and

$$\eta = \frac{4}{3} \tag{22.16}$$

for the case of diffuse scattering.

One should have in mind that Eq. (22.13) refers to the total cross-section measured at distances larger then the Fraunhofer length, $L_F = a^2/\lambda$, from the scatterer, where quantum small-angle scattering smears the classical shadow.

22.4 Discussions and conclusions

Although we used a two-step (classical-to-quantum) derivation of Eqs. (??) and (22.16), it should be emphasized that their validity range is wider than that of the intermediate expressions, which involve the classical total cross-section $\sigma_{\text{tot}}^{\text{cl}}$. In fact, the latter quantity is well-defined only for scatterers with a sharp boundary. If, on the contrary, the scattering potential falls off continuously with the distance, the classical total cross-section $\sigma_{\text{tot}}^{\text{cl}}$ is infinite, no matter how small the potential is away from the center [5]. This makes the very notion of the classical total cross-section very restricted. On the contrary, the true quantum total cross-section accounts for the weakness of scattering by the potential tail and remains finite if the potential decays sufficiently fast [5].

Experimentally, the transport scattering rate $1/\tau_{\text{tr}}$ is extracted from the conductivity which is determined just by the decay rate of the electron momentum projection on the direction of the current. The quantum decay rate, $1/\tau_{\text{q}}$, may be obtained, in principle, by measuring attenuation of an electron beam propagating through a disordered stripe, similar to light extinction coefficient measured in optical experiments. However, it is more practical to extract $1/\tau_{\text{q}}$ from the amplitude of the de Haas-van Alfen or Shubnikov-de Haas oscillations in relatively weak magnetic fields (to avoid significant changes of the decay rate caused by the field itself). In both types of experiments, attenuation of the measured quantities (the beam amplitude or the amplitude of magneto-oscillations) is due to deflecting particles from their original trajectories by scattering, no matter in what direction. Consequently, $1/\tau_q$ is related to the total cross-section.

To conclude, we have studied electron propagation through a random array of strong mesoscopic scatterers (e.g., quantum antidots) of a typical size a greater than the electron de Broglie wavelength λ. For a given type of a disorder, the relative strength of backward and forward scattering is characterized by parameter $\eta = \tau_{\mathrm{tr}}/\tau_q$–the ratio of the "quantum" (total), $1/\tau_q$, and transport, $1/\tau_{\mathrm{tr}}$, elastic scattering rates (associated with the total, σ_{tot}, and the transport, σ_{tr}, scattering cross-sections, respectively). For a wide class of scatterer's shapes, namely for convex ones, we have shown that η does not depend on the scatterer's shape and size. In particular, for specular scattering, η is a universal constant determined only by the dimensionality (D) of the system: $\eta = 2$ for $D = 3$ and $\eta = 3/2$ for $D = 2$. This universality emerges as a consequence of averaging over random orientations of convex scatterers and of the universal relation between the total classical and quantum scattering cross-sections of a mesoscopic opaque obstacle.

Acknowledgements

We acknowledge stimulating discussions with V. E. Kravtsov, A. Yu. Kuntsevich, S. P. Obukhov, and V. M. Pudalov.

This work was supported by RFBR Grants (No. 03-02-17285 and 06-02-16744) and "Nanostructures" program of Russian Academy of Sciences (V. I. Yu.) and NSF Grant No. DMR-0308377 (D. L. M.). We acknowledge the hospitality of the Abdus Salam International Center for Theoretical Physics (ICTP) where part of this work was done.

References

1. See, e.g., D. G. Polyakov, F. Evers. A. D. Mirlin, and P. Wölfle, Phys. Rev. B **64**, 205306 (2001) and references therein.
2. C. Nachtwei, Z. H. Liu, G. Lütjering, R. R. Gerhadts, D. Weiss, K. von Klitzing, and K. Eberl, Phys. Rev. B **57**, 9937 (1998).
3. T. Ando, A. B. Fowler, F. Stern, Rev. Mod. Phys. Phys. Rev. B **54**, 437 (1982).
4. H. C. van de Hulst, *Light Scattering by Small Particles*. (Dover Publications, New York, 1981), Section 8.
5. L. D. Landau and E. M. Lifshitz, *Quantum Mechanics*, 4th edn. (Pergamon, Oxford, 1977).

GEOMETRIC PHASES IN OPEN MULTI-LEVEL SYSTEMS

S. SYZRANOV[1] AND Y. MAKHLIN[2,3]

[1] *Theoretische Physik III, Ruhr-Universität Bochum, 44801 Bochum, Germany.* sergey.syzranov@ruhr-uni-bochum.de
[2] *Landau Institute for Theoretical Physics, Kosygin st. 2, 119334 Moscow, Russia.* makhlin@itp.ac.ru
[3] *Moscow Institute of Physics and Technology, Institutskii per. 9, Dolgoprudny, Russia*

Abstract. We analyze the geometric phases in quantum systems coupled to a dissipative environment, when the Hamiltonian of the system, and possibly its coupling to the environment, are slowly varied in time. We find that the coupling to the environment modifies the values of the geometric phases and also induces a geometric contribution to dephasing and relaxation. For a multi-level system with equal level splittings, coupling to the environment makes the dynamics more complex, and we analyze the interplay between various geometric phases in such situations.

Keywords: geometric phases, multi-level system, relaxation, dynamics

23.1 Introduction

According to the adiabatic theorem, a quantum system remains in its instantaneous non-degenerate eigenstate, when the Hamiltonian varies slowly enough. If the Hamiltonian is varied along a closed path and returns to its initial value, the initial and final wave functions can differ only by a phase factor. The phase, acquired by a non-degenerate state of the Hamiltonian in addition to the usual dynamical phase $-\int E(t)dt$, is called the Berry phase [1,2]. It is of geometric nature, i.e., it depends only on geometry of the path but not on the rate and the details of its traversal.

The widespread criterion of adiabaticity requires that the rate of changes of the Hamiltonian be small compared to the energy gap to the neighboring levels. At the same time, any system is coupled, however weakly, to the rest of the universe (which we will refer to as an environment, a bath, or a reservoir),

typically with continuous spectrum. This implies that 'true' adiabatic manipulations are impossible, and the adiabatic behavior cannot be approached.
On the other hand, a vanishingly weak coupling to the environment cannot
change the behavior dramatically. In this context an interesting, important,
and significant question is about the Berry phase in an open quantum system.
The Berry phases for open systems were analyzed in earlier theory work [3–8].
In some of this work the behavior of the Berry phases in systems subject to
(classical or quantum) noise was analyzed in a language, equivalent to the
master equation used below. In part of this work, the visibility of the geometric phases, masked by dephasing, was studied carefully but no modification of
the Berry phase was found; e.g., Refs. [3–6] neglected the effect of the finite
rate $\partial_t \hat{\mathcal{H}}_0$ on the dissipative rates and the Lamb shift and as a result they
found that the BP remains intact. A modification of the Berry phase was
found in Ref. [9] for a spin-half in a magnetic field varied along a specific
path. Ref. [10] analyzed the variance of the phase and the dephasing due to a
random Berry phase, but found no change in its mean value.

Here we use an 'operational' definition of the Berry phase for open systems
via measurable quantities. Specifically, we analyze the evolution of the density
matrix: for an isolated system the off-diagonal entries of the density matrix
acquire phase factors, which contain the dynamic phase and the geometric
phase. It can be shown that for an open two-level system the situation is
similar: under conditions specified below, the evolution of the off-diagonal
entry in the eigenbasis is decoupled from the rest of the density matrix, and
as a result of the evolution this entry acquires a factor with a phase, which
contains the dynamic and geometric contributions. This Berry phase may
differ from its value in an isolated system, and the modification was found in
Ref. [9] for a spin-half in an external magnetic field, manipulated in a specific
way. Later [11] this modification was calculated for an arbitrary loop, traversed
by the tip of the external field, and it was shown that the modification is also of
geometric nature, similar to the Berry phase itself. Moreover, it was observed
that the modification is complex, or in other words, not only the phase but
also the dephasing acquires a geometric contribution. Recently, we have shown
that one can also generate geometric phases via adiabatic manipulations of
the properties of the noise (that is of the bath of the system-bath coupling)
and found this contribution for a two-level system [12].

In this paper we consider the adiabatic dynamics of a multilevel quantum
system coupled to an environment. The situation in multi-level systems may
be more complex. Indeed, the density matrix has more entries and even in a
static field the dynamics of different entries may influence each other. Indeed,
the secular approximation, useful for the description of the dynamics and
derivation of the Bloch-Redfield equations of motion [13–15], is not applicable
to systems with Liouville degeneracies, i.e., when energy gaps between (at
least) two pairs of levels are very close [16, 17]. We first analyze the general
case without Liouville degeneracies and find the values of Berry phases, accumulated independently between various level pairs. We further analyze the
situation with degenerate level splittings and derive the coupled equations

of motion for the corresponding coherences, i.e., off-diagonal elements of the density matrix. Furthermore, we study the dynamics of the level occupations, whose evolution is governed by the rate equations. For all these cases we find the dynamical and geometric contributions to the dynamics, that is to the phases and amplitudes acquired during the evolution. In particular, we find geometric contributions to relaxation. In systems with degenerate levels cyclic adiabatic dynamics may result in more complicated evolutions, coherent unitary transformations of the degenerate subspace for an isolated quantum system. Coupling to an environment may modify this behavior. In the present paper, however, we analyze quantum systems without level degeneracies but with possible Liouville degeneracies, which can make the dynamics non-trivial.

The analysis of geometric phases is of special interest in view of the recent progress in the theoretical and experimental analysis of solid-state, especially superconducting quantum-bit nano-circuits. These systems combine the coherence of the superconducting state with control possibilities of single-electronic and squid devices. They are macroscopic quantum systems and their behavior can be observed with solid-state quantum detectors. On one hand, the level of coherence in the recent experiments is sufficiently high and allows to study even slow adiabatic processes; indeed, recently geometric phases have been demonstrated directly for the first time in solid-state systems [18, 19]. On the other hand, decoherence in these systems is strong enough and its effect on the geometric phases was observed in two-state quantum systems [18] (for earlier direct observations of the Berry phase in various systems see, e.g., Refs. [2, 20–23]). Of special interest are geometric phases and their interplay with decoherence in multi-qubit systems and systems, which include qubits and quantum resonators, i.e., multilevel systems, which are analyzed in the present paper.

The paper is organized as follows: First, we review the known results for the influence of noise on Berry phases in a two-level system. Then we analyze the general structure of geometric phases in multi-level systems. In the further sections we study the geometric phases and dephasing in multi-level system without and with equal level splittings in the energy spectrum. In Section 23.5 we analyze geometric contributions to relaxation.

23.2 Berry phase for a two-level system

Before considering the structure and values of geometric phases for a multi-level system we review the results of Refs. [11, 12, 24, 25] for the noise contribution to the Berry phase in a two-level system.

The full Hamiltonian of a two-level spin-half system and its noisy environment reads

$$\hat{\mathcal{H}} = -\frac{1}{2}\mathbf{B}\hat{\boldsymbol{\sigma}} - \frac{1}{2}\hat{X}\mathbf{n}\hat{\boldsymbol{\sigma}} + \hat{\mathcal{H}}_{\text{bath}}, \qquad (23.1)$$

where the three terms pertain correspondingly to the spin, spin-environment coupling, and the environment. The fast stationary fluctuating quantity $\hat{X}(t)$

represents noise, and **n** is an adiabatically varying dimensionless vector, indicating the direction and the power of fluctuations. The Berry phase can be measured as (a contribution to) the phase of rotation of the spin component orthogonal to **B**. In other words, the off-diagonal (in the eigenbasis) entry of the density matrix as a result of the evolution is multiplied by a factor:

$$\rho_{\uparrow\downarrow}(t) = \rho_{\uparrow\downarrow}(0)e^{i\int_0^t (B+i\Gamma)dt + i(\Phi^0 + \delta\Phi)}, \qquad (23.2)$$

where the "phase" in the exponent is complex, that is it also describes the change in amplitude. Here the first term in the exponent, $\int (B + i\Gamma)dt$, gives the dynamical phase, where

$$\Gamma = i\left(2|n_+|^2 \int \frac{d\Omega}{2\pi}\frac{S(\Omega)}{\Omega - B + i0} + n_z^2 \int \frac{d\Omega}{2\pi}\frac{S(\Omega)}{\Omega + i0}\right) \qquad (23.3)$$

gives the dephasing rate $(\operatorname{Re}\Gamma)$ and modification of the level splitting $(\operatorname{Im}\Gamma)$ by the environment. The second term in Eq. (23.2) gives the Berry phase, Φ_0 being the conventional Berry phase for an isolated system and the contribution of the noise to the Berry phase is

$$\delta\Phi = \int \left(i\frac{S(0)}{B} - \frac{1}{2}\int \frac{d\Omega}{2\pi}\frac{S(\Omega)(3B - 2\Omega)}{B(\Omega - B + i0)^2}\right)\frac{\mathbf{nB}}{B}\frac{\mathbf{n}(\mathbf{B} \times d\mathbf{B})}{B^2}$$
$$-\frac{1}{2}\int \left(\int \frac{d\Omega}{2\pi}\frac{S(\Omega)}{(\Omega - B + i0)^2}\right)\frac{\mathbf{B}(\mathbf{n} \times d\mathbf{n})}{B} \qquad (23.4)$$

with integration along the path of the varying fields **B** and **n**. Here $S(\Omega)$ is the (symmetrized) noise power spectrum of the fluctuating field $\hat{X}(t)$. In writing Eq. (23.4) we omitted the non-universal "boundary phase", determined by details of the initial preparation and the final read-out, and the integral of a full derivative, which vanishes for a closed loop [12].

The first term on the rhs of Eq. (23.4) describes the modification of the Berry phase by the noise [11] and arises, when the varying controlled field **B** rotates about the direction of the fluctuating field **n**. The second term gives the contribution due to rotation of the noise about the controlled field [12].

Below we find the generalization of Eq. (23.4) to the case of a multi-level system.

23.3 Structure of Berry phases in a multi-level system

Geometric phases in a multi-level system may be calculated by similar means. The Hamiltonian of a multilevel quantum system, weakly coupled to a reservoir, reads

$$\hat{\mathcal{H}} = \hat{\mathcal{H}}_0(t) + \hat{X}\hat{V}(t) + \hat{\mathcal{H}}_{\text{bath}}, \qquad (23.5)$$

where $\hat{\mathcal{H}}_0(t)$ is the Hamiltonian of the quantum system, which is varied in time via control parameters[4] $\hat{\mathcal{H}}_{\text{bath}}$ governs the dynamics of the bath. The second term describes the coupling between the system and the bath, we take in the form $\hat{X}\hat{V}$, where the field $\hat{X}$ of the bath describes the fluctuations and the operator $\hat{V}$ of the quantum system controls the coupling. We assume that $\hat{\mathcal{H}}_0$ and $\hat{V}$ are varied slowly in time. For later use we represent the Hamiltonian $\hat{\mathcal{H}}_0$ in terms of its (time-dependent) eigenstates and eigenenergies: $\hat{\mathcal{H}}_0 = \sum_k E_k(t) |k_t\rangle \langle k_t|$ (below we often omit the subindex t).

Let us recall that the concept of the Berry phase [1, 2, 26] is meaningful only for closed paths: During adiabatic evolution (without degeneracies and level crossings) the system, initially in an eigenstate $|k_0\rangle$, remains in an instantaneous eigenstate of the Hamiltonian $\hat{\mathcal{H}}_0(t)$. If finally, at $t = t_P$, the controlled Hamiltonian assumes its initial value[5] (i.e., the Hamiltonian $\hat{\mathcal{H}}_0(t)$ was varied over a closed path), the final state may differ from the initial state $|k_0\rangle$ only by a phase factor, and one can separate this phase into the dynamic and Berry's contributions. At intermediate times, when $\hat{\mathcal{H}}_0(t) \neq \hat{\mathcal{H}}_0(0)$, the notion of the relative phase is ambiguous.

Instead of the analysis of the relative phase of the components of the wave function one can equivalently follow the evolution of (the off-diagonal elements of) the density matrix. For quantum systems coupled to their environments the wave function is ill-defined, and one in fact has to analyze the density matrix. As we find below, under certain generic conditions (and, in particular, for weak noise) the evolution of an off-diagonal entry may be decoupled from the rest of the density matrix; however, during the evolution such an entry gets multiplied by a factor, which is not purely a phase factor, but also changes the amplitude (dephasing or decay of coherence). Let us remark that this suppression of amplitude (which also contains dynamic and geometric part in the adiabatic limit) is well-defined also at intermediate times, since it is not influenced by the phase uncertainty.

In connection with this discussion let us remark that there is certain freedom in the choice of the phase factors of the eigenstates $|k\rangle$, which however does not influence the final results. The gauge invariance w.r.t. this choice of the phases imposes constraints on the expressions for the geometric phases.

---

[4] Typically, in the discussion of geometric phases and adiabatic manipulations, one refers explicitly to a set of control parameters $\mathbf{R}$ and discusses paths and the geometry of loops in the parameter space. Here, for brevity of notation, we just discuss explicit time dependencies $\hat{\mathcal{H}}_0(t)$ and consider paths in the space of the Hamiltonians (the $\mathbf{B}$-space for two-level system); one can say that the space of Hamiltonians serves as the (natural) parameter space.

[5] It is sufficient that the eigenbasis of $\hat{\mathcal{H}}_0(t_P)$ is the same as at $t = 0$. By analogy with the spin-half case, one can say that the "direction" of the Hamiltonian should return to its initial value (cf. [12]).

23.3.1 BERRY PHASES IN AN ISOLATED SYSTEM

Let us first consider a closed coherent quantum system, that is a system, which does not interact with its environment. The analysis of the evolution is simplified by a transformation to the instantaneous eigenbasis: consider the transformation U, which maps a fixed basis (e.g., the eigenbasis of $\hat{\mathcal{H}}_0$ at $t = 0$) to the eigenbasis at time t, $U : |k_0\rangle \rightarrow |k_t\rangle$. Then we find that the evolution of the transformed wave function $\psi' = U^\dagger \psi$ is described by the Schrödinger equation with the Hamiltonian $\hat{\mathcal{H}}'_0 = U^\dagger \hat{\mathcal{H}}_0 U - i U^\dagger \dot{U}$. The first term is diagonal in the basis $|k_0\rangle$, and for non-degenerate levels in the adiabatic limit the second term is a weak perturbation. The matrix elements of this perturbations are $\langle k_0 | - i U^\dagger \dot{U} |l_0\rangle = -i \langle k| \partial_t |l\rangle$.

The effect of the perturbation for a quantum system without degenerate levels is two-fold: it modifies the energy levels:

$$\delta E_k = E'_k - E_k = -i \langle k| \partial_t |k\rangle , \tag{23.6}$$

and the eigenstates:

$$|k'\rangle = |k_0\rangle - i \sum_{l \neq k} \frac{\langle l| \partial_t |k\rangle}{E_k - E_l} |l_0\rangle . \tag{23.7}$$

This determines the evolution operator (for the transformed wave function ψ' and hence for ψ), which is diagonal in this basis with the eigenvalues $\exp[-i \int dt E'_k(t)]$. Note that the corrections to the eigenstates (23.7) modify the evolution operator only slightly, whereas the corrections to the eigenenergies (23.6) are multiplied by t in the exponents and result in much stronger changes of the evolution at growing t. (In particular, the possible difference between the primed and non-primed bases at the initial and final moments is a negligible effect.)

For a system with level degeneracies, the off-diagonal matrix elements of the perturbation $-i U^\dagger \dot{U}$ between degenerate states are also relevant. In fact, the evolution in each degenerate subspace is determined by the projection of this perturbation onto this subspace, and this may result in arbitrary unitary holonomic transformations in this subspace [26]. In the current paper we consider only systems without degenerate levels.

23.3.2 INFLUENCE OF FLUCTUATIONS

Our analysis of geometric phases and dephasing in an open system is based on the Bloch-Redfield approach; we derive a markovian master equation of motion for the reduced density matrix of the quantum system, which is coupled to a reservoir (see, e.g., Refs. [14–16, 27] for the derivation and the discussion). To take into account the slow variations of the Hamiltonian of the system, $\hat{\mathcal{H}}_0$, and the effect of the environment, $\hat{V}$, we perform the derivation in the primed representation (i.e., in the primed basis). We assume the following conditions for the time scales involved: $\tau_c, \Delta E^{-1} \ll t_P \ll T_2$, where τ_c is the noise correlation time, T_2 is the dephasing time scale (the decay time of 'coherences', i.e. off-diagonal elements of the density matrix), ΔE is energy gap in the spectrum of $\hat{\mathcal{H}}_0$. This implies, in particular, that the noise is weak and short-correlated [16], and that on the time scale t_P of the evolution the noise correlations are local and the coherence is not destroyed completely (and thus the phase information can be detected). In fact, such conditions should be specified independently for different matrix elements since the gaps to the neighboring levels, decay times, and correlation times of various noise components may differ. Moreover, for some matrix elements the decoherence may dominate over the coherent evolution, changing the character of the dynamics (cf. [28]). Here we want to understand the influence of the noise and do not aim at a full analysis of all possible regimes of behavior, and thus would limit ourselves to the case when the conditions specified hold uniformly for all relevant matrix elements. We further introduce the notation for the typical scale of variations, $\omega \sim 1/t_P$, and the typical adiabaticity parameter is $\omega/\Delta E$.

We analyze the phases accumulated by the system (and the dephasing) between times 0 and t_P. Detection of this phase may involve preparation of the initial state (e.g., a superposition of various eigenstates to facilitate observation of the relative phases) and the final direct or indirect measurement. We do not specify details of these events, and hence neglect their contributions to the evolution (boundary effects, cf. Refs. [9,12].

In general, the Bloch-Redfield equations couple all elements of the density matrix: $\partial \rho_{mn} = -i(E_m - E_n)\rho_{mn} - \sum_{kl} \Gamma_{mn}^{kl} \rho_{kl}$. If we consider the second term as a perturbation, we find that to the leading order the time dependence is $\rho_{mn} \propto \exp(-i(E_m - E_n)t)$. Then the rotating-wave approximation shows that only the coupling between matrix elements ρ_{mn} with equal level splittings $E_{mn} \equiv E_m - E_n$ is relevant, and the influence of ρ_{kl} onto ρ_{mn} with a different level splitting averages out on the times scale $|E_{kl} - E_{mn}|^{-1}$. In particular, when all the level splittings are different, each off-diagonal entry ρ_{mn} evolves on its own (Section 23.4.1). If two or more level splittings coincide, the corresponding matrix elements follow joint evolution (Section 23.4.2). In particular, the diagonal elements ρ_{nn} 'correspond' to the zero level splitting $(E_n - E_n = 0)$, and hence their relaxational dynamics are coupled (Section 23.5).

As indicated above, it is convenient to perform the analysis in the primed basis. In this basis the matrix elements of the fluctuating field are:

$$\delta V_{mn} = V_{n'm'} - V_{nm} = -i \sum_{k \neq n} \frac{\langle n| \partial_t |k\rangle}{E_n - E_k} V_{km} - i \sum_{k \neq m} \frac{\langle k| \partial_t |m\rangle}{E_m - E_k} V_{nk} . \quad (23.8)$$

23.4 Calculation of geometric phases and dephasing in multi-level systems

23.4.1 MULTI-LEVEL SYSTEM WITH NON-DEGENERATE ENERGY SPLITTINGS

Consider first a quantum system without Liouville degeneracies, that is such a system that all level splittings $E_{mn} = E_m - E_n$ are different throughout the evolution.[6] As we have discussed above, in this situation the dynamics of each off-diagonal element ρ_{mn} decouples from the dynamics of the other matrix elements. The Bloch-Redfield equation of motion for this element can be obtained from the following integro-differential master equation [14,15,27]:

$$(i\partial_t - E_{mn})\rho_{mn}(t) = -i \int_{-\infty}^{t} \left\langle \left[[|n_t\rangle \langle m_t|, \hat{\mathcal{V}}(t)], \hat{\mathcal{V}}(t_1) \right] \right\rangle dt_1. \quad (23.9)$$

Here $\mathcal{V}$ describes the effect of the environment on the quantum system (cf. Eq.(23.5)), and we assume that its average over noise realizations vanishes, when the bath is decoupled from the quantum system. The angle brackets in Eq. 23.9 stand for averaging over noise realizations and the state of the system: $\langle \ldots \rangle = \text{tr}(\ldots \rho)$. In our calculations, we take the perturbation in the form

$$\hat{\mathcal{V}} = \hat{X}\hat{V}, \quad (23.10)$$

where $\hat{X}$ is the fluctuating field of the bath, and $\hat{V}$ is an operator of the quantum system. For further analysis, we introduce the matrix elements of the perturbations in the instantaneous eigenbasis:

$$\hat{V} = \sum_{k,l} V_{kl} |k\rangle \langle l| . \quad (23.11)$$

Note that although a shift of $\hat{V}$ by a scalar operator $\propto \hat{1}$ does not influence the level splittings of the quantum system directly, it modifies the average value of $\hat{X}$ due to the response of the bath, and thus also contributes to the dynamics of the quantum system. $\hat{V}$ is hermitian and hence $V_{kl} = (V_{lk})^*$.

After using the Redfield and rotating-wave approximations, we find the markovian equation of evolution for ρ_{mn} in the form:

$$\partial_t \rho_{mn} = \left(-iE_{mn} - \Gamma_{mn} + i(\dot{\Phi}_{BP}^{0,mn} + \delta\dot{\Phi}_{BP}^{mn}) \right) \rho_{mn}. \quad (23.12)$$

where the first term describes the dynamic phase for a coherent system, Γ_{mn} describes the dynamical effects of the bath (the dephasing and the Lamb shift), and for brevity we refer to it as the dephasing rate; $\dot{\Phi}_{BP}^{0,mn} = -(\delta E_m - \delta E_n) = i(\langle m| \partial_t |m\rangle - \langle n| \partial_t |n\rangle)$ (cf. Eq. (23.6)) describes the contribution to the Berry

---

[6] In fact, we consider a pair of levels m, n, such that E_{mn} is different from all other level splittings.

phase in a closed system, and $\delta\dot{\Phi}_{\mathrm{BP}}$ describes the modification of the Berry phase by the environment. The latter contribution is complex and contains the modification of the geometric phase and also the geometric dephasing.

The dynamical dephasing rate is

$$\Gamma_{mn} = -i \int \frac{d\Omega}{2\pi} \left(\sum_k \left[\frac{S_c(\Omega)|V_{mk}|^2}{\Omega - E_{mk} - i0} + \frac{S_c(-\Omega)|V_{nk}|^2}{\Omega + E_{nk} - i0} \right] - V_{mm}V_{nn}\frac{2S(\Omega)}{\Omega - i0} \right).$$
(23.13)

Here $S_c(\Omega)$ is the Fourier image of the noise correlator $S_c(t-t_1) = \langle \hat{X}(t)\hat{X}(t_1)\rangle$, and the noise power $S(\Omega)$ is the symmetrized correlator, i.e., $S(\Omega) = (S_c(\Omega)+ S_c(-\Omega))/2$. Notice that $S_c(\Omega)$ is real, $(S_c(\Omega))^* = S_c(\Omega)$.

The noise-induced geometric phase emerges from the following expression:

$$i\delta\dot{\Phi}_{\mathrm{BP}}^{mn} = i \int \frac{d\Omega}{2\pi} \Bigg\{$$

$$\sum_k \left[\frac{S_c(\Omega)\,\delta(|V_{mk}|^2)}{\Omega - E_{mk} - i0} + \frac{S_c(-\Omega)\,\delta(|V_{nk}|^2)}{\Omega + E_{nk} - i0} \right] - \frac{2S(\Omega)}{\Omega - i0}\delta(V_{mm}V_{nn})$$

$$+ \sum_k \left[\frac{|V_{mk}|^2 S_c(\Omega)}{(\Omega - E_{mk} - i0)^2}\delta E_{mk} - \frac{|V_{nk}|^2 S_c(-\Omega)}{(\Omega + E_{nk} - i0)^2}\delta E_{nk} \right]$$

$$+ \sum_k \left[\frac{i|V_{mk}|^2 S_c(\Omega)}{(\Omega - E_{mk} - i0)^3}\dot{E}_{mk} - \frac{i|V_{nk}|^2 S_c(-\Omega)}{(\Omega + E_{nk} - i0)^3}\dot{E}_{nk} \right]$$

$$+ \sum_k \left[\frac{iS_c(\Omega)}{(\Omega - E_{mk} - i0)^2}V_{mk}\dot{V}_{km} + \frac{iS_c(-\Omega)}{(\Omega + E_{nk} - i0)^2}\dot{V}_{nk}V_{kn} \right]$$

$$- \frac{iS_c(\Omega)}{(\Omega - i0)^2}\dot{V}_{mm}V_{nn} - \frac{iS_c(-\Omega)}{(\Omega - i0)^2}V_{mm}\dot{V}_{nn} \Bigg\},$$
(23.14)

where $\dot{V}_{kl}$ is a notation for $\partial_t(V_{kl}) = \partial_t(\langle k| \hat{V} |l\rangle)$, and this time derivative includes the effects of the time dependence of the perturbation $\hat{V}$ ('rotation of noise') as well as of the basis states $|k\rangle$, $|l\rangle$ (variation of the Hamiltonian $\hat{\mathcal{H}}_0(t)$). Further, $\dot{E}_{kl}$ is the time derivative of the level splitting, i.e., of the difference between the kth and lth eigenenergies of $\hat{\mathcal{H}}_0(t)$. Finally, δV_{kl} and δE_{kl} arise due to the difference between the primed and non-primed frames: $\delta V_{kl} = V'_{kl} - V_{kl}$ as given by Eq. (23.8) and $\delta E_{kl} = \delta E_k - \delta E_l$, where $\delta E_k = -i\langle k| \partial_r |k\rangle$ as given by Eq. (23.6).

Let us discuss these results (23.13) and (23.14). Note that the (dynamical) dephasing rate can be expressed via the outgoing transition rates from levels m, n to other levels and the rate of pure dephasing (cf. Refs. [16, 17, 29]):

$$\mathrm{Re}\,\Gamma_{mn} = \frac{1}{2}\sum_{k\neq n}\Gamma^{\mathrm{rel}}_{k\leftarrow n} + \frac{1}{2}\sum_{k\neq m}\Gamma^{\mathrm{rel}}_{k\leftarrow m} + \Gamma^{\varphi}_{mn},$$
(23.15)

where $\Gamma^{\mathrm{rel}}_{j\leftarrow i} = |V_{ij}|^2 S_c(E_{ij})$ is the relaxation rate $|i\rangle \to |j\rangle$, and $\Gamma^{\varphi}_{mn} = \frac{1}{2}(V_{mm} - V_{nn})^2 S(0)$ is the pure-dephasing rate of ρ_{mn}.

As for the modification of the level splitting, the "Lamb shift", it can be presented as:

$$\text{Im}\, \Gamma_{mn} = \delta E_m - \delta E_n\,, \qquad (23.16)$$

where $\delta E_m = \sum_k \delta E_m^k$, and

$$\delta E_i^j = |V_{ij}|^2 P.V. \int \frac{d\Omega}{2\pi} \frac{S_c(\Omega)}{E_{ij} - \Omega} \qquad (23.17)$$

is the modification of the energy of level i by the coupling (virtual transitions) to level j. Note that for a two-level system the quantities $\delta E_1^2 - \delta E_2^1$ and $\text{Re}\,\Gamma_{12}$ depend only on the symmetrized noise correlator (cf. Ref. [11, 24]).

Now let us discuss the geometric contribution (23.14). The first two lines, which can be denoted as $-\delta \Gamma_{mn}$, show the difference of the expression (23.13) in the primed and non-rimed frames. In the fourth line one can replace $V_{mk}\dot{V}_{km}$ by $(\frac{1}{2}\partial_t + i\dot{\Phi}_{km})|V_{mk}|^2$, where Φ_{km} is the phase of $V_{km} = |V_{km}|e^{i\Phi_{km}}$, and similarly $\dot{V}_{nk}V_{kn} = (\frac{1}{2}\partial_t + i\dot{\Phi}_{nk})|V_{nk}|^2$. Then the first terms here, $\propto \partial_t|V|^2$, form a full derivative together with the third line (which contains $\dot{E}_{mk}$), and this contribution vanishes after integration over a closed loop; whereas the terms with $\dot{\Phi}$'s can be combined with the second line, and these time derivatives of phases and δE's enter only in combinations $\delta E_{mk} + i\dot{\Phi}_{mk}$. This fact is a consequence of the gauge invariance w.r.t. to multiplication of the basis states $|k_t\rangle$ by arbitrary time-dependent phase factors.

23.4.2 SYSTEM WITH EQUAL ENERGY SPLITTINGS

Consider now the simplest example of a system with a Liouville degeneracy, i.e., a system with two pairs of energy levels, $|n_1\rangle$, $|m_1\rangle$ and $|n_2\rangle$, $|m_2\rangle$, with equal splittings: $E_{n_1 m_1} = E_{n_2 m_2} \equiv E_{nm}$ (Fig. 23.1). This example allows one to understand also the evolution in more complicated situations, for instance, with a larger set of level pairs with the same energy splitting. We assume for simplicity that the splittings coincide at all times and that all other energy splittings differ from these two.

Using the Bloch-Redfield approach, we find the joint equation of motion for the off-diagonal matrix elements of the density matrix to the leading order in the adiabatic parameter:

$$\frac{d}{dt}\begin{pmatrix} \rho_{m_1 n_1} \\ \rho_{m_2 n_2} \end{pmatrix} = \Bigg[-iE_{m_1 n_1}$$

$$+ i\begin{pmatrix} C^1 & 0 \\ 0 & C^2 \end{pmatrix} + \begin{pmatrix} -\Gamma^{11} & \Gamma^{12} \\ \Gamma^{21} & -\Gamma^{22} \end{pmatrix} + i\begin{pmatrix} a^{11} & a^{12} \\ a^{21} & a^{22} \end{pmatrix} \Bigg] \begin{pmatrix} \rho_{m_1 n_1} \\ \rho_{m_2 n_2} \end{pmatrix} \quad (23.18)$$

Here the quantities $C^{1,2} = \dot{\Phi}_{\text{BP}}^{0,m_i n_i}$ give rise to the standard Berry phases for the two off-diagonal matrix elements; the rates Γ describe the contribution of

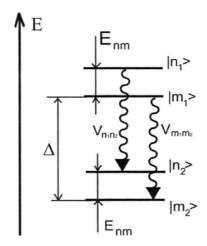

Fig. 23.1. A scheme of energy levels with two pairs having equal splittings. The relaxation processes from one pair to the other make the corresponding elements of the density matrix evolving dependently.

the noise to the dynamical phase (and dephasing), whereas a's are responsible for the geometric contributions. The diagonal entries Γ^{ii} of these matrices are given by the expressions $\Gamma_{m_i n_i}$ from the previous section, Eq. (23.13), and a^{ii} are given by $\dot{\delta\Phi}_{\mathrm{BP}}^{m_i n_i}$ from the previous section, Eq. (23.14). Note that since Eq. (23.18) is a matrix equation, we have to keep full time derivatives, e.g., in the a-matrix, even for evolution along a closed loop.

For the off-diagonal entries we find:

$$\Gamma^{12} = -iV_{m_1 m_2}V_{n_2 n_1}\int \frac{d\Omega}{2\pi}\left(\frac{S_c(\Omega)}{\Omega + \Delta - i0} + \frac{S_c(-\Omega)}{\Omega - \Delta - i0}\right) \quad (23.19)$$

$$= V_{m_1 m_2}V_{n_2 n_1}S_c(-\Delta). \quad (23.20)$$

Here we have introduced the notation $\Delta \equiv E_{m_1 m_2} = E_{n_1 n_2}$.

The geometric contributions are controlled by

$$a^{12} = -i\delta(V_{m_1 m_2}V_{n_2 n_1})S_c(-\Delta) \quad (23.21)$$

$$-\int \frac{d\Omega}{2\pi}\left\{V_{m_1 m_2}V_{n_2 n_1}\left[\frac{S_c(\Omega)}{(\Omega + \Delta - i0)^2}\delta E_{n_2 n_1} - \frac{S_c(-\Omega)}{(\Omega - \Delta - i0)^2}\delta E_{m_2 m_1}\right]\right.$$

$$-i\frac{S_c(\Omega)\dot{V}_{m_1 m_2}V_{n_2 n_1}}{(\Omega + \Delta - i0)^2} - i\frac{S_c(-\Omega)V_{m_1 m_2}\dot{V}_{n_2 n_1}}{(\Omega - \Delta - i0)^2}$$

$$\left.-i\frac{S_c(\Omega)V_{m_1 m_2}V_{n_2 n_1}}{(\Omega + \Delta - i0)^3}\dot{\Delta} - i\frac{S_c(-\Omega)V_{m_1 m_2}V_{n_2 n_1}}{(\Omega - \Delta - i0)^3}\dot{\Delta}\right\}. \quad (23.22)$$

Γ^{21} and a^{21} are given by similar expressions with the substitution $1 \leftrightarrow 2$. In the case of an arbitrary number N of equal splittings in the energy spectrum, we would arrive at non-diagonal $N \times N$ matrices Γ^{ij} and a^{ij}, given by similar expressions. Note also that the evolution in Eq. (23.18) is typically dominated by the coherent terms $C^{1,2}$; when these coherent terms dominate over the noise-induced contributions $\check{\Gamma}$ and $\check{a}$, the off-diagonal entries can be neglected as they provide only small corrections (unless C^1 and C^2 coincide or are very close during the evolution).

Notice that Γ^{21} (the 'transition rate between $\rho_{m_1 n_1}$ and $\rho_{m_2 n_2}$') is related to the transition rates between the levels m_1 and m_2 and between the levels n_1 nd n_2, both with the same splitting Δ:

$$|\Gamma^{21}| = \sqrt{\Gamma^{\mathrm{rel}}_{m_2 \leftarrow m_1} \Gamma^{\mathrm{rel}}_{n_2 \leftarrow n_1}} \tag{23.23}$$

and similarly $|\Gamma^{12}| = \sqrt{\Gamma^{\mathrm{rel}}_{m_1 \leftarrow m_2} \Gamma^{\mathrm{rel}}_{n_1 \leftarrow n_2}}$, in the lowest-order of the expansion in V. In particular, $|\Gamma^{21}| \leq \frac{1}{2}(\Gamma^{\mathrm{rel}}_{m_1 \leftarrow m_2} + \Gamma^{\mathrm{rel}}_{n_1 \leftarrow n_2}) \leq \mathrm{Re}\,\Gamma_{m_1 n_1} = \mathrm{Re}\,\Gamma^{11}$.

Since the noise couples the evolution of two off-diagonal entries, by measuring only one of them, one can deduce information about both Berry phases. Consider, for example, two level pairs, each of which, in the absence of fluctuations, acquires a relative geometric phase uniformly in time, with the rates $C^{1,2}$. Let us assume further that the relaxation induces transitions only in one direction, so that $\Gamma^{21} \neq 0$ but $\Gamma^{12} = 0$. Then one finds

$$\rho_{m_2 n_2}(t) = e^{i\Phi^2_{\mathrm{BP}} - \Gamma^{22}t} \left[\rho_{m_2 n_2}(0) + \frac{\Gamma^{21}}{i(C^2 - C^1) + (\Gamma^{11} - \Gamma^{22})} \rho_{m_1 n_1}(0) \right]$$

$$- e^{i\Phi^1_{\mathrm{BP}} - \Gamma^{11}t} \frac{\Gamma^{21}}{i(C^2 - C^1) + (\Gamma^{11} - \Gamma^{22})} \rho_{m_1 n_1}(0), \tag{23.24}$$

where $\Phi^i_{\mathrm{BP}} = C^i t$ is the Berry phase, which would be acquired by $\rho_{m_i n_i}$ in the absence of fluctuations ($i = 1, 2$). In particular, the effect of the Berry phase Φ^1_{BP} may be observed in $\rho_{m_2 n_2}$ at $t \sim 1/\Gamma \sim C^i \sim C^1 - C^2$.

23.5 Geometric relaxation

Let us now consider the dynamics of the diagonal entries of the density matrix, that is transitions between the levels (relaxation/excitation depending on the direction in energy). All these entries ρ_{nn} correspond to zero energy difference, $E_n - E_n = 0$, and their dynamics, in general, is coupled. Since we consider systems without degenerate levels, no other entries are coupled with the diagonal elements in the rotating-wave approximation. Thus, the dynamics of the diagonal entries is described by a rate equation. From the Bloch-Redfield formalism, one finds the transition rates between two distinct levels k and m (from k to m):

$$\Gamma^{\mathrm{rel}}_{m \leftarrow k} = |V_{mk}|^2 S_{\mathrm{c}}(E_{km}). \tag{23.25}$$

The first adiabatic correction to this expression is

$$\gamma_{m \leftarrow k}^{\mathrm{rel}} = \delta(|V_{mk}|^2) S_c(E_{km}) + |V_{mk}|^2 S_c'(E_{km}) \delta E_{km}$$
$$+ \int \frac{d\Omega}{2\pi} \left[\left(\frac{S_c(\Omega)|V_{mk}|^2}{(\Omega - E_{km} - i0)^3} - \frac{S_c(-\Omega)|V_{mk}|^2}{(\Omega + E_{km} - i0)^3} \right) \dot{E}_{km} \right.$$
$$\left. + \frac{S_c(\Omega)}{(\Omega - E_{km} - i0)^2} \dot{V}_{mk} V_{km} + \frac{S_c(-\Omega)}{(\Omega + E_{km} - i0)^2} V_{mk} \dot{V}_{km} \right],$$

$$(23.26)$$

and this correction is responsible for the geometric contributions to the relaxational dynamics.

Note that the relaxation rates are real and that the dynamic and geometric contributions to relaxation are gauge-invariant and well-defined for open paths. Similar to geometric phases and geometric dephasing [11] the geometric contribution to relaxation changes sign, when the same path is traversed in the opposite direction (i.e., when the Hamiltonian and/or the parameters of the fluctuations are varied backwards along the same path).

23.6 Conclusions

In this paper we have analyzed the influence of fluctuations on the adiabatic evolution in multi-level quantum systems. During adiabatic evolution the off-diagonal entries of the system's density matrix acquire dynamic and geometric phases. We found that the noise modifies the values of the geometric phases and also induces geometric contributions to dephasing and relaxation. In a multi-level system without degenerate levels and without equal energy splittings for different level pairs, the dynamics of different off-diagonal entries are decoupled, whereas in systems with equal level splittings (but no degenerate levels) the equations of motion for several off-diagonal entries may be coupled, which results in more complicated evolutions. Similarly, the dynamics of diagonal entries (occupations of the eigenstates) are coupled and are described by the rate equations, which also contain geometric contributions. We acknowledge useful discussions with A. Shnirman. This work was partially supported by the projects INTAS 05-1000008-7923, MD-4092.2007.2, and the Dynasty foundation.

References

1. M.V. Berry, Proc. R. Soc. Lond. **392**, 45 (1984)
2. A. Shapere, F. Wilczek (eds.), *Geometric phases in physics* (World Scientific, 1989)
3. D. Gamliel, J.H. Freed, Phys. Rev. A **39**, 3238 (1989)
4. D. Ellinas, S.M. Barnett, M.A. Dupertuis, Phys. Rev. A **39**, 3228 (1989)

5. A. Carollo, I. Fuentes-Guridi, M.F. Santos, V. Vedral, Phys. Rev. Lett. **90**, 160402 (2003)
6. K.M. Fonseca Romero, A.C. Aguiar Pinto, M.T. Thomaz, Physica A **307**, 142 (2002)
7. F. Gaitan, Phys. Rev. A **58**, 1665 (1998)
8. J.E. Avron, A. Elgart, Phys. Rev. A **58**, 4300 (1998)
9. R.S. Whitney, Y. Gefen, Phys. Rev. Lett. **90**, 190402 (2003)
10. G. De Chiara, G.M. Palma, Phys. Rev. Lett. **91**, 090404 (2003)
11. R.S. Whitney, Yu. Makhlin, A. Shnirman, Y. Gefen, Phys. Rev. Lett. **94**, 070407 (2005)
12. S. Syzranov, Yu. Makhlin. in preparation
13. F. Bloch, Phys. Rev. **70**, 460 (1946)
14. F. Bloch, Phys. Rev. **105**, 1206 (1957)
15. A.G. Redfield, IBM J. Res. Dev. **1**, 19 (1957)
16. Yu. Makhlin, G. Schön, A. Shnirman, in *New Directions in Mesoscopic Physics (Towards Nanoscience)*, ed. by R. Fazio, V.F. Gantmakher, Y. Imry (Springer, 2003), pp. 197–224
17. J. Schriefl, Dephasing in coupled qubits. Master's thesis, Universität Karlsruhe/ ENS Lyon (2002)
18. P.J. Leek, J.M. Fink, A. Blais, R. Bianchetti, M. Goppl, J.M. Gambetta, D.I. Schuster, L. Frunzio, R.J. Schoelkopf, A. Wallraff, Science **318**, 1889 (2007)
19. M. Möttönen, J.J. Vartiainen, J.P. Pekola, arXiv:0710.5623 (2007)
20. W. Wernsdorfer, R. Sessoli, Science **284**, 133 (1999)
21. J.A. Jones, V. Vedral, A. Ekert, G. Castagnoli, Nature **403**, 869 (2000)
22. W. Wernsdorfer, M. Soler, G. Christou, D.N. Hendrickson, J. Appl. Phys. **91**, 7164 (2002)
23. J.B. Yau, E.P. De Poortere, M. Shayegan, Phys. Rev. Lett. **88**, 146801 (2002)
24. R.S. Whitney, Yu. Makhlin, A. Shnirman, Y. Gefen, in *Theory of Quantum Transport in Metallic and Hybrid Nanostructures*, ed. by A. Glatz, V. Kozub, V. Vinokur (Springer, 2006), pp. 9–23
25. S. Syzranov, Influence of noise on Berry phases in multilevel systems. Master's thesis, MIPT (2007). (in Russian)
26. F. Wilczek, A. Zee, Phys. Rev. Lett. **52**, 2111 (1984)
27. C.W. Gardiner, P. Zoller, *Quantum Noise* (Springer, 2000)
28. Yu. Makhlin, G. Schön, A. Shnirman, Rev. Mod. Phys. **73**, 357 (2001)
29. C.P. Slichter, *Principles of magnetic resonance* (Harper & Row, 1963)

24

MAGNETIC AND TRANSPORT PROPERTIES
OF NANOCRYSTALLINE TITANIUM CARBIDE
IN CARBON MATRIX

N. GUSKOS[1,2], E.A. ANAGNOSTAKIS[1],
K.A. KARKAS[1], A. GUSKOS[2], A. BIEDUNKIEWICZ[3],
AND P. FIGIEL[3]

[1] *Solid State Physics, Department of Physics, University of Athens,
Panepistimiopolis, 15 784 Zografos, Athens, Greece.*
ngouskos@phys.uoa.gr
[2] *Institute of Physics, Szczecin University of Technology, Al. Piastow
17, 70-310 Szczecin, Poland*
[3] *Mechanical Department, Szczecin University of Technology, Poland*

Abstract. Samples of titanium carbide (TiC_x) in a carbon matrix have been prepared by the nonhydrolytic sol-gel process. The nanocomposite powder samples
containing one the TiC_x of average size ca. 30 nm encapsulated in carbon cages of 3
wt.% and other the TiC_x in carbon matrix (ca. 10 wt.%) have been obtained. The
temperature dependence of the EPR spectra of titanium carbide has shown coexistence of two different paramagnetic centers, one arising from conducting electrons
and the other from trivalent titanium ion complexes. Comparison with a similar titanium nitride (TiN_x) is made, where no EPR spectra of trivalent titanium ions exist.
The titanium nitride has shown only the EPR spectra arising from magnetic localized centers and not from trivalent titanium ion complexes. The magnetic ordering
and superconducting states are observed in titanium carbide in the low temperature
region, while in titanium nitride only the later state is recorded. The titanium nitride
is a good conducting material while the titanium carbide shows an extraordinary
behaviour, especially in the higher temperature region where a sharp jump in conductivity is recorded about 250 K. It is suggested that the disorder-order processes
are more intense in the nonstoichiometric titanium carbide.

Keywords: titanium carbide, magnetic, transport properties, EPR spectra

J. Bonča, S. Kruchinin (eds.), *Electron Transport in Nanosystems.*   315
© Springer Science + Business Media B.V. 2008

24.1 Introduction

Titanium carbide (TiC_x) and titanium nitride (TiN_x) belong to a class of important technological materials with very interesting physical properties [1–7]. As refractory materials they have gained much attention due to their extraordinary hardness (wear resistance and stability at very high temperatures), accentuating the promise of $Ti - C$ coating technologies [16], and electrical transport [8], and their electrical and thermal conductivities being close to those observed for pure metals. The physicochemical properties of materials containing titanium carbides are being found to be depending strongly upon preparation procedure, kind and degree of doping, and size and peculiarities of (nano)particles involved in their synthesis.

Titanium compounds, and especially titanium carbide, are used in the fabrication of cutting tools (as protective coatings) and, owing also to their high melting points (being over 3,300 K for titanium carbide), in the production of high-temperature alloys for aerospace engineering applications. TiC nanopowder has been fruitfully utilised as a reinforcement agent within a Mg host, through in situ formation in molten [24]. It has been observed [9–11] that it is difficult to produce a single phase material obeying its nominal chemical formula. These compounds are the most intensively studied, both experimentally and theoretically concerning their physical properties, especially the electronic structure, e.g. [5–7, 9, 10].

A few works have been published that are devoted to studying the electronic properties of titanium carbide, both separate and embedded in carbon matrix, by using the EPR method [11–15]. Extreme broadness of the EPR spectra was observed reflecting the high electrical conductivity of the former samples and suggesting that the electronic behaviour may be not dissimilar to the one of bulk metals [12] in which the electrons exhibit fairly short spin-lattice relaxation times tantamount to excessively wide spectral lines. The composites of nanocrystalline titanium carbide dispersed in carbon matrix synthesised by the nonhydrolytic sol-gel process have shown the coexistence of magnetic centres due to conducting electrons and trivalent titanium ion complexes. The possibility of a secondary contribution to EPR effect from localised defect spins coupled to conduction electrons has also been examined [15].

Recently [15], the temperature dependence of EPR spectra of TiC_x/C samples prepared by differing methods has pointed to an additional broad and intense resonance line attributable to trivalent Ti ions, with the respective resonance magnetic field shifting largely as temperature decreases, indicatively of the emanation of strong magnetic interactions. Also, the ageing effect has been studied for such samples, monitored to both suppress the conductivity electron narrow resonance and generate an additional, satellite resonance attributable to the oxidation process being liable to form new trivalent Ti ion complexes. The trivalent titanium ions could be influencing essentially the physical properties of titanium carbide and the existence of even a very low percentage of magnetic clusters or agglomerates into the nanocomposite system could be traceably modifying the physical properties of such functional materials, thus providing a positive change for their application in tribology and in metallurgy. Furthermore, the electric and magnetic properties of nanoporous carbon systems are expected [23] to be rather similar to those of nanohorns and nanotubes, as having in common in their major configuration a quasiamorphous net of randomly oriented nanometric graphite fragments. Interesting conducting properties of titanium carbide, in particular a semiconductor-to-metal transition, have been already identified in nanocrystalline samples [8] and related to disorder connected with the presence of trivalent Ti ions.

The aim of this review is to present the temperature dependence of the EPR spectra, magnetisation and electrical transport response of titanium carbide compounds dispersed in carbon matrix, as synthesised by using the method of nonhydrolytic sol-gel process. This provides additional data for a better understanding of complicated disorder-order processes in nonstoichiometric titanium carbide.

24.2 Experimental

The nanocomposite powder samples of the TiC_x encapsulated in carbon matrix were prepared using organotitanium gel precursor. The process preparation of the precursor was described elsewhere [17]. Average crystallites size of titanium carbide in these samplee was equal ca. 30 nm. The titanium carbide nanoparticles were released from carbon using ultrasounds after dispersion in the alcohol. Samples were characterized using SEM, TEM, XRD and chemical (TOC) analytical techniques.

DC electrical resistivity transport measurements for the nanocrystalline TiC/C composite were made in the 90–320 K temperature range for small pellets (about 6 mm in diameter, with a thickness of 3–4 mm) prepared under pressure of 70–80 bar and with the use of a special dielectric glue. The resistance was measured with a Keithley 181 electrometer according to the two-point geometry, with highest limit that of $2 \times 10^{11} \Omega$ for the circuitry of the experiment. The measurements of temperature dependence of magnetization were performed on a SQUID MPMS model by Quantum Design, operating in the 1.8–400 K temperature range, and in the 0–5.5 T range of magnetic fields. EPR measurement were performed on powder samples sealed in quartz tubes using a Bruker E 500 X-band spectrometer ($\nu = 9.4\,\mathrm{GHz}$) with 100 kHz field modulation and an Oxford flow cryostat for temperature-dependent measurements (4–300 K). Prior to the measurements, the samples had been magnetised by a steady magnetic field of 1.4 T for saturating any domain structure. The field was scaled with a NMR gaussmeter, while the g factor and EPR intensity were measured with respect to a standard calibrating sample.

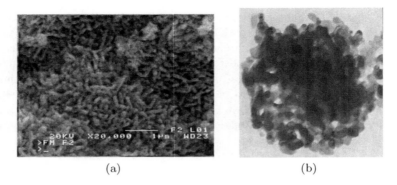

(a)   (b)

Fig. 24.1. The SEM picture of the titanium carbide dispersed in carbon cages (a) and the TEM picture of the titanium carbide released from carbon matrix (b).

24.3 Results and discussion

Figure 24.1a presents the scaning electron microscopy (SEM) picture of TiC_x/C nanocomposite, obtained by a JEOL JSM 6100 scanning electron microscope. The titanium carbide is surrounded by carbon and Fig. 24.1b is showing TEM micrographs of titanium carbide nanoparticles released from carbon matrix. The nanocrystallites in form of cubic-like of about 30 nm in average size are seen, the size of titanium carbide nanoparticles is essentially smaller. Modelling of the atomic structure [32] and electronic and energetics (especially as contrasted to "bulk" energetics [33]) of nanostructurised titanium carbides has begun rather recently and has up to now mostly concerned the quasi-zero-dimensional cluster forms of TiC, termed the nanocrystallites of titanium carbide [6]. The extensive family of TiC nanoclusters includes nanocrystallites with $[C]/[Ti] \approx 1$, their structure retaining the cubic, type B1 with space group Fm3m, symmetry of the condensed titanium monocarbide. Extended, quasi-unidimensional, TiC nanoparticles can be divided into hollow unidimensional cylindrical forms (tubes) and monolithic unidimensional nanoscale entities of various forms (nanorods, nanofibers, nanoribbons), configurations (from ideal crystalline to amorphous), and chemical compositions. Substances having layer phases regularly act as precursors of nanotubes, whilst nanowires derive from a broader compass of inorganic compounds possessing a variety of crystal structures, as the carbides, nitrides, oxides, and silicides of p- and d-orbital metals. Extended unidimensional TiC nanoforms, in particular, preserving the cubic structure of the original phase, have got faceting of the nanowalls and polygonal nanosections, capable of being visualized as extended nanocrystallites.

Figure 24.2 shows the temperature and magnetic-field dependence of the magnetisation of titanium carbide dispersed in a carbon matrix. A strong magnetic interaction could be inferred. The higher oxidation states of titanium ions are diamagnetic and recently the EPR measurements have shown

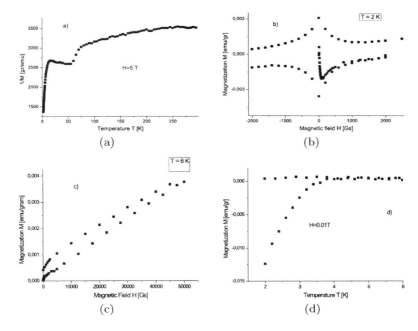

Fig. 24.2. The temperature dependence of the inverse magnetization of titanium carbide dispersed in carbon (a), magnetic field dependence of the magnetization at 2 K (b), magnetic field dependence of the magnetization at 6 K (c) and temperature dependence of the magnetization in ZFC and FC modes (d).

the appearance of the lower state of oxidation (trivalent titanium ions) which is paramagnetic [15]. The temperature dependence of the inverse dc magnetisation $M^{-1}(T)$ for TiC_x/C, obtained from dc-field-cooled (FC) mode (5 T) (Fig. 24.2a) shows an amazing behaviour: At higher temperatures the magnetisation follows the Curie-Weiss law yielding a positive Curie-Weiss temperature ($\theta = 42(1)K$), while two critical transitions have appeared at lower temperatures (below 60 K). Figure 24.2b shows the field dependence of the isothermal magnetisation M(H) at 2 K.

The observed magnetisation forms a hysteresis loop and the saturation is achieved in fields above $6T$. It could imply that the ferromagnetic ordering dominates at lower temperatures. The field dependence of the magnetisation at higher temperature (6 K) displays a similar character with saturation starting from $4.5T$ (Fig. 24.2c). A similar behaviour of magnetisation was recorded for samples with lower concentration of titanium carbide and with higher concentration of carbon, where the value of magnetisation was lower. Figure 24.2d presents the temperature dependence of magnetisation in zero-field-cooled (ZFC) and FC ($H = 100\,\mathrm{Gs}$) modes. A strong diamagnetic signal is observed which suggests that below 3.5 K a superconducting state is formed that was previously observed below 3 K [3].

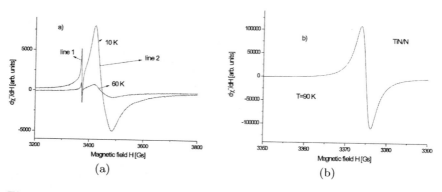

Fig. 24.3. The some EPR spectra of titanium carbide dispersed in carbon (a) and titanium nitride dispersed in carbon (b).

Figure 24.3 presents the EPR spectra of titanium carbide and nitride dispersed in carbon. The EPR spectra of the nanocrystalline titanium carbide/carbon nanocomposite at two different temperatures (of 10 and 60 K) show the presence of two paramagnetic centers, one from conducting electrons (line 1) and the second from trivalent titanium ions (line 2) (Fig. 24.3a). Nanocrystalline titanium nitride nanocomposite produced only the spectrum from magnetic localized centers (Fig. 24.3b) [13–15]. The titanium carbide EPR spectral line of titanium ions has appeared below 120 K and the increase of its intensity with decreasing temperatures is greater than for the resonance line arising from conduction electrons. The titanium ion-connected spectral line behaves versus temperature, in terms of locus and amplitude, in compliance with already monitored [25] trends for $BaTiO_3$ ceramics. The EPR line of trivalent titanium ion complexes is centered at $g_{eff} = 1.95(1)$ with linewidth $\Delta H_{pp} = 62.5(1)Gs$, whereas the very narrow line arising from conducting electrons is centered at $g_{eff} = 2.0026(3)$ with linewidth $\Delta H_{pp} = 2.6(1)Gs$, at 60 K. The g-factor of the narrow line remains very close to the free electron g-value ($g_e = 2.0023$).

The conductivity electron-related resonance of carbide titanium is especially believed to be regulated by free electron spin resonance in the graphite matrix hosting the TiC_x/C nanotubes; that is to be stemming from pure carbon contribution out of the overall synthetic nanocomposite. The conduction electron spin resonance of graphite was initially investigated by Wagoner in 1960 [27] for a natural single crystal specimen. Similar studies followed, reconfirming adequately his findings for a variety of well-defined graphite samples: The EPR absorption derivative spectrum is typically of Dysonian lineshape, indicating a large g-anisotropy and a narrow linewidth increasing with temperature lowering [28]. In highly oriented pyrolytic graphite (HOPG), proton irradiation appears through the resulting EPR spectrum (dominated by carbon electron spin resonance and pertaining to a Curie temperature well above

room temperature) to be inducing ferromagnetism [29], thus paving the way
of nanopatterned ferromagnets of light elements.

In the McClure - Yaret theory [30], the g-value anisotropy vanishes for two-
dimensional graphite, owing to complete mutual compensation of the electron
and hole contributions. This prediction has been corrected by including the
contribution of the spin - orbit splitting of the Landau levels [31]. Hence, the
g-parameter anisotropy may be notionally deduced on the basis of the spin -
orbit coupling parameter, the effective electron density at a given ambient
temperature and working field, and the electron orbital susceptibility. Through
such a perception, the large reduction in the g-parameter anisotropy after
proton irradiation of HOPG, coexistent with a preservation of the large orbital
susceptibility anisotropy, is interwoven with an increase in the effective carrier
density.

Figures 24.4 and 24.5 present the temperature dependence of the EPR
spectrum of conduction electrons of titanium carbide and magnetic localized

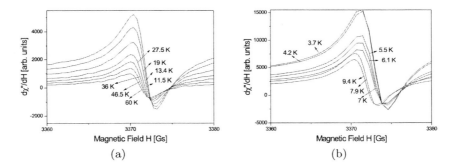

Fig. 24.4. The temperature dependence of the EPR spectrum of conducting
electrons in titanium carbide/carbon nanocomposite.

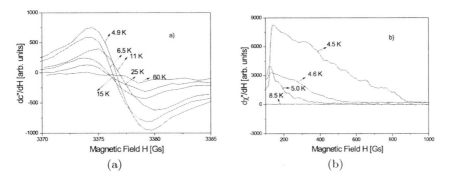

Fig. 24.5. The temperature dependence of the EPR spectrum of magnetic localized centers in titanium nitride/carbon nanocomposite (a) and the temperature
dependence of the zero-field absorption non-resonance line (b).

centers nitride carbon nanocomposites. The EPR lines could be fitted very well by using Dysonian lineshapes, which suggests that the system is strongly conducting for titanium carbide and Lorentzian for nitride carbide [13]. At room temperature, the average g-factors and linewidths are the following $g_{eff} = 2.0030(5)$ with $\Delta H = 2.40(2) Gs$ for titanium nitride. The resonance field shifts to higher values with temperature decrease in the high temperature range (Fig. 24.5a). For titanium carbide, a strongly anomalous behavior below 4.2 K is observed with opposite shift of the resonance magnetic field. For titanium nitride in the high temperature region a similar behavior is recorded. Additionally, for titanium nitride a zero-field non-resonance absorption is observed (Fig. 24.5b). The intensity of this signal increases strongly with decreasing temperature and above 5.5 K the signal disappears and this temperature is typical for TiN. In other two samples the superconducting state has not been observed down to 4 K.

Figure 24.6 presents the temperature dependence of the intensity, linewidth and resonance field for titanium carbide dispersed in carbon. The all these

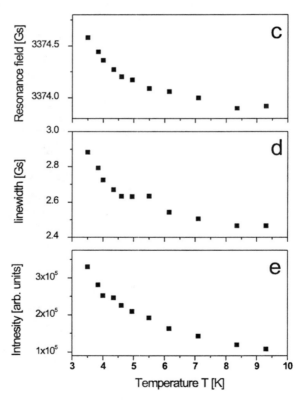

Fig. 24.6. The temperature dependence of the EPR parameters of the titanium carbide dispersed in carbon, (a) resonance field, (b) linewidth and (a) integrated intensity.

parameters has shown some essential temperature behavior which the resonance field displayed a temperature shift. Nitride carbide compounds have
shown some differences in the EPR spectra at low temperatures regions
which involves superposition of different localized magnetic centers. Usually,
the observed EPR line could arise from carbon being in different crystallographic states. The graphite or multiwall carbon nanotubes have shown similar
conduction-electron spin resonance (CESR) spectra [17]. The changing with
temperature of the g-factor could be well described in terms of the band model
of quasi-two-dimensional graphite by conduction carriers and an additional
effect of localized centers. It is observed that the temperature dependence of
g-parameter could change significantly after annealing processes. The temperature dependence of the intensity of CESR spectra at higher temperatures
was very weak and it was suggested that the Pauli processes corresponding
to the electron conduction were dominating. The linewidth has tendency for
fluctuation near 60 K and at lower temperatures a drastic increase of the
linewidth was observed [17].

According to the 2D graphite band model the Δg at higher temperatures
could be described well by the following relation [17]:

$$\Delta g = h\nu/\mu_B H_r = [\lambda_{eff}/k_B(T+\delta)]\times \tag{24.1}$$
$$\times \left\{ \left[3m_e a^2 \gamma_0^2/4\hbar^2 k_B(T+\delta)\cosh^2(\eta/2)\right]/2\ln\left[2\cosh(\eta/2)\right]\right\}$$

where $\lambda_{eff} \approx 2.4 \times 10 - 5eV$ is the effective spin-orbit energy interaction,
$\gamma_0 \approx 3eV$ is the 2D band parameter, $\eta = E_F/k_B(T+\delta)$ is the reduced
chemical potential determined from the conduction of charge neutrality, T is
the lattice temperature and δ is the additional temperature related to the
carrier relaxation time due to the effect of scattering [17].

It is well known that the carbon-cluster-based magnets are very interesting materials for studying [18] and ferromagnetism is observed in pure
carbon material [19]. Similar temperature dependence of the g-parameter was
observed at low temperature for the non-oxygenated Re123 compounds (high
temperature superconductors in oxygenated state) [20, 21], in which antiferromagnetic ordering is formed at low temperature. The localized magnetic
cluster could influence essentially the EPR spectra [22]. It is suggested that
the shift of resonance field at low temperatures with decreasing temperature is
arising from a change of the resonance condition by introduction of an internal
magnetic field, produced by magnetic ordering processes.

The decomposition of the titanium carbide nanocomposite EPR spectrum
of the broad line into two Lorentzian lineshapes and the values of g-parameters
are close to those expected for trivalent titanium ions and the character of
temperature evolution is similar to the one for the bulk. The values of g-
parameters are close to those expected for trivalent titanium ions and the
character of temperature evolution is similar to the one for the bulk. The
resonance field shifts to higher values with temperature decrease and two
main regimes differing in the value of resonance field temperature-gradient

$r = \Delta_H/\Delta_T$ are distinguishable: 100 down to 75 K with $r = 1.54(2)$Gs/K and 60 down to 20 K $r = 0.50$ Gs/K.

Such a shifting of the resonance locus would be connected with magnetic ordering incorporating an additional internal magnetic field, with the magnetic resonance condition being modified into

$$h\nu = g\mu_B(H_{ext} + H_{int}), \tag{24.2}$$

with h being Planck's constant, μ_B Bohr magnetron, H_{ext} the externally applied static magnetic field, and H_{int} the effective internal magnetic field. The exhibition of considerably greater resonance field temperature-gradient in some regime could be linked with the formation of a system of magnetic centre clusters, with the dipole - dipole interaction possibly permeating the internal magnetic field. The temperature evolution of the resonance field, also, is suggestive of a complicated character for the magnetic interactions, thus allowing for influential involvement on the part of disordering phenomena.

That above about 70 K the EPR spectrum integrated intensity could be satisfactorily fitted by a Curie - Weiss causality with a Curie - Weiss temperature of around 42 K, indicating in the high temperature range presence of strong ferromagnetic interactions, competing with other and thus not allowing formation of a magnetically ordered state. The integrated intensity temperature evolution in the low temperature region (below 18 K) by Curie - Weiss causality determines a Curie - Weiss temperature of about 1.5 K, hinting at the feasibility of magnetic ordering formation in that regime.

Additional insight into the functionality of magnetic centers active within titanium carbide has been gained by our studying of its ageing effect: Retracing the EPR spectrum of the sample against temperature after the elapse of one year from its synthesis and original EPR spectroscopy, we have observed the drastic diminishing of the narrow, conductivity electron-related resonance on the one hand and the emergence of a unprecedented side signal on the other. Indeed, whilst the narrow resonance has practically disappeared, a third-type line centered at $g_{eff} = 1.9(1)$ is now singled out, attributable to a new type of trivalent Ti ion complexes brought about by the aged sample having undergone oxidation processes. It is noteworthy that the temperature evolution of the spectral intensity of this new resonance turns out similar in causality to the one concerning (mainly) the trivalent titanium ion-related intensity of the fresh sample. Nonetheless, the significant difference of the new line effective g-factor from the free electron value pertains to a transition metal paramagnetic centre. Such oxidation processes interwoven with the ageing effect may have influenced the electrical transport properties of the nanocomposite system, thus being considerable as responsible for the suppression of the graphite conduction electron-connected narrow EPR resonance. According to Fig. 24.1 SEM picture, titanium carbide nanoparticles are amply surrounded by graphite, making the probability that internal oxidation processes would be

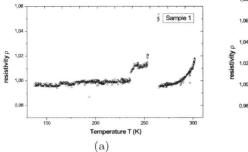
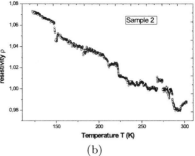

(a)   (b)

Fig. 24.7. The temperature dependence of the reduced resistivity for higher concentration of titanium carbide ($\rho_0 = 268.4\ \Omega m$ at 288 K) (a) and for lower concentration of titanium carbide ($\rho_0 = 168.3\ \Omega$ at 258 K) (b).

complicated worth mentioning. Therefore, ageing effect distortion occurrences in titanium carbide nanocomposites could believably induce drastic alterations of physical properties of theirs, such as application-critical hardness and corrosion resistance. And yet, the greater hardness of titanium carbide as compared to that of titanium nitrate may be stemming from the presence of trivalent Ti ions in the former, which could be deforming the surrounding lattice there in such a way as to shorten the bond length to the ligand.

Figure 24.7 presents the temperature dependence of the reduced resistivity of titanium carbide/carbon nanocomposite of two samples with different concentration. For higher concentration of titanium carbide the behaviour is similar to conducting materials (Fig. 24.7a), while for lower concentration it has got a semiconducting character (Fig. 24.7b). The major trend exhibited by the higher TiC concentration sample has been already recorded for bulk-like specimens, whereas for some nanocmposite Ti-systems a hysteretic resistivity temperature-variation has been reported [8,26]. The extraordinary behaviour observed at higher temperature could be connected with disordering processes involving trivalent titanium ions. Indeed; conduction processes might be emanating more importantly at higher temperatures through the activated skin depth effect.

In connection to conductivity peculiarities of monocarbides and mononitrides of the early transition metals, the lattice dynamics has been studied [5,7] by using first-principles density functional perturbation theory, soft pseudopotentials, and generalised gradient approximation to the exchange-correlation functional, with a view towards predicting superconductivity occurrence and enhancement: Calculated electron - phonon interaction constants foretell the possibility of superconductivity enhancement for substitutional transition metal carbo-nitrides.

24.4 Conclusions

Using sol gel method samples of titanium carbide embedded in carbon matrix in different concentrations were prepared and the samples have shown the coexistence of magnetic ordering (ferromagnetic) and superconducting state. The temperature dependence of the EPR spectra has shown different behaviour of their parameters, suggesting that some kinds of magnetic ordering process have appeared. A strong dependence on concentration of titanium carbide in carbon matrix of the conducting process has been registered.

Acknowledgments

This paper and the work it concerns were partially generated in the context of the MULTIPROTECT project, funded by the European Community as contract N NMP3-CT-2005-011783 under the 6th Framework Programme for Research and Technological Development.

References

1. L. E. Toth, Transition Metal Carbide and Nitrides, Academic, New York 1971.
2. B. C. Guo, K. P. Kerns, and A. W. Castleman Jr, Science **255**, 1411 (1992).
3. A. A. Rembel, Usp. Fiz. Nauk (Russia) **166**, 33 (1996).
4. A. I. Gusev and S. Z. Nazarov, Usp. Fiz. Nauk (Russia) **175**, 682 (2005).
5. E. I. Isaev, R. Ahuja, S. I. Simak, A. I. Lichtenstein, Yu. Kh. Vekilov, B. Johansson, and I. A. Abrikosov, Phys. Rev. B **72**, 064515 (2005).
6. A. I. Ivanovskii, Theor. Exper. Chem. **43**, 1 (2007).
7. E. I. Isaev, S. I. Simak, I. A. Abrikosov, R. Ahuja, Yu. Kh. Vekilov, M. I. Katsnelson, A. I. Lichtenstein, and B. Johansson, J. Appl. Phys. **101**, 123519 (2007).
8. N. Guskos, A. Biedunkiewicz, J. Typek, S. Patapis, M. Maryniak, and K. A. Karkas, Rev. Adv. Mater. Sci. **8**, 49 (2004).
9. J. Izquierdo, A. Vega, S. Bouarab, and M. A. Khan, Phys. Rev. B **58**, 3507 (1998).
10. C. Acha, M. Monteverde, M. Nunez-Requeiro, A. Kuhn, and M. A. A. Franco, Eur. Phys. J. B **34**, 421 (2003).
11. A. M. Hassib, A. A. S. Musmus, and M. A. A. Issa, Phys. Stat. Sol.(a) **89**, 147 (1985).
12. K. J. D. MacKenzie, R. H. Meinhold, D. G. McGavin, J. A. Ripmeester, and I. Moudrakovski, Solid State Nuclear Mag. Res. 4, 193 (1995).
13. T. Bodziony, N. Guskos, M. Maryniak, and A. Biedunkiewicz, Acta Phys. Pol. A (2005).
14. T. Bodziony N. Guskos, A. Biedunkiewicz, J. Typek, R. Wrobel, and M. Maryniak, Materials Science-Poland **23**, 899 (2005).
15. N.Guskos, T. Bodziony, M. Maryniak, J.Typek, and A. Biedunkiewicz, J. All. Comp. (2007).
16. A. Biedunkiewicz, Materials Science (Poland) **21**, 445 (2003).

17. Likodimos, S. Glenis, N. Guskos, and C. L. Lin, Phys. Rev. B 68, 045417 (2003); 76, 075420 (2007).
18. F. Beuneu, C. l'Huillier, J. -P. Salvetat, J. -M. Bonard, and L. Forro, Phys. Rev. B **59**, 5945 (1999).
19. V. A. S. Kotosov and D. V. Shilo, Carbon **11**, 1649 (1998).
20. M. Kosaka, T. W. Ebbesen, H. Hiura, and K. Tanigaki, Chem. Phys. Lett. **233**, 47 (1995).
21. H. Sato, N. Kawasatsu, T. Enoki, M. Endo, R. Kobori, S. Maruyama, and K. Kaneko, Solid State Comm. **125**, 641 (2003).
22. T. L. Makarova, D. Sundqist, R. Hohne, et al., Nature 413, 716 (2001).
23. R. N. Kyutt, E. A. Smorgonskaya, A. M. Danishevskii, and S. K. Gordeev, Solid State Phys. (in Russian) **41**, 1359 (1999).
24. H. Y. Wang, Q. C. Jiang, X. L. Li, J. G. Wang, Q. F. Guan, and H. Q. Liang, Materials Research Bulletin **38**, 1387 (2003).
25. T. Kolodiazhnyi and A. Petric, J. Phys. Chem. Solids **64**, 953 (2003).
26. S. Lakkis, C. Schienker, B. K. Chakraverty, and R. Buder, Phys. Rev. B **14**, 1429 (1976).
27. G. Wagoner, Phys. Rev. **118**, 647 (1960).
28. K. Matsubara and T. Tsuzuku, Phys. Rev. B **44**, 11 845 (1991).
29. K. W. Lee and Ch. Eu. Lee, Phys. Rev. Lett. **97**, 137 206 (2006).
30. J. W. McClure and Y. Yafet, Proceedings of the Fifth Conference on Carbon, (edited by S. Mirozowsky and P. L. Walker, Pergamon, New York, 1962), 1, 22.
31. A. S. Kotosonov, Carbon **26**, 735 (1988).
32. L. Largo, A. Cimas, P. Redondo, V. M. Rayon, C. Barrientos, Chem. Phys. **330**, 431 (2006).
33. M. Guemmaz, G. Moraitis, A. Mosser, M. A. Khan, and J. C. Parlebas J. All. Comp. **262–263**, 397 (1997).

Part IV

Sensors

25

RAPID METHODS FOR MULTIPLY DETERMINING POTENT XENOBIOTICS BASED ON THE OPTOELECTRONIC IMAGING

B. SNOPOK

V. Lashkaryov Institute of Semiconductor Physics, National Academy of Sciences, 41 Prospect Nauki, 03028 Kyiv, Ukraine.
snopok@isp.kiev.ua

Abstract. Some predictions concerning the technological expansion for the optoelectronic imaging systems for the screening potent xenobiotics are made based on the analysis of the state-of-the-art "multivariate" array technology. Emphasis is placed on the multiparameter aspect of such systems performance, in particular, the additional value of the scattered light under surface plasmon resonance conditions when forming chemical images for composite multicomponent media using the multisensor arrays.

Keywords: multivariate array, surface plasmon resonance, potent xenobiotics

25.1 Motivation

Smart sensors are vital elements of domestic life and industry. Although a great variety of sensors have become well established in industries, food production, medicine, and many other areas, development of new sensing techniques and elements proceeds at an unprecedented rate [1]. During recent years sensor markets in industrialized countries have been growing at an average rate in excess of 10% per year, and there has been a correspondingly high level of investment in sensor research and development. As a consequence, there is an increasing demand for sensors and sensing technologies appearing as a purchasing power on the sensor market. This is governed by various factors. On the one hand there is a growing concern for the protection of the environment, for the process optimization and for the enhancement of safety and security. In the last case, both the prevention of accidents and the minimizing the effects of potential xenobiotic pollutions on civilian populations (independent of their origins, e.g. terrorist acts, environmental security or food safety) are essential [2–5]. On the other hand the industrial environment has reached the level necessary for backing the production of highly sophisticated sensors.

Rapid identification of the chemical or biological agents involved in any hazardous material incident is vital for the effective treatment of casualties [7–10]. A wide variety of detection equipment is available commercially and through the different special organizations; most of the mobile detectors and monitors have been developed by the armed forces. However traditional military approaches to battlefield detection of chemical and biological weapons are not necessarily suitable or easily adapted for use by civilian health providers dealing with a heterogeneous population of casualties in a peacetime civilian setting. Moreover, since a xenobiotic "attack" of any sort (e.g. terrorist act as well as environmental pollution or food contaminations) is a *very low-probability event*, the ultimate goal of equipment for detection and measurement of chemical/biochemical agents is to rapidly (i) and inexpensively (ii) detect in the field presence of potent xenobiotics (iii) using portable (iv) instrument and transfer information (v) by standard operating procedures and communication for data analysis and to take a decision. However, a potential complication that can easily be overlooked is the possibility that a potential xenobiotics "attack" may involve the use of *a priori* unknown or more than a single toxic agent, – *simultaneous processing of multiple, different analyts* must be carry out.

To obtain information about the complex mixtures, the most efficient traditional techniques try to provide the highest *selectivity toward a given analyte* when detecting the useful signal against a background produced by attendant contaminants [11]. This approach to both individual chemical compounds and *strictly defined* mixtures analysis has gained wide usage. However the potentialities to create specific sensor sets that are optimized to detect complex xenobiotic mixtures have fundamental restrictions. This results from the fact that such multicomponent chemical media (*MCM*) can contain abundant quantities of *fractions differing in their chemical and physical nature*.

The concept of using cross-reactive sensor arrays for chemical analysis offer advantages over individual sensors concerning sensitivity to a wider range of analytes, improved selectivity and the capability for recognition of both single and complex analytes (*MCMs*). A relationship between the *MCM* chemical functionality (that determines its quantitative, as well as qualitative, composition) and its chemical image (*CI*) occurs through the stages of parametrization of array response and follows classification, identification and interpretation based on previous calibration. At that, eventually any problem in sensor array development can be reduced to a search for ways to *improve sensitivity and discrimination ability of a sensor system with maximally decreased time of analysis*.

25.2 Sensing technologies based on evanescent wave phenomena

Many successes achieved during detection on potent xenobiotics have become possible owing to the sensor instrument engineering for investigations of (bio)chemical processes at the molecular level that is actively developing. The most promising of them seem to be those based on transducer-based sensing elements [12–17]. However, only a few analytical techniques now exist that can be used to study the binding process of unlabeled components in real time based on evanescent wave phenomena (i.e. optical Surface Plasmon Resonance (SPR), waveguide spectroscopy (WG), and - gravimetric Quartz Crystal Microbalance (QCM) and Surface Acoustic Waves (SAW)).

Optoelectronic techniques and, in particular, SPR imaging has recently come to the fore as a valuable technology for the analysis of various, especially, biological samples, through its integration with existing analytical methods [18–20]. Indeed, just in the classical configuration SPR is sensitive to the presence of sub-monolayers of material of different dielectric constant and can therefore register the presence of very small quantities of materials. The label free nature of the technique enables rapid method development with reduced reagent preparation time and avoids the possibility of assay interference from the label. However, to date SPR has not been utilized to its full potential in various profiling or imaging technologies because of the using only SPR angle shift for generation an analyte specific information signal.

For the nanostructured sensitive layers optoelectronic techniques (e.g. SPR) can give additional analyt specific information since intensity, polarization, and Stokes factors of scattering from the sensor surface light (decoupling owing to interactions with interfacial architectures) have strong dependence on fluctuation of dielectric permittivity near or inside sensitive films and, as a consequence, on the presence of extrinsic molecules [21]. Thus, the detection of scattered light under the SPR conditions give a more information about the interfacial process, namely, (1) SPR reflection with overall information about the composition changes on the surface; (2) scattering light from the surface represents some specific interfacial processes; (3) fluorescence signal can be detected for labeled analytes or competitors. In multiple assay formats the technology will be used with varying sample types, such as nucleic acids, proteins, peptides, etc. Developing chips can be as good examples of relatively universal approach. From the one side, ones can be used only to monitor a molecular patterns (since identifying changing biomarker expression patterns more finely and quickly); from the other side ones is the basis for the identification the actual biomolecules associated with the xenobiotic "attack", - in both cases biochips present an best approach. The optoelectronic imaging sensor is about few square mm, and can consists of a hundreds of independent sports (each of the spot is 150 μm square, produced by commercially available spotters). When the array is introduced in the mixture, the both reflection and scattering (angle or wavelength dependent process) are measured for each sport which corresponds to the concentration change of analyte bound to the sport. Additional advantages of the scattered based SPR arrays are a simple optical system and that the basic single sensor configuration is already developed, and in fact is commercially available as SPR scanners. Moreover, portable biohazard detection instrument can be built based on the array sensor that will be rapid, and inexpensive relative to existing devices and methods. However in spite of high sensitivity of light scattering to small changing of thin film optical parameters, the scattering techniques are still not extensively used to creation of SPR based sensor devices.

25.3 Surface functionality specific to the nanometer scale

Sophisticated devices of the future like mobile personal assistance will rely heavily on microelectronic and micro/nano-sensor components. However, the more essential rather (bio)chemical than microelectronic aspect of the problem - identification ability of the instrument depends mainly by the stability, sensitivity and selectivity properties of recognition system. Despite the wide spectrum of different artificial receptors, the bioreceptors based on immunospecific proteins and DNA fragments combined with nanostructured elements remain the more promising objects for such architectures [22, 23]. The main reason for that is that it is possible in short time to create the affinity type receptor with selectivity against specific analytes.

Difficulties with practical realization of the systems for the analysis of multicomponent mixtures are caused first of all by problems with formation the nanostructured sensitive layers. Key to the success in the R&D of biohazard detection instrument is the attachment of capture agents at the correct density and orientation to assure rapid and selective capture, even of low concentrations of xenobiotic macromolecules. Indeed, the ability to manipulate and orient macro and biomolecules may improve sensitivity and selectivity for many already known systems that can be perfected using the full control of location and orientation in preparation of films. The technologies to manipulate and to position the macromolecules or aggregates are still in its infancy and many new tools have to be devised. Only few approaches based on the self-assembling and tractable/self-limiting layer-by-layer deposition techniques can provide the required arrangement of receptor centers with desired chemical functionality for layer formation on various substrates. In particular, immobilization approaches using the natural binding agent for immunoglobulin, protein A as well as mixed self-assembled thiol films can be used efficiently for gold surfaces of SPR transducers. The advantage of the approaches lee in fact that both procedures permit to fabricate the interfacial architectures with controlled density of immobilized receptors and thickness of buffer layer and provide oriented immobilization of bioreceptors.

Finally, the following aspects will be taken into account for a design of nanostructured sensing architectures: (i) stability and availability of receptors;

(ii) nondestructive, self-limiting and scalable procedures for immobilization of receptors on transducer; (iii) optimal spatial distribution of the receptors at the surface with (iv) specifically organized interfacial structure; (v) specific chemical functionality of the whole array (based on optimal number and kind of sensing spots).

25.4 Nanosensors: effects of nanostructural features on chemical properties

Specific chemical functionality (e.g. molecular recognition) may be only realized within a definitive length scale of structured matter. Particularly, a certain level of functional complexity is a result of combining smaller subunits that are spatially arranged in a specific manner - new properties and functions are realised by hierarchically organisation of materials in discrete steps ranging from the atomic to the macroscopic scale. This rather broad concept encompasses a wide variety of issues of current technological interest [23]. The focus of the modern activity in sensor sector it is not the mechanism of organization at the nanoscale, but the influence on properties that may be observed due to these ways of organization. There is particular interest in answering questions such as "How are chemical reactivity or selectivity influenced by the size, shape or ordering of the surface elements?" (structural nanotechnology) or "How does the ordering of molecules or macromolecules on the surface influence these properties?" (bionanotechnology). However, there is little understanding of the interrelations between structure and properties of such complex systems in which the chemical reactivity, molecular recognition, catalytic efficiency are affected by a nano-structured arrangement.

The main objective of investigations in the bionanotechnology is the both development and characterization of local structure (density, distribution, orientation, etc.) and prediction the effect of nanostructural features on (bio)chemical properties [24–26]. This opens a way to the proper design of a sensitive surface for the construction of any desired sensor because the local environment should be a crucial factor determining advanced properties of both artificial and natural receptors.

The combination of nanoparticles with evanescent wave transducers represents the structural nanotechnology aspect in sensor science. One of the possible ways is the control of the selectivity of adsorption-desorption processes by variation the dimension of functionalized nanoparticles [27]. Indeed, if the typical dimension of the nanoparticles is less than de-Broglie wavelength of electron the quantum confinement of charge carriers occurs and carrier energy is quantized. In this case the electron levels on the particle surface (and correspondingly chemical functionality) is dependent on nanoparticle volume/surface area ratio. In the frame of the electronic theory of adsorption this phenomenon leads to ability of analyte to change the adsorption status; this process can be connected with localization/delocalization of hole or electron. It means that nanoparticles with different dimensions can realize different form of interfacial complexes with the same analyte (i.e., charged or neutral) depending on the dimensions of nanoparticles. So, it is possible to control the reactivity and selectivity profile of nanomaterials by appropriate selection of corresponding size distribution of adsorbent nanoparticles. This fact can be used as basis for creation of novel sensor architectures with size-controlled selectivity profile. Finally, the system of sensitive elements based on the same chemical compounds, each element has unique selectivity profile, can be created,-this is the basis for multichannel analyzers.

25.5 Biochemical fingerprint technologies

The reliability of array applications is mainly based on the possibility to extract the xenobiotic relevant specific information from sensor response for multiply analytes. Typically, methods based on a "univariate" arrays which have completely independent channels with no significant cross-reactivity overlap [28]. Critical components of the system in this case are specific receptors as well as a sophisticated surface chemistry that efficiently couples the capture ligands and preserves their activity, while minimizing nonspecific adsorption of all other biomolecules.

Alternative "multivariate" approach based on biological receptors that show differences in their affinity patterns and can improve the efficiency of "univariate" immunoassay limited by the cross-reactivity [29–31]. Indeed, the *MCM* can be described not by a sum of their individual components (proteins, etc.) but by some abstract representation – a *chemical image (CI)* – a virtual fingerprint with a set of intrinsic parameters for a given *MCM*. An arrays of low-selective/cross-reactive sensors ("multivariate" sensing systems) typically used to form a *CI* of a *MCM* with further classification and identification based on the image recognition techniques [32, 33]. The signals from sensors have to represent result of the interaction between the *MCM* components and a sensitive (bio)chemical layer with desired chemical functionality: arrays of sensors mapping the MCM variety in the biochemical space onto the m-dimensional response space (m is the number of sensors). The discrimination ability of patterns is determined by the number of clusters of specific sampling point's arrangement in a multidimensional space, classified according to the property we are interested in. This is determined by the type of mathematical description of the classification problem. The process of sensor array *calibration* enables to establish correlation between the *CI* clusters in the response space and qualitative/quantitative *MCM* characteristics using supervised learning methods.

The actual challenge is to extend the using the "multivariate" sensing array in diagnostic tools for identification of various *MCM* and, especially, in security sector. Thus, when used with complex mixtures, analysis methods are intended to identify the set of features (which are linear or nonlinear combinations of response characteristics) that can segregate identifiable *MCM*. The classification algorithm based on the overall trends of molecular profile changes will be more suitable for fast prediction in the field.

25.6 Concluding remarks

The organization and functioning of living organisms serve as inexhaustible source of ideas for development and production of various technical facilities. This is particularly true when developing smart systems for the security sector are that imitate such high level functions of living organisms as taste and olfaction (both determined by the chemical composition of the ambience). Beyond any question, that this principles (molecular recognition, cross-reactivity, pattern recognition, etc.) could be applied in engineering of nanostructured artificial systems as intelligent multichannel analyzers [34].

Development the "multivariate" sensing array concerns solution of the following topics: (i) selection of receptor centers with required (bio)chemical functionality; (ii) integration of these receptors with physical transducer; (iii) parameterization of sensor array response using dynamically resolved data processing algorithms; (iv) integration of individual system components for better performance; (v) optimization the sample preparation steps, measuring procedure, and database formation algorithms for targeted chemical compounds.

The key element in "multivariate" array related R&D is a signal-generation technology [35]. Light scattering on the surface of nanostructured architectures in the conditions of resonance excitation of surface plasmon-polaritones seems to be optimal to a wide variety of molecular assay formats (for protein, immunohistochemical, immunocytological, DNA-based, or cellular assays) and can be easy combined with fluorescence technology, if necessary. One exploits variations of interfacial surface organized architectures that scatter in the near field to generate light depending of their dimension, optical properties and environment. The promising approach is the combination of advantages of SPR and light scattering approaches in the same chip; it will open the new way for multiparameter detection of biospecific interactions using different analyt specific responses.

Short analysis of the state of the art in the area of imaging optoelectronic systems indicate the presence of advanced physical transducers that can facilitate the development in near future smart instruments for detection and measurement of xenobiotics in aqueous and gaseous environments. New concepts based on multiparameter "multivariate" sensing arrays will result cheaper equipment and sensors with higher accuracy, better selectivity and moderate price, suitable as a disposable compact smart sensors for domestic life, industry and, especially, for security sector. Key applications of the new technology are foreseen in such applications as DNA screening, protein profiling, safety assessments (e.g. personal safety monitors, territory survey systems, emission control devices, etc.) as well as for medical purposes.

References

1. R. Dewhurst, G.Y. Tian, Sensors and sensing systems, *Meas. Sci. Technol.* **19**, 020101 (2008).
2. G.K. Mostéfaoui, P. Brézillon, Context-Based Constraints in Security: Motivations and First Approach, *Electronic Notes in Theoretical Computer Science* **14**, 685–100 (2006).
3. E.E. Schultz, Risks due to convergence of physical security systems and information technology environments, *Information Security Technical Report* **12**, 80–84 (2007).
4. M. Burgess, Biology, immunology and information security, *Information Security Technical Report* **12**, 192–199 (2007).
5. M. Friedewald, E. Vildjiounaite, Y. Punie, D. Wright, Privacy, identity and security in ambient intelligence: A scenario analysis, *Telematics and Informatics* **24**, 15–29 (2007).
6. N. Boudriga, M.S. Obaidat, Mobility, sensing, and security management in wireless ad hoc sensor systems, *Computers and Electrical Engineering* **32**, 266–276(2006).
7. S. Singh, Sensors-An effective approach for the detection of explosives, *Journal of Hazardous Materials* **144**, 15–28 (2007).
8. S.J. Toal, W.C. Trogler, Polymer sensors for nitroaromatic explosives detection, *J. Mater. Chem.* **16**, 2871–2883 (2006).
9. P.M. Boltovets, B.A. Snopok, T.P. Shevchenko, N.S. Dyachenko, Yu.M. Shirshov. Optoelectronic transducers for detection of biologically hazard agents, *Saint-Petersburg Journal of Electronics* **1** (38), 231–237 (2004).
10. Bioterrorism. Mathematical modeling applications in homeland security *Frontiers in Applied Mathematics, 28, Society for Industrial and Applied Mathematics (SIAM), Philadelphia*, PA Edited by: Banks HT, Carlos Castillo-Chavez. 2003.
11. G.J. Wegner, H.J. Lee, R.M. Corn, Surface Plasmon Resonance Imaging Measurements of DNA, RNA, and Protein Interactions to Biomolecular Arrays, Protein Microarray Technology, Wiley, ISBN: 3-527-30597-1, pp 107–129 (2004).
12. R. Rella, J. Spadavecchia, M.G. Manera, P. Siciliano, A. Santino, G. Mita, "Liquid phase SPR imaging experiments for biosensors applications", *Biosensors and Bioelectronics* **20**, 1140–1148 (2004).
13. A.M. Rossi, L. Wang, V. Reipa, T.E. Murphy, Porous silicon biosensor for detection of viruses, *Biosensors and Bioelectronics* **23**, 741–745 (2007).
14. J. Dostálek, J. Homola, Surface plasmon resonance sensor based on an array of diffraction gratings for highly parallelized observation of biomolecular interactions, *Sensors and Actuators B* **129**, 303–310 (2008).
15. A.D. Taylor, J. Ladd, Q. Yu, S. Chen, J. Homola, S. Jiang, Quantitative and simultaneous detection of four foodborne bacterial pathogens with a multichannel SPR sensor, *Biosensors and Bioelectronics* **22**, 752–758 (2006).
16. K. Arshak, C. Adley, E. Moore, C. Cunniffe, M. Campion, J. Harris, Characterisation of polymer nanocomposite sensors for quantification of bacterial cultures, *Sensors and Actuators B* **126**, 226–231(2007).
17. J.E. Valentine, T.M. Przybycien, S. Hauan, Design of acoustic wave biochemical sensors using micro-electro-mechanical systems, *J. Appl. Phys.* **101**, 064508 (2007).

18. J. Homola, Present and future of surface plasmon resonance biosensors, *Anal Bioanal Chem.* **377(3)**, p. 528–39 (2003).
19. P.T. Kissinger, Biosensors-a perspective, *Biosensors and Bioelectronics* **20**, Issue 12, P. 2512–2516 (2005).
20. M. Piliarik, H. Vaisocherová, J. Homola, A new surface plasmon resonance sensor for high-throughput screening applications, *Biosensors and Bioelectronics* **20**, 2104–2110 (2005).
21. A. Savchenko, E. Kashuba, V. Kahuba, B. Snopok. (2007): A novel imaging technique for the screening of protein-protein interactions using scattered light under surface plasmon resonance conditions, *Analytical Chemistry* **79**, 1349–1355.
22. Y.-S. Lee, M. Mrksich, Protein chips: from concept to practice, *Trends in Biotechnology* Vol. **20** No. 12, p. S14–S18 (2002).
23. X.-J. Huang, Y.-K. Choi, Chemical sensors based on nanostructured materials, *Sensors and Actuators B* **122**, 659–671 (2007).
24. K.K. Jain, Nanotechnology in clinical laboratory diagnostics, *Clinica Chimica Acta* **358**, P. 37–54 (2005).
25. B.A. Snopok, P.N. Boltovets, F.G. Rowell, Effect of Microenvironment and Conformation of Immobilized Ligand on Its Interaction with Macromolecular Receptor. *Theoretical and Experimental Chemistry* **42 (4)**, 210–216 (2006).
26. A.S. Pavluchenko, B.A. Snopok, An Inhibitor-type Competitive Analysis Model for Sensors with Small Sensitive Surface Area, *Sensor Letters* **5**, 380–386 (2007).
27. S.I. Lysenko, E.B. Kaganovich, I.M. Kizyak, B.A. Snopok, "Multiparametric Chemical Sensor Based on Nanocrystalline Silicon Waveguide", *Sensor Letters* **3** (2), 117–125 (2005).
28. M. Piliarik, H. Vaisocherová, J. Homola Towards parallelized surface plasmon resonance sensor platform for sensitive detection of oligonucleotides, *Sensors and Actuators B* **121**, 187–193 (2007).
29. J. Pavón, J.Gómez-Sanz, A. Fernández-Caballero, J.J. Valencia-Jiménez, Development of intelligent multisensor surveillance systems with agents, *Robotics and Autonomous Systems* **55**, 892–903 (2007).
30. J.H. Lambert, M.W. Farrington, Cost-benefit functions for the allocation of security sensors for air contaminants, *Reliability Engineering and System Safety* **92**, 930–946 (2007).
31. B. Barshan, T. Aytaç, Ç.Yüzba Target differentiation with simple infrared sensors using statistical pattern recognition techniques, *Pattern Recognition* **40**, 2607–2620 (2007).
32. B.A. Snopok, I.V. Kruglenko, Multisensor systems for chemical analysis: state-of-the-art in electronic nose technology and new trends in machine olfaction, *Thin Solid Films* **418**, 21–41 (2002).
33. B.A. Snopok, I.V. Kruglenko, Nonexponential relaxations in sensor arrays: forecasting strategy for electronic nose performance, *Sensors and Actuators B: Chemical* **106 (1)**, 101–113 (2005).
34. M.A. Gubrud, Nanotechnology and International Security, http://www.foresight.org/Conferences/MNT05/Papers/Gubrud/index.html
35. G. Boas, Taking It to the Streets with SPR, Biophotonics International April, 40 (2007).

ELECTRONIC NANOSENSORS BASED ON NANOTRANSISTOR WITH BISTABILITY BEHAVIOUR

V.N. ERMAKOV[1], S.P. KRUCHININ[2], AND A. FUJIWARA[3]

[1] *Bogolyubov Institute for Theoretical Physics, Metrologichna str. 14 b, 03143, Kiev-143 Kyiv, Ukraine.* vlerm@bitp.kiev.ua
[2] *Bogolyubov Institute for Theoretical Physics, Metrologichna str. 14 b, 03143, Kiev-143, Kyiv, Ukraine.* skruchin@i.com.ua
[3] *School of Materials Science Japan Advanced Institute of Science and Technology 1-1, Asahidai, Tatsunokuchi, Ishikawa, 923-1292 Japan.* fujiwara@jaist.ac.jp

Abstract. The creation of a nanosensor aimed at its multiple usage is one of the most important problems in nanotechnologies. In the present work, we demonstrate a possibility of its creation on the basis of a nanotransistor with bistable characteristics. Such a transistor can be a nanostructure including a quantum dot. The reason for the appearance of bistability can consist in the correlation phenomena between current carriers or in their interaction with atoms of the quantum dot.

Keywords: nanodevice, nanosensor, bistability, quantum dot

26.1 Introduction

The problem of the usage of electronic nanounits as sensors for the control over the running of various physical processes in small spatial volumes under a variation of their characteristic parameters on the level of noise is one of the central problems of nanotechnologies. From the basic viewpoint, an electronic nanounit intended for the sensor purposes must consist of two components: the sensor part proper which reads out the information from an object under study and the electronic part for its processing. As for the first component, the situation is sufficiently clear. Indeed, according to the quantum-mechanical description, any change in a state of the molecular aggregate causes necessarily a change of the electrical conduction of the system. Thus, by recording a change in the electrical conduction of the testing part of the device, we can get the information about a state of the system under study. The main problem arises if we try to process such information. Here, two basically different approaches are possible. The first approach is related to the output of the information received from the sensor part through nanowires to microelectronic units, where it is processed. The second approach is based on the processing of information directly near the sensor part with the subsequent transfer of the processed information to a registering unit. It is easy to see that, in this case, it is necessary to possess a logical nanounit. A drawback of the first approach is the obligatory presence of nanowires that can and will introduce uncontrolled disturbances into output signals. Ways to solve this problem depend on a specific architecture of a nanodevice, and their general consideration meets difficulties. The usage of the second approach is restricted by the absence of a suitable reliable logical nanounit aimed at the processing of information. As such units, we can take those including quantum dots or quantum wells. But, in this case, it is very important that they will be able to operate in a nonlinear mode. In particular, this mode can be related to the appearance of a bistability in the system due to, for example, the correlation effects between current carriers in a quantum dot [1]. As is shown experimentally, such units can perform logical functions [2]. The second, very important property of nanosensors should consist in the possibility of their multiple use. This property can be decisive for practice. Indeed, the introduction of a nanosensor into an object under study can be, in a number of cases, a complicated task that can become unsolvable if the procedure of introduction must be repeated many times. Within the second approach with the use of nanochips for the direct diagnostics of a state of the sensor part of the device (a probe), such a task seems to be simple.

26.2 Intrinsic bistability of the resonance tunneling

Resonant electron tunneling of particles through a system of double potential barriers is very sensitive to a position of electronic states in a quantum well. This circumstance can be used for effective governing of the tunneling process. For example, it is possible to change potential field in the well by accumulation of electric charge in it under tunneling. This process supposes the existence of a large number of electronic states in the interbarrier space. Actually such a condition requires that the system has a macroscopic size for which a concept of electric capacitance can be introduced. In the case of a small-area quantum well one should consider electron-electron interaction using the quantum mechanical description with an account of its influence on the tunneling. For a limited number of electrons this problem has been considered in Ref. [2]. However, the influence of Coulomb interaction can be increased when the states in the quantum well are degenerated. We consider this problem for the case of N-fold degenerate electronic state when the accumulation up to N electrons in the well is possible. Taking into account the interaction between them one can derive a number of properties, typical of nonlinear tunneling. For instance, that can be an appearance of steplike form of current-voltage characteristics, tunneling bistability and others. We confine ourselves by the consideration of one-dimensional case of tunneling. Studying of fluctuations in these systems shows that they can be virtually suppressed [5]. The latter is typical of double-level systems.

26.2.1 HAMILTONIAN OF THE SYSTEM

As a model of double-barrier tunneling system we take a structure with the energy profile shown in Fig. 26.1. Hamiltonian describing tunneling of electrons through such a structure, can be chosen in the form

$$H = H_0 + H_W + H_T \tag{26.1}$$

The first term of this Hamiltonian is

$$H_0 = \sum_{k\sigma} \varepsilon_L(k)a^+_{k\sigma}a^+_{k\sigma} \sum_{k\sigma} \varepsilon_R(k)a^+_{p\sigma}a_{p\sigma}$$

One describes electrons in the left electrode (source) and in the right electrode (drain). Here $a^+_{k\sigma}(a_{k\sigma})$ and $a^+_{p\sigma}(a_{p\sigma})$ are the creation (annihilation) operators for the electrons in the source and the drain, respectively. $\varepsilon_L(k) = \varepsilon_L + \hbar^2k^2/2m_L$ is the energy of electrons in the source. $\hbar k$ and m_L are their quasimomentum and effective mass, respectively, σ is the electron spin. For the drain with an account of external potential V, applied across the system we have $\varepsilon_R(k) = \varepsilon_R + \hbar^2k^2/2m_R - V$, $\hbar p$ being the momentum and m_R the effective mass.

Hamiltonian H_W describes electronic states in the quantum well. We consider the case when there is a N-fold degenerate state in the quantum well. Then H_W can be written in the form

$$H_W = \sum_{\alpha} E_0 a^+_\alpha a_\alpha + \frac{1}{2} \sum_{\alpha_1 \neq \alpha_2} V_{\alpha_1\alpha_2} a^+_{\alpha_1} a^+_{\alpha_2} a_{\alpha_1} a_{\alpha_2}$$

where $\alpha = (l, \sigma)$, σ is the spin number, l is a number of the quantum state, which takes values from 1 to N. An energy of the degenerate state in the well with account of the applied bias is written as follows $E_0 = \varepsilon_0 - \gamma V$, where ε_0 is the energy of the resonant state in the quantum well, and γ is the factor depending on a profile of potential barriers (for identical barriers $\gamma = 0.5$), and $V_{\alpha_1\alpha_2}$ is a matrix element of the electron-electron interaction in the interbarrier space (quantum well). For simplicity, we approximate it a positive constant $V_{\alpha_1\alpha_2} = U$ corresponding to repulsion.

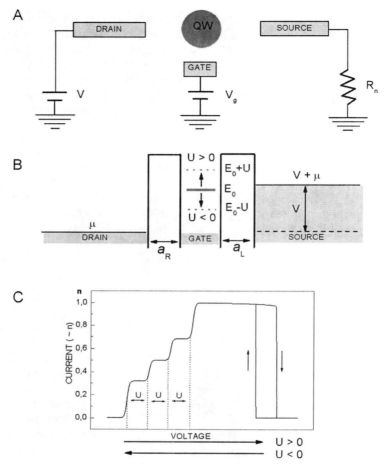

Fig. 26.1. The example of nanotransistor has nonlinear volt-ampere characteristic. (A) Schematic drawing of a nanotransistor. (B) The energy level diagram of the nanodevice showing resonant tunnelling phenomena through degenerated energy states with electron correlation U. (C) Theoretical prediction of current-voltage (I-V) curves. I-V characteristics depend on the sign of U. Applied voltage for positive (negative) U affects the right (left) direction of the shift in I-V curves in (C).

Hamiltonian H_T describing the tunneling of electrons through the barriers has the conventional form:

$$H_T = \sum_{k\alpha} T_{k\alpha} a_{k\sigma}^+ a_\alpha + \sum_{p\alpha} T_{p\alpha} a_{p\sigma}^+ a_\alpha + e.c.$$

where $T_{k\alpha}$ and $T_{p\alpha}$ are matrix elements of tunneling through the emitter and the collector, respectively. In general case, they depend on the applied bias.

26.2.2 OCCUPATION NUMBERS FOR QUANTUM STATES IN THE WELL

When we apply constant external bias across the system, nonequilibrium steady-state electron distribution sets in. We assume that the electron distribution functions in electrodes (source, drain) are equilibrium by virtue of their large volumes, but their chemical potentials change. They are connected by the relation $\mu_L - \mu_R = V$ (where μ_L and μ_R are the chemical potentials of the source and the drain, respectively). The electron distribution function $g(E)$ in the quantum well is essentially nonequilibrium. It can be determined from the condition of equality of the tunneling current through the source and the drain. Then distribution function has the form:

$$g(E) = \frac{1}{\Gamma(E)}[\Gamma_L(E)f_L(E) + \Gamma_R(E)f_R(E)], \quad \Gamma(E) = \Gamma_L(E) + \Gamma_R(E)$$

Γ_L and Γ_R are rates of electron transmission source-QW and QW-drain, respectively. f_L and f_R are electron distribution functions in the source and the drain, respectively. They have Dirac's forms. The occupancy of the QW states $(n_\alpha = \langle a_\alpha^+ a_\alpha \rangle)$ in the quantum well can be determined using the following expression

$$n_\alpha = -\frac{1}{\pi} \int dE\, g(E)\, ImG(\alpha, E) \tag{26.2}$$

Where $G(\alpha, E)$ is the Fourier transform from the retarded Green's function. Using the Hamiltonian H_W the Green's function can be calculated exactly. For example, for the state of number N one can obtain

$$G(\alpha) = \frac{1}{(E' - E)} \left\{ 1 + \sum_{m=1}^{2N-1} \sum_{\substack{\alpha_1, ..., \alpha_m \neq \alpha \\ \alpha_1 \neq \alpha_2 \neq ...\alpha_m}} \prod_{m_1=1}^{m} n_{\alpha_{m_1}} \frac{U}{(E' - E - m_1 U)} \right\} \tag{26.3}$$

with $E' = E + i\eta$ for $\eta \to +0$. Green's function has poles at $E_m = E_0 + mU$, where $m = 0, 1, 2, ..., 2N - 1$. Thus, the electron-electron interaction leads to splitting of states in the quantum well. New states are separated by the gap U. Using a cyclic indices permutation in the formula (26.3) we can obtain Green's functions $G(\alpha, E)$ for all the states of the quantum well.

As it follows from (26.3), the expression for n_α does not depend on the index α. Therefore, mean values of occupation numbers are also independent on the number of the quantum state, and we can assume that $n_\alpha = n$. Thus, finally, we get for n

$$n = F(n) \tag{26.4}$$

Where

$$F(n) = \sum_{m=0}^{2N-1} C_{2N-1}^m g_m (1-n)^{2N-1-m} n^m, \qquad C_{2N-1}^m = \frac{(2N-1)!}{m!(2N-1-m)!}$$

g_m is $g_m = g(E_m)$. Thus we have obtained the algorithmic equation of power $2N-1$ for occupation numbers n. In general case, this equation can have several solutions in the interval $0 \leqslant n \leqslant 1$. In the case when all $g_m = g$ it follows from (11) that $n = g$, i.e. occupation of electronic states will be defined only by the distribution function $g(E)$.

26.2.3 THE EXAMPLE: DOUBLE DEGENERATE STATE

An analysis of Eq. (26.4) shows that for $N = 2$ in the interval $0 \leqslant n \leqslant 1$ it has three solutions under condition $g_0 = g_1 = 0$. These solutions are as follows:

$$n_1 = 0,$$

$$n_{2,3} = -\frac{2}{3} \frac{g_2}{g_3 - 3g_2} \pm \sqrt{\frac{9g_2^2 + 4(g_3 - 3g_2)}{4(g_3 - 3g_2)^2}} \qquad (26.5)$$

Accordingly to the condition $0 \leqslant n \leqslant 1$, expression (26.5) leads to

$$0 < \frac{3g_2}{3g_2 - g_3} < 2$$

$$9g_2^2 - 4(3g_2 - g_3) > 0 \qquad (26.6)$$

Inequalities (26.6) are compatible when

$$g_2 \geqslant \frac{2}{3}(1 + \sqrt{1 - g_3}), \qquad g_3 > \frac{3}{4}$$

Therefore, Eq. (26.4) has three solutions when the values of g_2 and g_3 are close to unity. The two solutions n_1 and n_3 are stable, while the third one, n_2 is unstable. The stable states correspond to the cases, when there are no electrons in the well or there are four electrons occupying levels. The latter is possible since the system is essentially nonequilibrium.

26.3 Bistability-based nanotransistor

Consider an example demonstrating the possibility to use the process of resonance tunneling through degenerate states in a logical unit processing the information obtained from a sensor nanoprobe. Such a unit is, by its structure, a nanotransistor (Fig. 26.1A), whose component is a quantum dot with degenerate energy spectrum. Both the profile of the potential field along the motion

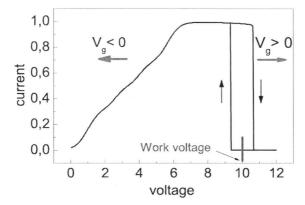

Fig. 26.2. Picture to demonstrate an influence on conductance by the gate voltage.

direction of current carriers and the current-voltage characteristic (CVC) are shown in Figs. 26.1A and 1C.

Depending on the nature of the correlation between current carriers, namely the Coulomb or electron-phonon one, the sections of the hysteresis curve presenting the conduction of a unit have the different directednesses. In the figure, this property is revealed in the behavior of the curve under the increase in the applied voltage. The step-like form of CVC corresponds to the process of elimination of the degeneracy of energy states at the quantum dot due to the correlation. The value of U determines the energy of interaction between two current carriers. Varying the voltage on a gate, one can control the conduction of the system. If the voltage on a gate depends on the potential of the probe, the conduction of such a device will be determined by a state of the probe. In Fig. 26.3, we display the peculiarities of using the property of bistability of the conduction of a nanotransistor for the processing of the information obtained from a probe. When the work voltage is fixed, and its value is in the region of bistability (blue line), a change in the voltage on a gate induces a shift of the profile of CVC. The profile shifts to the right (right red arrow) as the voltage on a gate increases and to the left (left red arrow) as it decreases. If a value of the work voltage leaves the region of bistability on such a shift, the conduction of a nanotransistor changes jumpwise. Thus, if the voltage on a probe is used as the voltage on a gate, such a construction allows one to realize the diagnostics of a state of the probe part of a nanosensor device.

26.4 Nanosensor with a bistability-based transistor

In Fig. 26.3, we give the basic diagram of a nanosensor device, in which the above-presented scheme of the processing of information can be realized. The device is a sandwich consisting of dielectric and conducting layers. The work

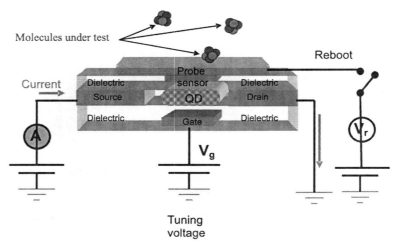

Fig. 26.3. The construction diagram for a electronic nanosensor devices. The function base is a nanotransistor with bistability conduction.

current passes from an emitter to a collector through a quantum dot so that the system is positioned in the conduction bistability region. The voltage supplied in the lower part of the device to the gate V_g allows one to shift a state of the quantum dot to the boundaries of the bistability region. In other words, this voltage is intended for the tuning of the device into the work state. The upper part of the device contains a testing probe that is some small conducting platform joined with an electrode analogous to a gate. On the right part of the device, a system ("reboot") restoring the probe into the working mode is positioned. It allows the necessary voltage to be supplied to the probe in a controlled way. How can such a nanosensor work?

As an example, let us consider its operation in the mode of chemical sensor. In this case, the attachment of tested molecules to the probe can change its potential. The electronic device under consideration does allow one to register such a change. Indeed, while the work voltage is supplied to the quantum dot being in the conduction bistability region, the conduction of the system will depend on the way, in which this voltage is supplied. For definiteness, we choose zero as the initial value of the conduction. In this case, the voltage on a reboot is switched-off, $V_r = 0$. Then, by varying the voltage on a gate V_g, we can shift the current-voltage profile shown in Fig. 26.3 to the right so that the value of work voltage turns out to be near the left boundary of the bistability region. Let now the device be positioned in the gaseous or liquid medium, where some number of tested molecules is present. Then these molecules will touch the probe and, hence, change its potential. In such a way, the potential field around the quantum dot is changed. Therefore, the current-voltage profile will be additionally shifted, which will lead to a jump-like variation of the conduction of the system. There appears the current in the work circuit, and it can be registered with an ammeter. The value of additional shift related to the touching of the probe by a molecule is an individual characteristic of the probe-molecule contact. The voltage on a gate can be selected with regard for this circumstance. Further, even if a molecule leaves the probe, the system remains in the current-conducting state. In other words, the sensor device has remembered the presence of tested molecules in the medium. In order to restore the device in the initial state, it is necessary to switch-on, for some time interval, the recharging voltage V_r, whose value must exceed the width of the bistability region. This will lead to two effects. The current-voltage profile will be shifted to the left by a value exceeding the bistability interval, which transfers the quantum dot into the nonconducting state. In addition, the recharging voltage will repel adhered tested molecules from the probe. That is, there occurs the purification of the probe. After the switching-off of the recharging voltage, the sensor device is in the initial state and is ready again to the testing process.

26.5 Conclusions

Thus, a device consisting of a nanotransistor operating in the bistability interval of voltages and a nanoprobe can successfully fulfill the sensor functions. The above-considered device can be also applied to the testing of pressure variations, acoustic oscillations, etc. To this end, it is sufficient to coat a probe by a substance possessing the piezoelectric properties. However, as a significant obstacle on the way to the fabrication of a similar device, we mention the absence of a suitable nanotransistor with the necessary properties. Eligible can be a transistor using either molecules of rotaxane as a quantum dot or short carbon nanotubes [3, 4]. But, in all the cases, it is basically important to understand the mechanism of the appearance of the property of bistability in nanotransisters [5–8].

Acknowledgment

S.K is grateful for a financial support of the MUES(project M/145-2007).

References

1. V.N. Ermakov, S.P. Kruchinin, H. Hori, A. Fujiwara, Int. J. Mod. Phys. B, **21** 1827 (2007)
2. V.N. Ermakov, E.A. Ponezha, J. Phys. Condens. Mater **10**, 2993 (1998)
3. C.P. Collier et al., Science 285, 391 (1999)
4. M.S. Fuhrer et al., Nano Lett., **2**, 755 (2002)
5. M. Radosavljevic et al., Nano Lett., **2**, 761 (2002)
6. V.N. Ermakov, Physyca E **8**, 99 (2000)
7. A.S. Alexandrov et al., Phys. Rev. B **65**, 155209 (2002)
8. A.S. Alexandrov, A.M. Bratkovsky, Phys. Rev. B **67**, 235312 (2003)

27

A BIO-INSPIRED ELECTROMECHANICAL SYSTEM: ARTIFICIAL HAIR CELL

KANG-HUN AHN

Department of Physics, Chungnam National University, Daejeon 305-764, Republic of Korea. ahnkh@cnu.ac.kr

Abstract. Inspired by recent biophysical study on the auditory sensory organs, we study electromechanical system which functions similar to the hair cell of the ear. One of the important mechanisms of hair cells, *adaptation*, is mimicked by an electromechanical feedback loop. The proposed artificial hair cell functions similar to a living sensory organ in the sense that it senses input force signal in spite of the relatively strong noise. Numerical simulation of the proposed system shows *otoacoustic sound emission*, which was observed in the experiments on the hair cells of the bullfrog. This spontaneous motion is noise-induced periodic motion which is controlled by the time scale of adaptation process and the mechanical damping.

Keywords: electromechanical system, emission, hair cell

27.1 Introduction

Humans communicate with each other mainly through speech in daily life. Therefore the hearing organs require precise sound detection in spite of the background noise. Humans can detect auditory stimuli and acceleration at the level of thermal energy [1,2]. The amazing function of the ear is from its peculiar signal amplification process which is done through heavily damped resonator. Although the complete understanding of the cochlear amplification, where mechanoelectric transduction arises, still remains uncertain, recent progress of biophysical study of the hair cells reveals the underlying strategy in hearing.

Understanding how we hear is important not only in life science, but also important because it opens the possibility of utilizing the novel mechanisms as a strategy of bio-mimetic sensor technology. Weak force detection by electromechanical systems is currently important topics in many fields of physics including gravitational wave detection [3] and searching for quantum behavior of mechanical systems [4].

Here I will introduce our recent effort on developing new scheme of force measurement which is currently under theoretical and experimental study [5, 6]. The basic idea in this approach is designing ultra sensitive force sensor based on the experimental findings of the novel mechanisms in hair cells of the ear [7]. Hair cells are complex micro-mechanical systems where displacement of hair bundles generates electric response via mechanically sensitive ion channels in the cell membrane. We designed an electronic feedback circuit which functions similar to the biochemical processes in the hair cell. The numerical simulation of the mechanical systems with the designed feedback loop shows that ultra sensitive detection of the sound, force, and inertia might be possible through this scheme. The rich nonlinear dynamics will be discussed which might help us understand the hearing mechanism and design noise-robust force sensor.

In sections 27.2 and 27.3, I will introduce the physics of hearing and the physiology of the ear and hair cell. The electromechanical feedback circuit model for the adaptation process is introduced in section 27.4. The summary and future work will be described in section 27.5.

27.2 Physics of the ear

When the sound waves enter the ear, they strike tympanum where the resulting displacement is about 10^{-10} m. The vibration of the tympanum is delivered to the oval window through small bones called ossicles. They are maleus, incus, and stapes (see Fig. 27.1). The sound amplification already arises here in the middle ear as the three ossicles play the roles of hammer, anvil, and stirrup, respectively. The much larger area of the tympanum compared to the oval window produces amplification; the pressure at oval window is greater.

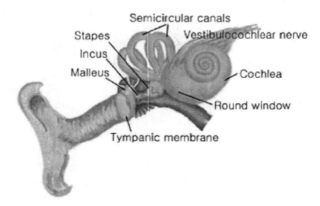

Fig. 27.1. The anatomy of the ear.

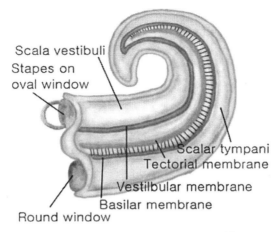

Scala vestibuli
Stapes on
oval window

Scalar tympani
Tectorial membrane
Vestilbular membrane
Basilar membrane
Round window

Fig. 27.2. The structure of the cochlea.

The cochlea is the organ where the sound is transformed into electric signal (see Fig. 27.2). The sound wave propagates from the oval window to the round window through scalar vestibuli and scala tympani. The sound receptors of the ear are in the organ of Corti which lies on the basilar membrane within the scalar media. The receptors are called hair cells because they end in hairlike structures called stereocilia. Hair cells are amazingly sensitive and can detect mechanical displacement as small as the diameter of a hydrogen atom.

The stereocilia are linked together by protein filaments. When the stereocilia are bent in the direction of the tallest stereocilium, more tension is applied in the protein filament and the ion channel opens. The stereocilia is in endolymph which has high concentration of K^+. The body of the hair cell is located in perilymph which is low in K^+ and high in Na^+. When the potassium channel is opened by the protein filament, more potassium enter the hair cell producing a less polarized state.

Loud sounds cause larger changes in the number of open potassium channels. The frequency of the sound is encoded relying on the fact that the sound of different frequency causes deflection of the basilar membrane of different region. The basilar membrane is stiff and narrow near the oval windows, so the peak displacement position of the basilar membrane is closer to the stapes at the higher frequency.

The acceleration is detected in the vestibular apparatus in the ear which consists of the semicircular canals, the utricle, and the saccule. The semicircular canals detect the rotational acceleration and the utricle and the saccule detect the linear acceleration in the direction of the forward-backward (utricle) and up-down (saccule). The acceleration is transformed into the electric signal again by the hair cells in the vestibular systems. As in hearing, mechanical bending of stereocilia causes the ion channel opening. The mechanical bending arises easier with the aid of additional mass in the vestibular system. In the utricle and the saccule, hair cells with stereocilia extends up into gelatinous material. On top of the material, there are small calcium crystals called otoliths which add mass to the gelatinous material.

27.3 Adaptation in hair cells

Biophysical experiments on the stereocilia of hair bundles show that hair cells contain an active force generating mechanism [8], which gives rise to the region of negative stiffness in force-displacement relation [7]. The amplification by hair cell relies on the existence of the unstable region of the negative stiffness. The instability, however, is not enough to explain the ear's high sensitivity. There is so called *adaptation* process [9, 10] which plays very important role in sensory organs.

'Sensory adaptation' in general means the decay of a sensory response to constant stimulus. Such response decay arises in diverse sensory system from diverse mechanisms, including decreases synaptic gain, change in the sensitivity of a transduction process. In hair cells, the adaptation arises via ion-dependent channel opening and closure (Fig. 27.3). The adaptation is considered to be mediated by an adaptation motor which may be a cluster of myosin molecules [11, 12]. The adaptation motor moves the hair bundles close to the range of negative stiffness. By doing so, the hair cells are ready to detect mechanical stimulus.

The negative stiffness and the adaptation process together lead to a spontaneous oscillation of stereocilia, which is called otoacoustic emission [7, 13]. The ear of many vertebrates can emit sound. The spontaneous otoacoustic

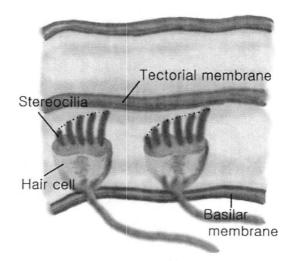

Fig. 27.3. The structure of the sound receptor in the ear.

emission arises as a natural results of amplification process in the ear similar to an electrical oscillator with a tunnel diode. It is not clear what is the role of the otoacoustic emission itself in the human ear. Perhaps, the oscillation might amplify sound signal using certain resonance. It has been proposed that non-mammalian vertebrates may use the resonance between the sound and the otoacoustic oscillation as an amplification mechanism [14].

27.4 Electromechanical model for hair cell

In previous sections, I reviewed on how the ear senses the auditory stimuli and the balance. In this section, we introduce an electromechanical model [5] for noise-robust artificial sensory organs by mimicking the mechanisms of the adaptation and negative stiffness.

The first step toward this task can be done by studying an electromechanical system which has a feedback loop functioning as an adaptation motor. Let us consider a system where external force is measured through displacement x of a cantilever. The cantilever is a mechanical oscillator of mass m and the resonance frequency ω_0.

In the setup shown in Fig. 27.4, the output voltage of the position sensor V_x varies proportional to the displacement of the cantilever,

$$V_x = ax. \tag{27.1}$$

In the feedback circuit, a lossy integrator gives output voltage V_s which is the average value of V_x over a time scale τ_a (let us call this 'adaptation time');

$$V_s(t) = \frac{1}{\tau_a} \int_0^\infty \exp(-\frac{s}{\tau_a})V_x(t-s)ds. \tag{27.2}$$

The output voltage of the integrator contains the information of the averaged displacement near time t, $V_s(t) = a\bar{x}(t)$;

$$\bar{x}(t) = \frac{1}{\tau_a} \int_0^\infty \exp(-\frac{s}{\tau_a})x(t-s)ds \tag{27.3}$$

The adaptation force F_{ad} is exerted by the adaptation force generator after the displacement x is compared to its mean value $\bar{x}$ in the comparator.

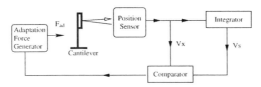

Fig. 27.4. Schematic figure for the electromechanical system which mimics the adaptation process and the negative stiffness.

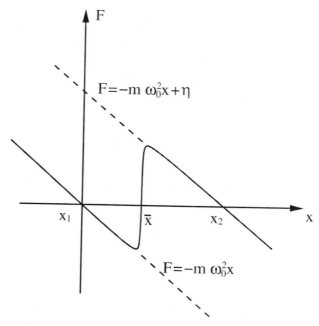

Fig. 27.5. The force-displacement relation of the cantilever where the adaptation feedback force F_{ad} is applied.

The adaptation force F_{ad} is designed to have two different values depending on the sign of $x - \bar{x}$ which proves the system operates similar to hair cell's adaptation [5, 6].

The effective force-displacement relation defined by $F_{\text{eff}}(x) = -m\omega_0^2 x + F_{ad}$ is shown in Fig. 27.5. There are two stable equilibrium positions x_1 and x_2. The averaged position $x = \bar{x}$ is an unstable equilibrium point, where the negative stiffness region is formed around. Note that this system is not passive. When the cantilever is located at x_1 initially, then $\bar{x}$ spontaneously moves to x_1 in time scale of τ_a. This movement of $\bar{x}$ is essentially necessary to play the role of the adaptation motor in hair cells. The past history of the constant stimuli is memorized in $\bar{x}$, so the force-displacement relation is adapted to the past stimulus.

The governing equations of motion for the proposed electromechanical system including external signal and noise are given by

$$m\ddot{x} + m\gamma\dot{x} = -m\omega_0^2 x + F_{ad} + F_{ext} + F_{noise} \tag{27.4}$$

$$\tau_a \dot{V}_s = V_x - V_s \tag{27.5}$$

$$F_{ad} = \eta\Theta(V_x - V_s), \tag{27.6}$$

where γ is the damping parameter, $F_{noise} = \xi(t)$ is the (thermal) noise force exerted on the cantilever which can be modeled as white noise

$$< \xi(t) > = 0 \tag{27.7}$$

$$< \xi(t)\xi(s) > = 2D\delta(t - s). \tag{27.8}$$

The numerical simulation based on the above equations shows interesting dynamics which do not appear in passive systems. The white noise is simulated using Box-Mueller algorithm [15] and the time-step for the white noise is small enough to ensure that the bandwidth is larger enough than any physically meaningful frequency in this system.

Figure 27.6 shows the numerical simulation for the cantilever positions as a function of time in the absence of any periodic signal. Interestingly, the spontaneous oscillation similar to the otoacoustic sound emission from bullfrog's ear [7] is found in the numerical results. Depending on the adaptation time τ_a and the damping parameter γ, the cantilever shows two different types of motion. When $\gamma\tau_a$ is large enough, the system shows *bistable(BS) oscillation* similar to otoacoustic sound emission. The BS oscillation becomes very periodic as the noise strength becomes weaker but the BS oscillation itself does not exist in the absence of the noise. The BS oscillation is noise-induced periodic motion. When $\gamma\tau_a$ is small enough, the system shows periodic *sinusoidal oscillation*. The amplitude of this oscillation has non-monotonic dependence of $\omega_0\tau_a$.

When the cantilever is located at x_1 (Fig. 27.5), $\bar{x}$ continuously approaches to x_1 letting $x - \bar{x}$ as small as possible. The system stays long time around x_1 as can be seen in Figs. 27.6a and b. A small noise force eventually kicks

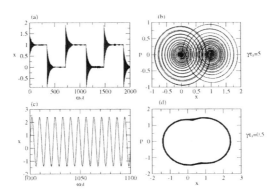

Fig. 27.6. Numerical simulation of the displacement of the cantilever with the adaptation feedback loop. Time series of the displacement (a) and (c) and the trajectory in the displacement-momentum phase space (b) and (d). Given length scale l of the system, the parameter in use are $D = 10^{-5}m^2\omega_0^2 l$, $\eta = m\omega_0^2 l$, $\gamma/\omega_0 = 0.05$ and $\omega_0\tau_a = 100$ (for (a) and (b)), $\omega_0\tau_a = 10$ (for (c) and (d)).

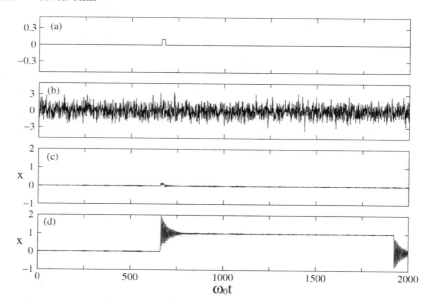

Fig. 27.7. The simulated response of the single artificial hair cell to the external force. (a) input signal force (b) the noise plus input signal force (c) the output displacement signal without adaption process ($\eta = 0$) (d) the output displacement signal with adaptation process with $\eta = m\omega_0^2 l$. l is the length scale of the system, $D = 10^{-5}m^2\omega_0^2 l$, $\gamma/\omega_0 = 0.05$ and $\omega_0\tau_a = 100$, $\omega_0\tau_a = 400$, and the units of the force is $m\omega_0^2 l$.

the cantilever so that $x - \bar{x}$ changes its sign. The cantilever escaped out of the local minimum x_1 moves to new local equilibrium point x_2. This process is repeated again also at x_2 and gives rise to the spontaneous oscillation shown in Figs. 27.6a and b.

When the cantilever oscillates near one of the equilibrium points in sufficient time, the system becomes more and more sensitive to the external force and it becomes more and more probable for the cantilever to jump to the other stable equilibrium point. The release of the mechanical energy in this process is the origin of the signal amplification in the proposed system. Fig. 27.7 shows that even the small input force signal (Fig. 27.7a), which is completely invisible because of the force noise, will be clearly detected through the proposed scheme. While the passive damped cantilever ($\eta = 0$) vaguely detects the input force signal (Fig. 27.7c), the simulated response of the artificial hair cell shows dramatic response to the external force signal (see Fig. 27.7d).

27.5 Summary and outlook

The mechanisms working in the proposed electromechanical system are very similar to the light emission by two-level atom. The spontaneous emission of light due to vacuum fluctuation of electromagnetic field is similar to the otoacoustic sound emission induced by thermal noise. The sensitive response to the weak external force signal can be understood as like the stimulated emission of light by excited two-level atoms. The adaptation then functions similar to pumping in laser system.

The proposed single electromechanical force sensor is an artificial hair cell and it might be a building block for the future artificial cochlea. The experiments following the proposed scheme based on opto-electromechanical systems will be reported elsewhere [6]. The material properties and the noise related issue will be important in the experimental study. Theoretical analysis on collective response of many artificial hair cells is now necessary for the improved bio-mimicking force (or inertia) sensor.

I thank Sukyung Park for helpful discussion and thank Hyery Kim for drawing the first three figures. This work was supported by the Korea Science and Engineering Foundation(KOSEF) grant funded by the Korea government(MOST) (No. R01-2007-000-10837-0).

References

1. H. De Vries, Physica **14**, 48 (1948).
2. H. De Vries, Acta Otolaryngol, **37**, 218 (1949).
3. V. Braginsky, S.P. Vyarchanin, Phys. Lett. A **293**, 228 (2002).
4. M.D. LaHaye, O. Buu, B. Camarota, K.C. Schwab, Science **304**, 74 (2004).
5. K.-H. Ahn, T. Song, Y.D. Park, S. Park, (submitted).
6. K.-H. Ahn, T. Song, J. Lee, (preprint).
7. P. Martin, A.D. Mehta, and A.J. Hudspeth, Proc. Natl. Acad. Sci. **97**, 12026 (2000).
8. A.C. Crawford and R. Fettiplace, J. Physiol. **364**, 359 (1985).
9. R.A. Eatock, D.P. Corey, and A.J. Hudspeth, J. Neurosci. **7**, 2821 (1987).
10. R.A. Eatock, Annu. Rev. Neurosci. **23**, 285 (2000).
11. P.G. Gillespie and R.G. Walker, Nature **413**, 194 (2001).
12. P.G. Gillespie and D.P. Corey, Neuron **19**, 955 (1997).
13. R. Probst, *Otoacoustic emissions: an overview*, in New Aspects of Cochlear Mechanics and Inner Ear Pathophysiology, edited by C.R. Pfaltz, Basel, Karger, p1–p91, (1990).
14. P. Martin, A.J. Hudspeth, Proc. Natl. Acad. Sci. **96**, 14306 (1999).
15. R.F. Fox, I.R. Gatland, R. Roy, and G. Vemuri, Phys. Rev. A **38**, 5938 (1988).

Si CIRCUITS, OPTICAL FIBERS, AND AlInGaP AND InGaN LIGHT-EMITTERS

J.D. DOW

Department of Physics, Arizona State University, Tempe, AZ 85287-1504 U.S.A. `catsc@cox.dat`

Abstract. Fifty years ago, there were three major questions facing the physics of semiconductors: (i) Would Si, having replaced Ge only a few years before, maintain its position as the dominant semiconductor, or would Si be replaced by faster semiconductors, such as III-V semiconductors (e.g., GaAs); (ii) would the direct-gap III-V semiconductors produce efficient light-emitters; and (iii) would optical fibers become sufficiently transparent to make light-wave communications possible?

Si is for circuits. The biggest surprise of the intervening fifty years is that Si became, and still is, the dominant semiconductor for electrical circuits in modern semiconductors. The first transistor was fabricated at Bell Laboratories out of Ge by Bardeen and Brattain, but Shockley showed that Si was better for transistor technology, and Si has remained dominant ever since. Even the future seems likely to be dominated by Si, which is currently being used for three-dimensional circuits. Hopes for faster useful transistors made out of III-V semiconductors, have dimmed in recent years, and so, all but a few of the III-V fabrication lines for *electronic* devices have been abandoned.

Optical fibers and Urbach's Rule. In the late 1960s there was a question of whether optical fibers were possible, and how long a fiber is compatible with transmission of visible light. At the time, fibers capable of transmitting light about a meter were in existence, and the question was whether this distance could be increased to over one km. The increase has occurred.

III-V light emitters. Si is not a candidate for light emission, since it is an indirect-gap semiconductor. The preferred light-emitting semiconductors are currently $A\ell_{1-x-y}In_xGa_yP$ or $In_{1-x}Ga_xN$, and cover the visible spectrum from red to violet and white. So now it is possible to compute with Si, to transmit the computed information over optical fibers, and to fabricate light-emitting diodes of most visible colors.

Keywords: optical fibers, light emitters

J. Bonča, S. Kruchinin (eds.), *Electron Transport in Nanosystems.*
© Springer Science + Business Media B.V. 2008

28.1 Introduction

In the 1960s, three problems were dominant in semiconductor physics. (i) Would Si, having replaced Ge only a few years earlier, be the dominant semiconductor of the future, or would III-V semiconductors take over with their visible light emission and their higher electronic mobilities? (ii) Would optical fibers of length exceeding one km become possible? (iii) Would III-V light emitters evolve into high intensity visible light emitters?

(i) The problem of Si's replacement seems to have been resolved, at least for the short term; it now appears that three-dimensional Si, not any III-V semiconductor, will form the electronic devices of the future. (ii) Ultratransparent optical fibers now exist, with absorption lengths less than 10^{-4}m^{-1}. Indeed, they have become so widespread that optical fibers have supplemented (and often replaced) Cu wires for information-transfer. (iii) Anyone who has examined a traffic light recently, or the back-up lights of an automobile, knows that III-V light-emitters are playing a growing role in this area, having replaced the incandescent lamps of old. Indeed, we can look forward to the next decade in which incandescent household lights are likely to be supplanted by solid state light emitters at all relevant wavelengths.

So of the three outstanding problems of the 1960s, two have proven to be important for modern semiconductor devices (optical fibers and colored lamps) and one (the determination of a III-V replacement for Si) appears to have been a failure.

28.2 Si versus III-V semiconductors

Once Ge [1–4] was replaced by Si [5], the early expectation was that Si would subsequently be replaced by some III-V semiconductor with higher mobility. But the current answer to the question about Si versus III-V semiconductors is that Si is still the dominant semiconductor for conduction today, and in the future, Si appears to be likely to remain dominant for some time. The III-V semiconductors expected to replace Si as an electronic material have failed to do so, and the current efforts to develop Si three-dimensional chips suggest that Si circuits will remain dominant for the next many years.

28.3 Optical fibers

Optical fibers were initially limited by exponential absorption edges (Urbach's Rule [6, 7]), but as originally hoped, the absorption edges did not remain constant in time, but became reduced and sharpened as the quality of the glass fibers improved. Optical fibers have replaced Cu wires as the dominant means of information transfer in our society.

28.3.1 URBACH'S RULE

The first question one must answer when considering the limits of optical fibers is "What kind of disorder is responsible for the exponential (Urbach) absorption edge?" The answer that appears to be best is that the hole of the electron-hole pair is relatively unaffected, because it is so heavy, but the electron is in the presence of both the hole and an electric field associated with the disorder.

Approximating the electrostatic potential felt by the electron due to the hole and the nearly random potential of the disordered internal fields (due to impurities or phonons) looks like the potential in Fig. 28.1a, which can be approximated by a Coulomb potential plus a uniform field (Fig. 28.1b). In three dimensions the potential felt by the electron is somewhat like that of Fig. 28.2, with the Urbach edge being associated with (i) the Coulomb

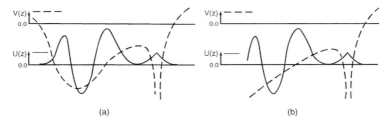

(a)   (b)

Fig. 28.1. Tunneling wavefunction U(z) and potential V(z) (a) versus z, and (b) in a uniform field approximation, versus z, after Ref. [6].

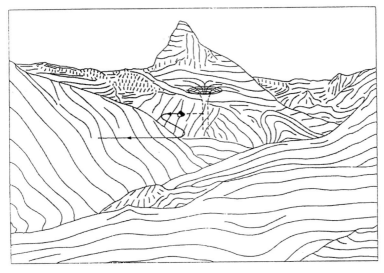

Fig. 28.2. Random potential with the Coulomb potential of the hole, illustrating the physics of absorption in a random potential, after Ref. [6].

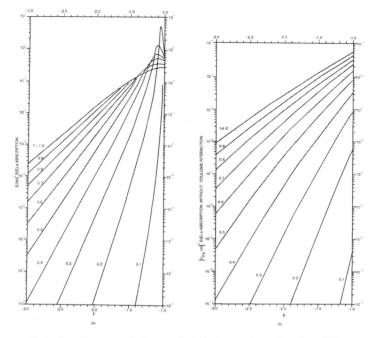

Fig. 28.3. Calculated spectral shape of the absorption edge for different uniform field strengths f. Absorption $|U(0)|^2 S(E)$ versus E (a) with the electron-hole interaction included, and (b) with no electron-hole interaction, after Ref. [8]. The energy E is measured from the energy gap, in units of the binding energy R of the unperturbed exciton.

potential of the hole, plus (ii) the random potential of the internal fields or phonons. This makes a uniform field approximation reasonable. Figure 28.3 shows the predicted absorption of an exciton in a uniform field, and with it, the corresponding results obtained if the electron-hole interaction is neglected [8,9]. Clearly the electron-hole interaction makes a significant contribution to the optical absorption strength.

Figure 28.4 has the predicted absorption versus average field.

It is now generally accepted that, in most cases, Urbach's Rule is due to internal electric fields, and causes broadening of the fundamental absorption edge.

28.4 III-V light emission

For light emission, III-V semiconducting light emitters [10] seem to be carving out a niche even today, especially for red and amber emission, and more recently for colors from the yellow to the ultraviolet (UV) [11–24]. These

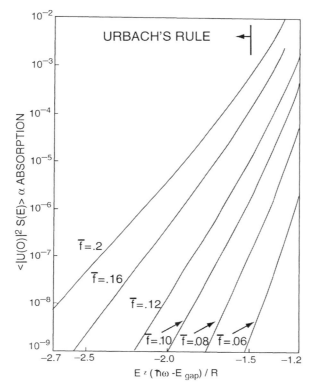

Fig. 28.4. Absorption versus E for various average fields $\overline{f}$.

devices appear likely to dominate the markets of the next several years. We already have light emitters that appear in tail-lights of cars and in traffic-lights, and there is an indication that these devices will continue to be important for some time. Solid-state lamps appear likely to eclipse incandescent lamps in the next decade, both for white lamps and for colored lamps: from the infrared to the ultraviolet.

28.4.1 IMPURITY-ASSISTED EMISSION: DEEP LEVEL

Central to the issue of light emission is whether the emission is facilitated by impurities which produce deep levels or shallow levels, because the deep levels in particular often facilitate light emission. (Many of the early light-emitters featured impurity-assisted emission). Originally a deep level was thought to be more than 0.1 eV within the fundamental band gap, but it is now known that a defect that has a different charge state from the atom it replaces produces *both* deep and shallow levels.

The deep levels originate from the central-cell potential of the defect, and are typically of order 10 eV deep with a potential of radius of order 0.3 Å.

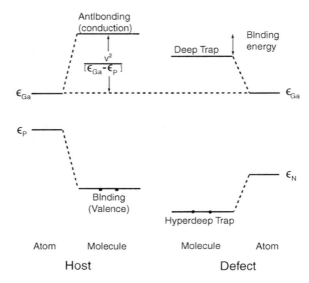

Fig. 28.5. Schematic bonding–anti-bonding diagram for (a) a GaP host and for (b) a N defect replacing P in that host. The deep level lies below the conduction band edge, and the hyperdeep trap lies below the valence band maximum, after Ref. [34].

The deep levels lie between the bonding valence-band states and the anti-bonding conduction-band states. So the "10 eV" binding energy is not actually a binding energy, but the difference between the anti-bonding energy of the host conduction band and the energy level of the defect. The shallow levels are hydrogenic in character and are bound relative to a band edge by the Coulomb potential, $-e^2/(\epsilon r)$, between the electron and the central-cell ion (e.g., P in bulk Si) with binding energy of order 10 meV. It is now generally accepted that, in most cases, Urbach's Rule is due to internal electric fields, and causes broadening of the fundamental absorption edge. To understand the physics of deep levels, consider GaP doped with N on a P site (Fig. 28.5). Since P is isoelectronic to N, there is no Coulombic force associated with P, and hence there are no shallow levels. Schematically the Ga and P energy levels (mostly their p-states) form bonding and anti-bonding states, with the bonding state becoming the valence band of GaP (fully occupied by electrons), and the anti-bonding state becoming the conduction band. The anti-bonding state is raised above the Ga energy level by about $v^2/(\epsilon_{Ga}-\epsilon_P)$, where v is the Ga-P matrix element and ϵ_{Ga} and ϵ_P are the energy levels of the atomic p-states. When a N impurity substitutes on a P site, its p-symmetric energy level ϵ_N is much more negative ($\sim$5 eV [25]) than the P level ϵ_P it replaces, forming an occupied *hyperdeep* trap state in or below the valence band and a *deep* trap state that either lies in the conduction band or below it, in the gap (see Fig. 28.6). As a result of this physics, the N deep state lies in the fundamental band gap of $GaAs_{1-x}P_x$ for most compositions $x > 0.4$. (It is

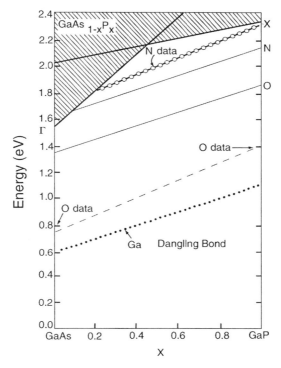

Fig. 28.6. Alloy diagram of the energies of GaAs$_{1-x}$P$_x$ versus alloy composition x (showing the Γ and X band edges). The N and O luminescence lines are shown, with their predicted energies and the Ga dangling bond energy, after Ref. [34].

below the Γ_1, X_1, and L_1 band edges). The data for N and O impurities on P or As sites in GaAs$_{1-x}$P$_x$ are presented in Fig. 28.7. Note that the N-line, N$_X$ (also called the A-line in GaP) is not attached to the near-by band-edge, but varies almost linearly with alloy composition x, unlike the shallow donor line, which is "attached" to the nearby conduction band edge with nearly constant binding energy. The donor lines D$^\Gamma$ and D$_0^X$ follow the band edges Γ_1 and X_1, as does the free exciton line.

28.4.2 N-N PAIRS AND (CATION, O) PAIRS

Figure 28.7 shows the luminescence in GaAs$_{1-x}$P$_x$ for N$_\Gamma$ (which becomes N$_{\Gamma-X}$ and N$_X$) as well as the phonon sidebands and the pair states NN$_1$ (nearest-neighbor) and NN$_3$ (third nearest-neighbor).

The pair states for a Ga-vacancy (V$_{Ga}$) paired with an O ion and for a (Zn,O) pair are given in Fig. 28.8. The chemical trends for pair states, including (V$_{Ga}$, O), (Mg, O), (Cd, O), (Ga, O), and (Si, O), are in Fig. 28.9.

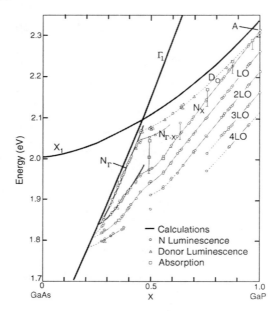

Fig. 28.7. Energies of N-doped $GaAs_{1-x}P_x$ versus x, showing the band edges Γ_1 and X_1, the N deep level N_X (called A in GaP), the N level $N_{\Gamma-X}$ as the material becomes indirect-gap, and the level N_Γ in indirect-gap material. The shallow donor state is D_0, and the longitudinal optical phonon sidebands of N_X are shown as LO 2LO, 3LO, and 4LO, after Ref. [34].

28.4.3 CURRENT STATUS: DEFECTS

All of the deep levels associated with point defects, such as the N deep level and the paired-defect states, are candidates for light emitters, but the developments of the past five years suggest that the best light-emitters will not be defects, but will be host materials, such as $GaAs_{1-x}P_x$ or $A\ell_x Ga_{1-x}N$.

28.4.4 BULK HOST MATERIALS

In recent years, the best light-emitters have evolved toward bulk host materials, not defects or impurity states. The first infrared light emitters were made out of the host bulk GaAs [26, 27]. Red light-emitting diodes were fabricated frombulk $GaAs_{1-x}P_x$ [10, 14, 28–30]. The favored yellow light emitter was InGaN [16] or N-doped GaP [19, 20]. $A\ell_x Ga_{1-x}N$ and $In_x Ga_{1-x}N$ are materials that produced blue light [11, 31–33].

Currently, the most promising light-emitters [39, 41] appear to be bulk $In_x Ga_{1-x}N$ or $A\ell_{1-x-y} In_x Ga_y P$. Both emit light over much of the visible spectrum, from red to violet (See Table 28.1).

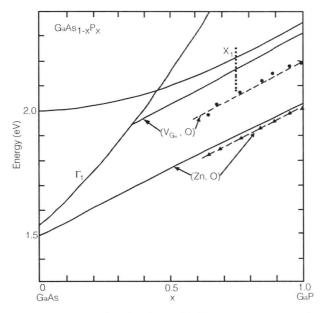

Fig. 28.8. Theory and data for the (V_{Ga},O) [Ga-vacancy, oxygen] pair, and for (Zn,O) paired levels in $GaAs_{1-x}P_x$ versus x.

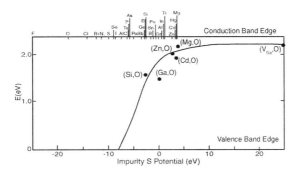

Fig. 28.9. Energies versus impurity s potential for paired defects (solid line), with data for (V_{Ga}, O), (Mg, O), (Cd, O), (Zn, O), (Ga, O), and (Si, O) in GaP. The oxygen is on a P-site, while the impurity is on the Ga-site. The dots correspond to data, while the solid line is the theory. After Ref. [35].

28.5 Summary

The original hope that Si computation would be replaced by III-V computation has not been realized. Optical fibers have had an enormous impact on the area of light transmission, and Cu electrical transmission lines are being replaced almost everywhere with optical fibers, most notably for under-ocean telephony. The problem of developing solid-state light-emitters for most of the

Table 28.1. Summary of bright light-emitting diodes available, after Refs. [12, 37–41]. Growth techniques include: MOCVD or metal-organic chemical vapor deposition; LPE or liquid phase epitaxy; MBE or molecular beam epitaxy; VPE or vapor-phasee epitaxy; and CVD or chemical vapor deposition. Structures include: DH or double heterostructure; QW or quantum well; and H or homojunction device.

Material	Peak emission	Color	External quantum efficiency	Growth technique	Kind of device	
Recent materials						
AℓGaInP		Red to white		MOCVD	DH	[39–41]
InGaN		Red to white		MOCVD	DH	[39–41]
Older materials						
AℓGaInP	636 nm	Red	24%	MOCVD	DH	[39]
AℓGaAs	650 nm	Red	16%	LPE	H	[13]
GaP:(Zn,O)	650 nm	Red	5%	MOCVD	DH	[14, 15]
AℓGaInP	590 nm	Amber	10%	MOCVD	DH	[39]
InGaN	590 nm	Amber	3.5%	MOCVD	QW	[16]
GaAsP:N	590 nm	Yellow	0.3%	VPE	H	[17]
GaP:N	565 nm	Yellow-green	0.3%	LPE	H	[18–20]
InGaN	520 nm	Green	10%	MOCVD	QW	[16, 21, 36, 37]
InGaN	470 nm	Blue	11%	MOCVD	QW	[16, 21]
InGaN	372 nm	UV	7.5%	MOCVD	DH	[22, 23]

colors of the visible spectrum has been largely (but not completely) solved. Most new cars are equipped with red lights that are solid state in character. In the next decade we expect that most incandescent lamps will be replaced by solid-state light emitters, and that the nature of home lighting will change.

Acknowledgments

We thank the U.S. Army Research Office (Contract W911NF-05-1-0346) for their support. We also want to express our gratitude to D. J. Wolford and M. G. Craford for their assistance.

References

1. J. Bardeen and W. H. Brattain, Phys. Rev. **74**, 230 (1948).
2. G. L. Pearson, J. Bardeen, Phys. Rev. **75**, 865 (1949).
3. J. Bardeen and W. H. Brattain, Phys. Rev. **75**, 1208 (1949).
4. W. Schockley and G. L. Pearson, Phys. Rev. **74**, 232 (1948).
5. W. Schockley, Bell Sys. Tech. J. **28**, 435 (1949).

6. J. D. Dow and D. Redfield, Phys. Rev. **B 5**, 594 (1972)
7. J. D. Dow, in *Optical properties of highly transparent solids*, ed. by S. S. Mitra and B. Bendow (Plenum Press, New York, 1975) pp. 131–143.
8. J. D. Dow and D. Redfield, Phys. Rev. **B 1**, 3358 (1970).
9. F. C. Weinstein, J. D. Dow, and B. Y. Lao, Phys. Rev. **B 4**, 3502 (1971).
10. N. Holonyak, Jr. and S. F. Bevacqua, Appl. Phys. Lett. **1**, 82 (1962).
11. S. Nakamura, Jpn. J. Appl. Phys. **30**, L1705–L1707 (1991).
12. S. P. DenBaars, p. 1, in *Introduction to nitride semiconductor blue lasers and light emitting diodes*, ed. by S. Nakamura and S. F. Chichibu, Taylor & Francis, London and New York, (2000).
13. D. J. Wolford, W. Y. Hsu, J. D. Dow, and B. G. Streetman, J. Luminescence **18/19**, 863–867 (1979).
14. M. Pilkuhn and H. Rupprecht, J. Appl. Phys. **36**, 684 (1965).
15. A. H. Herzog, W. O. Groves, and M. G. Craford, J. Appl. Phys. **40**, 1830–1838 (1969).
16. T. Mukai, H. Narimatsu, and S. Nakamura, Jpn. J. Appl. Phys. **37**, L479–481 (1998).
17. T. Mukai, M. Yamada, and S. Nakamura, Jpn. J. Appl. Phys. **38**, 3976–81 (1999).
18. D. J. Wolford, B. G. Streetman, W. Y.Hsu, J. D. Dow, R. J. Nelson, and N. Holonyak, Jr., Phys. Rev. Letters **36**, 1400–1403 (1976).
19. P. B. Hart, Proc. IEEE **61**, 880 (1973).
20. D. G. Thomas and J. J. Hopfield, Phys. Rev. bf 150, 680 (1966).
21. S. Chichibu, T. Azuhata, T. Sota, and S. Nakamura, Appl. Phys. Lett. **69**, 4188–4190 (1996).
22. S. D. Lester, M. J. Ludowise, K. P. Killeen, B. H. Perez, J. N. Miller, and S. J. Rosner, J. Crystal Growth **189–190**, 786–789 (1998).
23. T. Mukai, D. Morita, and S. Nakamura, J. Crystal Growth **189–190**, 778–781 (1998).
24. T. Mukai and S. Nakamura, Jpn. J. Appl. Phys. **38**, 5735–5739 (1999).
25. P. Vogl, H. P. Hjalmarson, and J. D. Dow, J. Phys. Chem. Solids **44**, 365 (1983).
26. R. N. Hall, G. E. Fenner, J. D. Kingsley, T. J. Soltys, and R. O. Carlson, Phys. Rev. Lett. **9**, 366 (1962).
27. M. I. Nathan, W. P. Dumke, and G. Burns, Appl. Phys. Lett. **1**, 62 (1962).
28. M. G. Craford, W. O. Groves, A. H. Herzog, and D. E. Hill, J. Appl. Phys. **42**, 2751 (1971).
29. M. G. Craford, R. W. Shaw, A. H. Herzog, and W. O. Groves, J. Appl. Phys. **43**, 4075 (1972).
30. M. G. Craford and W. O. Groves, Proc. IEEE **61**, 862 (1973).
31. J. I. Pankove, IEEE Trans. Electron. Devices **22**, 721–724 (1975).
32. S. Nakamura and S. F. Chichibu, Taylor & Francis, 11 New Fetter Lane, London EC4P–4EE (2000).
33. S. Nakamura and G. Fosol, "The blue laser diode," (Springer-Verlag, Heidelberg, 1997).
34. H. P. Hjalmarson, P. Vogl, D. J. Wolford, and J. D. Dow, Phys. Rev. Lett. **44**, 810–813 (1980).
35. O. F. Sankey, H. P. Hjalmarson, J. D. Dow, D. J. Wolford, and B. G. Streetman, Phys. Rev. Lett. **45**, 1656 (1980).

36. T. Mukai, S. Nagahama, T. Yanamoto, M. Sano, D. Morita, M. Yamamoto, M. Nonaka, K. Vasutomo, and K. Akaishi, Phys. Stat. Sol. (c) **2**, 3884–3886 (2005).
37. S. Nakamura, M. Senoh, and T. Mukai, Jpn. J. Appl. Phys. **32,** L8–L11 (1993).
38. S. Nakamura, T. Mukai, and M. Senoh, Jpn. J. Appl. Phys. **30,** L1998–L2001 (1991).
39. M. Krames, O. B. Shchekin, R. Mueller-Mach, G. O. Mueller, L. Zhou, G. Harbers, and M. G. Craford, J. Display Technol. **3,** 160 (2007).
40. Y. Narukawa, J. Narita, T. Sakamoto, T. Yamada, H. Narimatsu, M. Sano, and T. Mukai, Phys. Stat. Sol. (a) **204,** 2087–2093 (2007).
41. Y. Narukawa, M. Sano, M. Ichikawa, S. Minato, T. Sakamoto, T. Yamada, and T. Mukai, Jpn. J. Appl. Phys. **46,** L963–L965 (2007).

29

SENSING EFFECTS IN THE NANOSTRUCTURED SYSTEMS

V.G. LITOVCHENKO AND V.S. SOLNTSEV
Institute of Semiconductor Physics, National Academy of Sciences of Ukraine, 41 prospect Nauki, 03028, Kiev, Ukraine.
textttLvg@isp.kiev.ua

Abstract. The theoretical analysis and experimental data on the adsorption and catalysis of H - containing molecules: dipole-like H_2, H_2O and H_2S with slight dipole moment. The feature of the spectra of electron states in nanostructures demonstrates mechanisms responsible for the enhanced gas sensitivity and catalysis. Comparison of experimental results on the electro-adsorption phenomena with theory allows to estimate the characteristic energy bands on the surface E_a for different molecules.

Keywords: electro-adsorption effect, spillover effect, nanostructured systems, MIS sensor, porous silicon, gas sensitivity, gas catalysis

29.1 Introduction

The mechanisms of adsorption and catalysis that take place in the nanostructured systems are different from those observed for the bulk systems. In particular, the reconstruction of electronic configuration and chemical bonds in semiconductors and catalytically active transition metals upon the size decrease (e.g. in Pd/Cu *nanoclusters*) realises a new possibility for the dissociative adsorption of complex molecules and leads to an enhancement of the sensitivity of nanostructures to various gases [1–6]. In this paper, the catalytic-adsorption effects in nanostructured systems have been analysed. We have considered the dissociative catalysis of H-containing molecules (H_2, H_2O, H_2S) and the spillover effect in the Cu nanoclusters in the surface layer of a $Pd + Cu$ *structure*. The activation energies of dissociative catalysis have been estimated from the experimental data. The selectivity of the Pd/Cu sensor structure has been demonstrated.

J. Bonča, S. Kruchinin (eds.), *Electron Transport in Nanosystems.*
© Springer Science + Business Media B.V. 2008

29.2 Boundary transformation in the quantum-size structures

Quantum-size nanostructures (in particular, quantum dots QD) are the special type of material structure, which appears for the ultra-small sample sizes, when due to narrowing of the potential wall of the boundaries, remarkable linkage of the electron clouds take peace, i.e., delocalisation of the electrons , which create the chemical bonds of materials. For QDs, the reconstruction of electron spectrum is particularly large, which leads to its transformation to atomic-like with quasi-discrete levels (Fig. 29.1). In addition, the reconstruction of the spectra of other quasi-particles such as phonons, plasmons, exitons, mixed modes etc. -takes place also [1]. Figure 29.2 illustrates the shift of energy levels in Pd and $PdCu$ nanoclusters estimated by the method of crystal lattice field [3–5].

All these factors lead to a large change of adsorption and catalytic processes, which are the basis for the gas sensors and are important for the ecology gas monitoring. Below we consider a number of effects, which are responsible for the enhanced sensitivity of MIS sensor devices, based on the

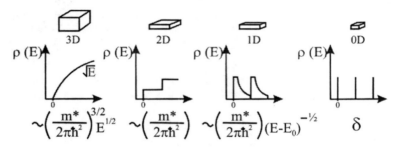

Fig. 29.1. Transformation of electron state spectrum for 3D (bulk) 2,1 a quantum dot.

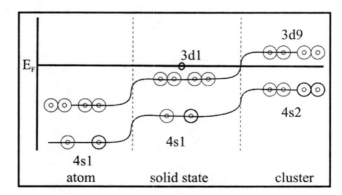

Fig. 29.2. Energy level shift under in nanoclusters [5].

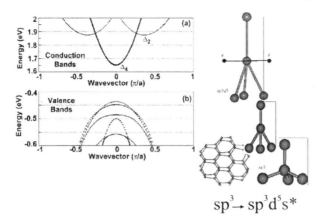

$$sp^3 \rightarrow sp^3d^5s^*$$

Fig. 29.3. The electron spectra and chemical bonds for ID nanostructured systems of silicon [4].

nanostructured Si with upper metallic layers (Pd, Cu). For figures use the reason for that is that the degeneration of levels in the nanoclusters is removed by a crystal lattice field and they become splitted [5], Fig. 29.2. In particular, in the Pd- octahedral field the energy of E_g orbital becomes larger and t_{rg} smaller. The values of splitting are different for different nanosized samples, which leads to the increase of the energy of chemical bonds for clusters (in comparison to the bulk) as well as to the appearance of $4d^9$ bonding orbital of Pd, transformed from the completely filled $4d^{10}$ orbital in the bulk. The same happens for other transition metals (such as Ni, Cu, and Pt with half-empty $5d^9$ orbital). Hence, for the nanoclusters, additional electron empty states above the Fermi surface and $1/2$ filling for $6s^1$ are predicted. Due to this, the catalytic activity (strength of adsorption molecules) of such nanoclusters becomes substantially larger. For silicon nano-wires, two principal results have been predicted in addition [2]:

1. Transformation of indirect band structure to a direct one, which leads to appearance of efficient photoluminescence.
2. Transformation of the character of the hybridisation of electron chemical bonds, distorted on the surface, and even change of hybridisation. In particular, the transformation of chemical bonds from sp^3 to sp^2 or to a complex bond structure spnd with changing of the spatial orientation and strength was demonstrated in [4], Fig. 29.3.

   For transition metal nanostructured layers (which are the main adsorbtion-catalytic materials in gas sensors) such electron bond transformations are important also. Namely, for the nanoclusters, the distribution of electron energy states transforms so that the states occupied in the bulk become at least partly free for some types of nanoclusters (like Cu_3Pd clusters on the surface of Pd) [4–6]. In particular, as can be seen from Fig. 29.4, some of

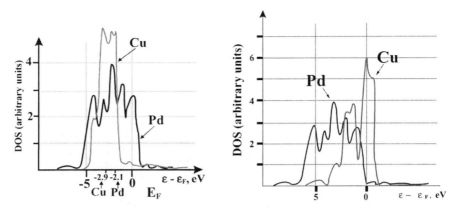

Fig. 29.4. Spectra of states for Pd and PdCux clusters [6].

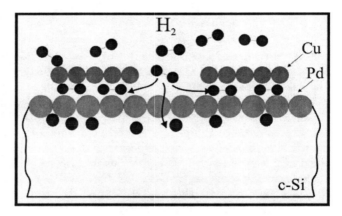

Fig. 29.5. Mechanism of spillover effect.

Cu states for nanostructures become non-occupied. Hence, a nanostructured cluster layer becomes catalytically very active.

One more important catalytic phenomenon is a **spillover** effect, which is perspective for the enhancement of gas sensitivity. It consists in an accumulation of gas molecules at the nanocluster/substrate interface. Figure 29.5 demonstrate the mechanism of spillover effect for H_2 adsorption: the increasing value of this effect is accompanied by an increase of the time of charge accumulation.

In our previous analysis [13], we calculated the values of energy levels and surface barriers for nanoclusters of different materials and found remarkable shift up of the potential energy level of upper band. The same happens for inter-metallic nanoclusters. That means more unoccupied bonds for smaller nanostructured composites. In the case of nanostructures, not only Pt, Pd, but also typically non-catalytic materials can show catalytic effects.

29.3 Experimental results

Below we analyse the experimental results, which demonstrate the important role of nanostructured systems.

Figure 29.6 shows the thermodesorption spectra of H_2 for ultrathin Pd and Pd/Cu (for different number of the deposited atomic layers of Cu) electrodes deposited on a W substrate. A remarkable change of thermodesorption spectra is observed when the nanosized spots of Cu (1 monolayer of Cu) form on a 2 monolayer thick Pd. The three observed peaks are related to the surface adsorption (1-st peak), H_2 absorbed by bulk Pd film (2-nd peak), and to H atoms couples at differences interfaces. The *first peak is related to the* surface, the second – to the adsorption at the $Pd - W$ interface (it is nearly the same for both systems), and a new third peak is related to the spillover effect realised by Cu nanoclusters. This peak becomes dominating over all peaks. This shows the substantial increase of the sensing efficiency of Pd covered with Cu nanoparticles. The other experiments that are concerned to the spillover effect are to study the kinetics of H_2 adsorption by the MIS-sensor with the porous Si-interlayer and covered with Pd activated by nanoclustered Cu [7–12]. Figure 29.7 demonstrates the increase of the values of electric charge effect under influence of H_2 adsorption. O_2 adsorption was used for the quick compensation of induced charge after removing H_2 from the ambient by Ar flow.

Next set of experiments is devoted to adsorption on the nanoporous Si layers. The attractive properties of this material are very large characteristic surface ($\sim 10^3 \mathrm{M}^2/\mathrm{g}$) and capillary structure, both of which promote both the increase of gas sensitivity and the spillover effect. The electro-chemical etching of standard $p - Si(100)$ allows to prepare layers $1\,\mu\mathrm{m}$ thick with a porosity degree up to 70%, the size of pores appears 2 to 10 nm. The Fig. 29.8 demonstrates the topology of MIS structure with porous Si layers. In Fig. 29.9, the isotherm of H_2 adsorbtion measured on a MIS structure is demonstrated. The isoterm of H_2S adsorbtion in N_0 a.d O_2 gas ambient is shown in Fig. 29.10.

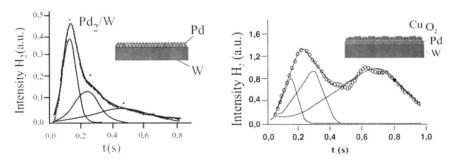

Fig. 29.6. Thermodesorption spectrum of H_2 for ultrathin Pd (2 monolayers) (a) and $PdCu$ (2+1 monolayers) (b) [6].

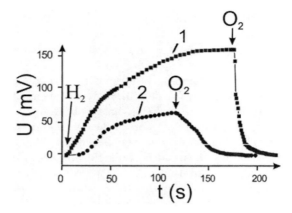

Fig. 29.7. Kinetic dependence of MIS capacitor sensor with thin (50 nm) Cu/Pd (1) and Pd (2) electrode layers under adsorption. The arrow indicates the beginning of gas inlet.

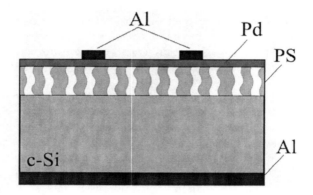

Fig. 29.8. Topology of MIS structure with porous Si lalyer.

The following conclusions can be made:

1. The use of the structure described enables to increase the sensitivity threshold about five times, from ~ 10 ppm (typical for previous investigations) to about $(1 \div 2)$ ppm.
2. Sensitivity to H_2S even at room temperatures (not observed before).
3. There are two regions with different adsorption laws. At low pressures (1–10^2 ppm), the square root dependence of adsorption centre concentration on pressure takes place:

$$\Delta n \sim A/(a + p/p^x)^{1/2} \tag{29.1}$$

At larger *pressures, this dependence turns to* inverse: $\Delta n \sim A^x/p$. The $1/p^{1/2}$ law is predicted for the dissociative atomic adsorption on the heterogeneous surface (Freindich law [13, 14]):

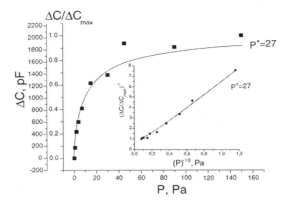

Fig. 29.9. Isotherm of H_2 adsorption on MIS sensor with porous layers.

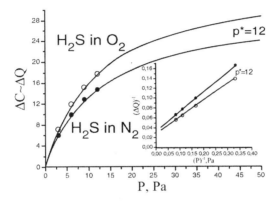

Fig. 29.10. Isotherm of H_2S adsorption in N_2 and O_2 ambients.

$$n = A \cdot p^{1/2} / (p^x + p)^{1/2} \sim AP^{-1/2} \qquad\qquad \text{at low p} \qquad (29.2)$$

and

$$n \approx A \cdot (p^x)^{-1/2} \approx Const \qquad\qquad \text{at large p,}$$

which corresponds to the monolayer covering. For multi-layer adsorption, the step-like dependence on pressure was observed: For figures use

$$n = \sum_{i=1}^{n} AiP^{-1/2} / (P_i^x + P)^{1/2}, \qquad (29.3)$$

which allows to determine characteristic values of P_i^x, indicateded on the creation of the each next mono layer. Besides, these value directly determine such important parameter as the characteristic energy E_{ai} of adsorption band for each type of molecules with definite substrate, calculated in [14, 15]:

$$p^x = \frac{B}{A}e^{-E_a/kT} \tag{29.4}$$

with $A = \nu \cdot C_a/4kT$ being the effective area of adsorption centre, $B = 1/\tau_a = \nu \cdot C_a N_{ta}$ the probability of adsorption, and E_a the energy position of adsorbtion

$$\frac{n_a}{N_t} = p^m/(p + p^*)^m, \quad m > 1/2 \quad or \quad m < 1/2, \tag{29.5}$$

and for many-center adsorbtion

$$n_a/N_{ti} = \sum_i^k (1 + p_i^*/p)^m \tag{29.6}$$

Here:

$$p^* = 4kTN_{ta}e^{-E_a/kT} \tag{29.7}$$

$$E_a = kT \cdot \ln \cdot \left(\frac{4kT \cdot N_{ia}}{p^*}\right) \tag{29.8}$$

The density of the electrically active adsorbtion centers N_{ai} can be estimated from the saturation parts of isotherms, obtained by capacity method: $\Delta Q = V_\Delta \cdot C$. For the saturation part of isoterm:

$$Q_{sat} = eN_a \cdot f_a \approx eN_a \tag{29.9}$$

The latter has place in the case of fully occupied centres, or

$$Q_{sat} \sim eN_a e^{-(E_a-E)/kT}. \tag{29.10}$$

(for small occupation).

Energy E_a indicates the effectiveness of catalytic processes by each of nanomaterials (nanocomposites of $PdCu$, $PtCu$). For the nanostructured surfaces and clusters, it acts as a marker for different adsorbed molecules and, hence, is responsible for the selectivity of sensors.

On Fig. 29.9 we illustrate typical isotherms of hydrogen electroadsorption effect for por-Si surface. The dependence follow the square-roof low (29.2) with saturation at large. The electro-adsorption effect for H_2S molecules is realised already at room temperatures completely due to the nanostructured substrate (nanoporous silicon layer, covered with nanocluster catalytically activ Pd and Cu layers). Here the isotherms of adsorption at small pressure are more or less linear (Fig. 29.10), which is reasonable for such strong inter-bonding molecule as H_2S [16]

$$\Delta n = \frac{Bp}{(p + p^*)} \sim B/p. \tag{29.11}$$

Table 29.1. The physical adsorption for nanosystems

Gas,	$\Delta C_{max},$ $\times 10^{-12} F$	$\Delta U,$ V	$\Delta Q,$ $\times 10^{-11} C$	$N,$ $\times 10^{8}$	$N_{ts},$ $\times 10^{13} m^{-2}$	$N_{t\nu}^{*},$ $\times 10^{22} m^{-3}$	$p^{*},$ Pa	$E_a,$ eV
H_2	2035	0,2	38,7	24,2	19,2	6,7	27	0,096
$H_2S \cdot (O_2)$	3540	0,03	10,62	6,63	5,26	1,8	12	0,084
$H_2S \cdot (N_2)$	2970	0,04	11,88	7,42	5,89	2,1	12	0,086

The energy E_a has twice or more smaller value than in previous case (H_2, H_2O). The analysis of H_2 gas adsorption isoterms has demonstrated its linear character in coordinates $P^{-1/2}$. That is typical for hydrogen dissociate adsorption. Whereas the analysis of hydrogen sulphide (H_2S) adsorption isotherm has demonstrated that corresponds to the P^{-1} dependence. This form of isotherm is typical for the molecular gas adsorption. However, in both cases the value of activation energy is rather small close to to 0,1 eV, which is typical for the physical adsorption, see Table 29.1.

29.4 Conclusion

In this paper, we have presented the theoretical analysis and experimental data on the adsorption and catalysis of H – containing molecules: dipole-like H_2, H_2O, and H_2S with slight dipole moment, which have however a strong energy of dissociation and small adsorption energy. The theory of the spectra of electron states in nanostructures demonstrates mechanisms responsible for the enhanced gas sensitivity and catalysis. The comparison of experimental results on the electro-adsorption phenomena with theory allows to estimate the characteristic energy bands on the surface, E_a, for different molecules. The latter can be regarded as the base for the development of selective sensors.

Acknowledgments

This work was supported by the following projects: STSU - 3819 (2007–2009), MON - grant M/175 - 2007 (Ukraine - Russia), and NASU - grant 25 - 2007.

References

1. Physics base of micro and nanoelectronics, *Unoversity, P.H. "BGC"*, Kiev (2005).
2. Y. Zhong, C. Rivas, R. Lake, K. Alam, T. Boyken, Y. Klimeck, Electronic properties of Silicon nanowires, *IEEE Transaction on Electron Devices* 52, 1097 (2005).
3. A. Zangwill, Physics of Surfaces, *Cambridge University Press* (1988).
4. M.O. Castro Gooyen, Complex Pd-Cu adsorption properties, *Surf. Sci. 307–309*, 387 (1994).

5. V.G. Litovchenko, Models of the adsorption-catalitic canters on transiton metals, *Condensed Matter Phys.* *1-2*, 388 (1998).
6. V.G. Litovchenko, A.A. Efremov, Ar. Kiv et al., Adsorption-catalytic properties of thin Pd and PdCux films, *Phys. Low-Dim. Struct.* *3-4,* 17 (2004).
7. V.G. Litovchenko, T.I. Gorbanyuk, A.A. Efremov, A.A. Evtukh, Effect of macrostructure and composition of the top metal electrode on properties of MIC gas sensors, *Microelectron. Realibility 40*, 821 (2000).
8. G. Bozzolo, J. Garces, R. Noeble, P. Abel, H. Mosca, Atomic modeling of surface and bulc properties of Cu, Pd and the Cu-Pd system, *Progress in Surface Science 73*, pp. 79 (2003).
9. V. Litovchenko, T. Gorbanyuk, O. Efremov et al., Catalytic proporties of ultrathin layers of Pd and alloys, *Ukr. Phys. J. 98*, 565 (2003).
10. V.G. Litovchenko, A.A. Efremov, Analyses of ageing mechanisms of gas MIS sensors with Pd(CuxPd) gate, *Sensor Electron. Microsyst. Technol. 1-2*, 5 (2004).
11. V.G. Litovchenko, T.I. Gorbanyuk, V.S. Solntsev, A.A. Evtukh, Mechanism of Hydrogen containing and oxygen molecules sensing by Pd- or Cu/Pd-porous Si-Si structures, *Appl. Surf. Sci. 234*, 262 (2004).
12. T.I. Gorbanyuk, A.A. Evtukh, V.G. Litovchenko and V.S. Solntsev, Nanoporous silicon doped by Cu for gas-sensing applications, *Physica E: Low-dimens. Syst. Nanostruct. 38*, 211 (2007).
13. V.G. Litovchenko, A.O. Grigorev, Calculation of the barriers for QW Structures, *Ukr. Phys. J. 52*, 17 (2007).
14. V.G. Litovchenko, Adsorbo-electrical effects in layered structures Insulator-Semiconductors, *J. Phys-Chem. 52*, 3063 (1978).
15. V.G. Litovchenko, Changing of the electronic parameters of surface by adsorption, *in Collection "Semiconductor techniques and microelectronics, Naukova Dumka,* Kiev, 3 (1972).
16. I. Lungstrem, Adsorption gas sensing, *Sensor Actuators 2*, 403 (1981).

BAND STRUCTURE AND ELECTRONIC TRANSPORT PROPERTIES IN II–VI NANO-SEMICONDUCTORS. APPLICATION TO INFRARED DETECTORS SUPERLATTICES AND ALLOYS

A. EL ABIDI, A. NAFIDI, A. EL KAAOUACHI, AND H. CHAIB
Group of Condensed Matter Physics, Physics Department, Faculty of Sciences, B.P. 8106 Hay Dakhla, University Ibn Zohr, 80000 Agadir, Morocco. nafidi21@yahoo.fr

Abstract. We report here electronic transport properties measurements, scattering mechanisms and theoretical results on Fermi energy, donor state energy and modeling carrier charge mobility in the ternary alloys $Hg_{1-x}Cd_xTe$ (x = 0.22) medium-infrared detector. We report again the band structure, magneto-transport results, X-ray diffraction, Seebeck and Shubnikov-de Haas effects (SDH) in the medium-infrared detector, narrow gap and two-dimensional p-type semiconductor HgTe(5.6 nm)/CdTe(3 nm) superlattice. The later sample is an alternative to the investigated alloy.

Keywords: band structure, magneto-transport, two-dimensional, nano- semiconductor, HgTe/CdTe superlattices

30.1 Introduction

The level of development achieved in the growth techniques and molecular beam epitaxy (MBE) of semiconductors has allowed the fabrication of different alloys, quantum wells and superlattices, and the observation of a number of fine aspects present in optical and transport properties of these structures. Among them The II–VI ternary random alloys $Hg_{1-x}Cd_xTe$ has been predicted as a stable alternative for application in infrared optoelectronic devices. Especially in the region of second atmospheric window (around 10 μm) witch is of great interest for communication. A number of papers have been published devoted to the band structure of this system as well as its

J. Bonča, S. Kruchinin (eds.), *Electron Transport in Nanosystems.*
© Springer Science + Business Media B.V. 2008

magneto optical properties [1]. The aim of a part of this work is to study the transport properties for Hg $_{1-x}$Cd$_x$Te (x = 0.22) and the contribution of various scattering mechanisms by doing theoretical curve fitting on the experimental temperature-mobility plot. The work of Essaki and Tsu in 1970 [2] caused a big interest to the study of superlattices made from alternating layers of two semiconductors. A number of papers have been published devoted to the band structure of this system as well as its magnetooptical and transport properties [3]. We report here the band structure, Landau levels (LL), new magneto-transport results, X-ray diffraction, Hall, Seebeck and SDH effects for a HgTe/CdTe SL grown by MBE.

30.2 Experimental techniques

The Hg$_{1-x}$Cd$_x$Te sample, with thickness of 5 μm, was grown by MBE on [111] CdTe substrate at 180°C. The nominal composition (x = 0.22) was determined by density measurements. The Hall Effect measurements and conductivity were done in the temperature range 4.2–300K. A weak current (I = 0.1 μA) flows along the sample under a transversal weak magnetic field (B = 0.1 T).

The HgTe/CdTe superlattice was grown by MBE on a [111] CdTe substrate at 180°C. The sample (90 layers) had a period d = d$_1$ + d$_2$ where d$_1$(HgTe) = 5.6 nm and d$_2$(CdTe) = 3 nm. It was cut from the epitaxial wafer, had typical sizes of 5 × 1 × 1mm^3. Carriers transport properties were studied in the temperature range 1.5–300 K in magnetic field up to 8 Tesla. Conductivity, Hall effect, angular dependence of the transverse magnetoresistance and the Hall voltage with respect to the magnetic field were measured. The measurements at weak magnetic fields (up to 1.2 T) were performed into standard cryostat equipment. The measurements of the magnetoresistance were done under a higher magnetic field (up to 7.7 T), the samples were immersed in a liquid helium bath, in the center of a superconducting coil. Rotating samples with respect to the magnetic field direction allowed one to study the angular dependence of the magnetoresistance and Hall voltage. The ohmic contacts of the two samples were obtained by chemical deposition of gold from a solution of tetrachloroauric acid in methanol after a proper masking to form the Hall crossbar.

30.3 The Hg$_{1-x}$Cd$_x$Te (x = 0.22) medium-infrared detector alloys

30.3.1 THEORY AND GENERAL FORMULATION

The calculation of the Fermi energy E_F at a temperature T have been done with the formula (30.1) giving the expression of the density n of electrons in the conduction band. We iterated on the value of E_F, between 0 and $E_g + 10k_BT$, that gives a density n near of the measured one n_{mes}.

$$n = \frac{2}{\sqrt{\pi}} N_c \int_0^\infty \frac{(E)^{1/2}}{1 + e^{(E-\eta)}} \, dE \qquad (30.1)$$

where: $\eta = (E_F - E_c)/k_BT$ and $N_c = 2\left(2\pi m^* k_BT/h^2\right)^{3/2}$. The energy gap was calculated by the empiric formula [4]:

$$E_g(x, T) = -0.302 + 1.93x - 0.810x^2 + 0.832x^3 + 5.03510^{-4}(1 - 2x)T \qquad (30.2)$$

Here T is the temperature in Kelvin, E_g is the material band gap in eV and x is the fractional composition value. The electron effective mass of m^* was obtained via Kane model band [4].

$$m^* = m_0/\left[-0.6 + 6.333(2/E_g + 1/(E_g + 1))\right] \qquad (30.3)$$

The effective mass of hole was fixed at $m_h = 0.55m_0$ [5] (m_0 is the free-electron mass). The origin of energies is taken at the top of the valence band ($E_v = 0$ and $E_c = E_g$). The carrier concentration n_{mes} was calculated from the standard expression: $n_{mes} = r_H/(|R_H| \cdot e)$ For convenience r_H was assumed to be unity. In order to determine the energy of the donor level E_d, we iterated in the formula (30.4) – giving the expression of E_F as a function of the temperature [6] – on values of E_d that vary between 0 and $E_g + 5k_BT$. The E_d energy is therefore the one that gives the E_F calculated previously. The calculation of E_d has been done in extrinsic regime (4.2K-77K). The density of donor impurities in the sample was $N_d \simeq n_{mes} \simeq 5.24 \times 10^{14}$ cm^{-3} in the regime of carriers freeze-out. The Hall mobility $\mu_H = |R_H.\sigma_0|$, where R_H is the Hall coefficient and σ_0 the conductivity without field, will be compared to the calculated electron mobility μ.

$$\begin{cases} \text{for } k_BT < E_g/10 : E_F = E_d + k_BT \ln\left\{1/4\left[-1 + (8 + (N_d/N_c)e^{(E_c - E_d)/k_BT})^{1/2}\right]\right\} \\ \text{for } k_BT > E_g/10 : E_F = \frac{E_c + E_v}{2} + \frac{k_BT}{2} \ln \frac{N_v}{N_c} \end{cases}$$
$$(30.4)$$

without field, will be compared to the calculated electron mobility μ.

In our calculations of the electron mobility, we included four types of scattering processes namely, acoustic phonons scattering, piezoelectric scattering, polar optical phonons scattering and ionized impurity scattering. The relations between electron mobility and temperature, for various scattering processes, are given as follows:

The acoustic phonons scattering mobility is given by the relation [7]:

$$\mu_{ac} = (3 \times 10^{-5}C_1)/(m^*/m_0)^{5/2} T^{3/2} E_{ac}^2 \qquad (30.5)$$

Where: $C_1 = 6.97 \times 10^{10} N \cdot m^{-2}$ [8] is the longitudinal elastic constant, $E_{ac} = 9.5$ eV [8] is the deformation potential for acoustic phonons.

The piezoelectric scattering mobility is given by the relation [9]:

$$\mu_{pz} = 2.6 \frac{\varepsilon_s}{(m/m_0)^{3/2} \, k'^2 \, (T/100)^{1/2}} \qquad (30.6)$$

ε_s is the dielectric static constant in the semiconductor and is given by the relation [5]:

$$\varepsilon_s = 20.5 - 15.5x + 5.7x^2 \qquad (30.7)$$

ε_0 is the dielectric constant in vacuum and k' is given by:

$$k' = \frac{e_{pz}}{\sqrt{e_{pz}^2 + \varepsilon_0 \cdot \varepsilon_s \cdot C_1}} \qquad (30.8)$$

Where e_{pz} is the piezoelectric constant. The polar optical phonons scattering mobility is given by the relation [9]:

$$\mu_{op} = 2.6 \times 10^5 \frac{\exp(\theta_D/T)}{\alpha \cdot (m^*/m_0) \cdot (\theta_D/k')} \qquad (30.9)$$

Where θ_D is the Debye's temperature and α is the coupling constant without dimension given by:

$$\alpha = \frac{1}{137} \sqrt{\frac{mc^2}{2k_B\theta}} \left(\frac{1}{\varepsilon_{op}} - \frac{1}{\varepsilon_s} \right) \qquad (30.10)$$

ε_{op} is the dielectric optical constant. α is 0.39 for CdTe and 0.05 for HgTe [10]. The Debye's temperatures of HgTe and CdTe are respectively 147.5 K [11] and 160 K [9]. When the alloy $Hg_{1-x}Cd_xTe$ is a two mode system in which the optical modes associated with CdTe and with HgTe are both observed, the mobility was calculated from Eq. (30.10) by first substituting the CdTe values for the appropriate parameters, then using the values for HgTe in the expression and averaging the two results [10] with the expression:

$$\frac{1}{\mu_{op}} = \frac{x}{\mu(CdTe)} + \frac{1-x}{\mu(HgTe)} \qquad (30.11)$$

The ionized impurity scattering mobility μ_{imp} is given by the Brooks-Herring equation [10]:

$$\mu_{imp} = \frac{3.28 \, 10^{15} \varepsilon_s^2 T^{3/2}}{(N_a + N_d)(m^*/m_0)^{1/2} \left[\ln(b+1) - \frac{b}{b+1} \right]} \qquad (30.12)$$

$$\text{where: } b = \frac{1.29 10^{14} \, (m^*/m_0) \, \varepsilon_s T^2}{N^*}$$

and N^* is the effective screening density given by:

$$N^* = n + \frac{(n + N_a)(N_d - N_a - n)}{N_d}$$

N_a and N_d are, respectively, the concentration of acceptors and donors. The total mobility μ is calculated using the relation: $1/\mu = \sum_i 1/\mu_i$ where μ_i refers to the individual calculated mobility from various scattering processes and i is the number of scattering process included.

30.4 Results and discussion

The measurement of R_H at different temperatures shows that the sample is an n-type semiconductor in Fig. 30.1a. At low temperature (in extrinsic regime), we are in presence of a carrier freeze-out and the density of the carriers is nearly independent of the temperature. After R_H increases slightly according to $10^3/T$ following the diminution of the density of carriers as described by: $n \simeq A \exp - (\Delta E_d / 2k_B T)$ [9], $\Delta E_d = E_c - E_d$ is the full activation energy. The slope of $|R_H| = f(10^3/T)$ curve allow us to determinate the donor ionisation energy $\Delta E_d = -0.67$ meV. That shows the existence of a donor state at 0.67 meV above the conduction band. When the temperature increases, in the intrinsic regime, the electrons go thermally from the donor level (caused by Te rather than Hg interstitials [13]) and valence band to the conduction band. So the carrier's concentration increases and Hall constant decreases. The slope of $|R_H| = f(10^3/T)$ gives $E_g = 178$ meV witch agree well with $E_g(300\text{K}, x = 0.22) = 184$ meV, given by the formula of Hansen et al. The corresponding detection wave length to this small gap is $\lambda = 6.89\,\mu$m. This situates the sample as medium-infrared detector (MIR). The measured lectrons concentration n_{mes} at 4.2 K, 77 K and at the ambient temperature 300 K and their corresponding calculated Fermi energy are given in Table 30.1.

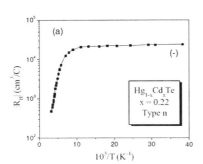
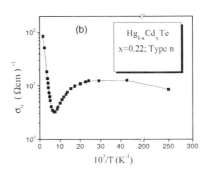

Fig. 30.1. Temperature dependence of the Hall constant (a) and the conductivity (b) in Hg $_{1-x}$Cd$_x$Te ($x = 0.22$).

Table 30.1. The concentration of electrons and calculated Fermi energy E_F. The gap E_g and the donors energy level's E_d at different temperatures

T(K)	n_{mes}(cm^{-3})	E_F(meV)	E_g(meV)	E_d (meV)
4	$(2.5 \pm 0.3) \times 10^{14}$	97	95.2	98
77	$(3.01 \pm 0.36) \times 10^{14}$	103	116.9	111
300	$(1.15 \pm 0.01) \times ^{16}$	154	183	–

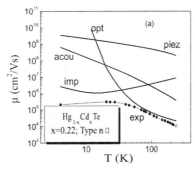 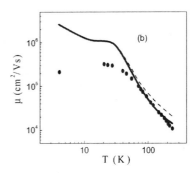

Fig. 30.2. Electron mobility as a function of temperature. The closed circles represent the experimental data. (a) The various other curves refer to calculated contributions of various scattering mechanisms and (b) The dashed curve represent the sum of all contributions of various scattering mechanisms while the solid line represent the sum taken in account the additional term.

At low temperatures (25–100 K), the thermal energy is sufficient to ionize all donors and one reaches a plateau of concentration of carriers. This allows us to extract the density $N_D : n \simeq N_D - N_A \simeq N_D \simeq 5.24 \times 10^{14}\,\mathrm{cm}^{-3}$. And to determine the donors energy level's E_d in Table 30.1.

In our calculations we found an electronic donor state approximately at 3 meV above the conduction band at T = 4.2 K. A resonant electronic state 8 meV above the conduction band at 4.2 K was observed in Shubnikov de Hass experiments in Hg $_{0.8}$Cd $_{0.2}$Te [14] witch is different to 0.67 meV from our Hall experiments and the calculated 3 meV. The intrinsic conduction is $\sigma_0 = |e|\,n_i\mu_h(1 + b)$ where $b = \mu_e/\mu_h$. Since the intrinsic concentration n_i rises exponentially with temperature as $n_i = \sqrt{N_cN_v} \cdot \exp - (E_g/2k_BT)$. The temperature dependence of σ_0 is shown in Fig. 30.1b. When the temperature is lowered, σ_0 enters the extrinsic region, where the carrier concentration is constant, and σ_0 rises the simplest case $\sigma_0 \, \alpha \, \tau_0 \, \alpha T^{-3/2}$. At lower temperatures, there is a carrier freeze-out.

In Fig. 30.2a we see the effect of single-scattering mechanisms on the total electron mobility. Ionized impurity scattering is seen to dominate below 25 K, optical phonons scattering at high temperatures (>70 K)and the mobility is generated by the both mechanisms of scattering in the intermediate temperatures. At lower temperatures the calculated mobility is about a factor of 10 greater than the measured value. This disagreement can be due to the impurities type acceptors and the non-parabolicity of the conduction band not taken into account in our calculations. The discrepancy is reduced at higher temperatures to a factor of 1.3. In order to improve the agreement between theory and experiment, in the intrinsic regime, we added a term which replaces additional scattering mechanisms not discussed in this paper such as electron-hole scattering and scattering by optical transversal phonons. We tried to estimate such a term by a trial function: $\mu_a = \mu_1 \exp(\mu_2/T)$ [12] where $\mu_1 = 15{,}835 \, \text{cm}^2\text{V}^{-1}\text{s}^{-1}$ and $\mu_2 = 239 \, \text{K}$ are fitting parameters. These values have been determined while minimizing the sum of squares which characterizes the quality of the fit. The additional term within these values gives a good agreement for temperatures higher than 70 K in Fig. 30.2b.

30.5 The HgTe($d_1 = 5.6$ nm)/CdTe($d_2 = 3$ nm) medium-infrared detector superlattice

30.5.1 THEORY OF STRUCTURAL BANDS

HgTe is a zero gap semiconductor (due to the inversion of relative positions of Γ_6 and Γ_8 band edges) when it is sandwitched between the wide gap semiconductor CdTe (1.6 eV at 4.2K) layers yield to a small gap HgTe/CdTe superlattice witch is the key of an infrared detector. Calculations of the specters of energy bands $E(k_z)$ and $E(k_p)$, respectively, in the direction of growth and in plane of the superlattice; were performed in the envelope function formalism [16]. The rapport d_1/d_2 governs the width of superlattice subbands (i.e. the electron effective mass). A big d_1/d_2 moves away the material from the bidimensional behaviour. In Fig. 30.3b, c we can see the specters of energy $E(k_z)$ and $E(k_p)$, respectively, in the direction of growth and in plane of the superlattice at 4.2 K. Along k_z the subbands E_1 and h_1 are large and non-parabolic. Along k_p, E_1 and h_1 increase with k_p whereas HH_n decreases. This yield to an anti-crossing of HH_1 and h_1 at $k_p = 0.01 \text{Å}^{-1}$ corresponding to a magnetic field of 4 Teslas. The gap is $E_g(\Gamma, 4.2 \, \text{K}) = 111$ meV. Using the value of ε_1 and ε_2 at different temperatures between 4.2 K and 300 K [17] and taking P temperature independent, this is supported by the fact that from Eq. (30.4) in [18] P $\approx \varepsilon_G(T)/m^*(T) \approx$ cte, we get the temperature dependence of the band gap E_g, in the center Γ of the first Brillouin zone in Fig. 30.4. Note that E_g increases from 111 meV at 4.2 K to 178 meV at 300 K.

30.5.2 EXPERIMENTAL RESULTS AND DISCUSSION

At low temperatures, the sample exhibits p type conductivity (confirmed by thermoelectric power measurements) with a concentration $p = 1.84 \times$

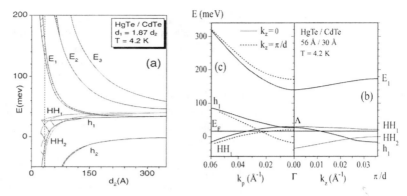

Fig. 30.3. (a) Energy position and width of the conduction (E_i), heavy-hole (HH_i), and the first light-hole (h_1) subbands calculated at 4.2 K in the center Γ of the first Brillion zone as a function of layer thickness for HgTe/CdTe superlattice with $d_1 = 1.87\,d_2$, where d_1 and d_2 are the thicknesses of the HgTe and CdTe layers, respectively. Calculated band along the wave vector k_z (b) and in plane along $k_p(k_x, k_y)$ (c) of the HgTe/CdTe superlatice at 4.2 K. E_F is the energy of Fermi level.

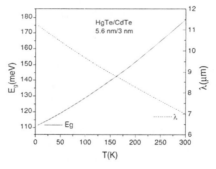

Fig. 30.4. Temperature dependence of the band gap E_g and cut-off wavelength λ, at the center Γ of the first Brillion zone.

$10^{12}\,\mathrm{cm}^{-2}$ and a Hall mobility $\mu_p = 8200\,\mathrm{cm}^2/\mathrm{V.s}$ in Fig. 30.5b [19]. The decrease of R_H (1/T) at 40K shown in "Fig. 30.5c" can be due to coupling between HgTe well (small d_1/d_2 and d_2), to the widening of carrier sub-bands under the influence of the magnetic field and/or to the overlap between involved carriers sub-bands (HH_1) and (h_1) at (k_z; k_p) = (π/d; $0{,}023\,\mathring{A}^{-1}$) along $E(k_p)$. In intrinsic regime, the measure of the slope of the curve R_H $T^{3/2}$ indicates a $E_g = E_1$-$HH_1 = 190$ meV witch agree well with calculated $E_g\,(\Gamma, 300\,\mathrm{K}) = 178$ meV. This relatively high mobility allowed us to observe the Shubnikov-de Haas effect until 18 Tesla. The measured period of the magnetoresistance oscillations (2D) gives $p = 1.80 \times 10^{12}\,\mathrm{cm}^{-2}$ in good agreement with weak field Hall Effect. At low temperature, the superlattice heavy holes dominate the conduction in plane. The HH_1 (and h_1)

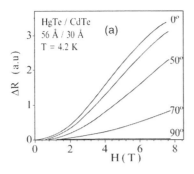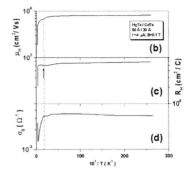

Fig. 30.5. Variation of magnetoresistance of the sample with various angles between the magnetic field and the normal to the HgTe/CdTe superlattice surface. Temperature dependence of the Hall mobility (b), weak-field Hall coefficient (c) and conductivity (d) in the investigated HgTe/CdTe superlattice.

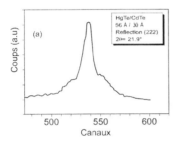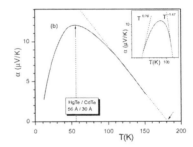

Fig. 30.6. (a) Room temperature X-ray diffraction profile around the (222) Bragg reflection of the HgTe/CdTe SL. (b) measured thermoelectrique power as a function of temperature of the HgTe/CdTe superlattice.

band is parabolic with respect to k_p^2. That permits us to estimate effectives masses $m_{HH1}^* = 0.297\, m_0$, $m_{h1}^* = 0.122\, m_0$ and the Fermi energy (2D) at 4.2K $|E_F - E_{HH1}| = |p\pi\hbar^2/m_{HH1}^*| = 14$ meV. Around the (222) Bragg reflection in Fig. 30.6a, a series of steps are observed corresponding to the oscillating counterparts described by Arch [20].

The difference with our results comes from the monocromacity and the power of their synchrotron source. Our SL is a modulated structure with high quality and interfaces. In Fig. 30.5a we can see that the angular dependence of the transverse magnetoresistance vanishes, when the field is parallel to the plane of the SL, indicating a two dimensional (2D) behaviour confirmed by the observation of SDH oscillations in Fig. 30.7a. The thermoelectric power, α, measurements shown in Fig. 30.6b indicate a p-type conductivity, confirmed by Hall effect measurements in Fig. 30.5c. At low temperature, $\alpha \sim T^{0.8}$ (in the top insert of Fig. 30.6b) is in agreement with theory where

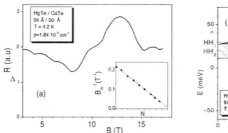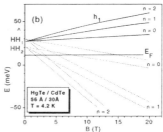

Fig. 30.7. (a) Variation of transverse magnetoresistance, with magnetic field of HgTe/CdTe superlattice, at 4.2 K (b) calculated Landau levels as a function of applied magnetic field of the HgTe/CdTe SL at 4,2K. E_F is the Fermi level energy.

$\alpha = [(\pi k_B^2 T(s + 1)]/3eE_F$ and the collision time $\tau \sim E^{s-(1/2)}$. This permits us to estimate the Fermi energy at $E_F = 12$ meV (in Fig. 30.7b) in agreement with the calculated $|E_F - E_{HH1}| = 14$ meV with s = 2.06 corresponding to holes diffusion by ionized impurities. It is relevant here to signal that the maximum of α at T = 55.2 K correspond to the shift of Hall mobility in Fig. 30.4b. Here $\mu_{HH1} = 0.411 \times \mu_{h1}$ at E_F. At T = 190 K, a reversal of α sign will occur corresponding exactly to the minimum of the conductivity σ_0 in Fig. 30.4d. In the intrinsic regime, the measure of the slope of the curve $R_H T^{3/2}$ indicates a gap $E_g = E_1 - HH_1 = 190$ meV witch agree well with the calculated E_g $(\Gamma, 300 \text{ K}) = 178$ meV. Here, $\alpha \sim T^{-3/2}$ indicates carrier scattering by phonons.

The relatively high Hall mobility ($\mu_p = 8,200 \text{ cm}^2/\text{Vs}$) allowed us to observe the SDH effect until 18 Tesla in Fig. 30.7a. The measured period of the magnetoresistance oscillations gives a concentration $p = 1.80 \times 10^{12} \text{ cm}^{-2}$ (in good agreement with $1.84 \times 10^{12} \text{cm}^{-2}$ of weak field Hall effect from Fig. 30.5c). Here we have deduced the effective masse of the degenerate heavy holes to be $m_{HH1}^* = \hbar^2 \pi p / E_F$ (in agreement with the theoretical $m_{HH1}^* = 0.297 m_0$). We calculated the energy of the Landau levels (LL) by transposing the quantification rule of the wave vector in the plane of the SL: $k_p^2 = (2n + 1)eB/\hbar$ where n are the quantum orders of LL. The crossing of E_F with LL inm Fig. 30.7b indicated the same magnetic field positions as those of the observed SDH oscillations minima in Fig. 30.7a.

The HgTe/CdTe superlattice is a stable alternative for application in infrared optoelectronic devices than the alloys $Hg_{1-x}Cd_xTe$ because the small composition x = 0.22 is difficult to obtain while growing the ternary alloys and the transverse effective masse in superlattice is two orders higher than in the alloy. So the tunnel length is small in the superlattice [21].

30.6 Conclusions

We have presented transport properties measurements and theoretical results of Fermi energy, donor state energy and mobility of Hg $_{1-x}$Cd$_x$Te (x = 0.22). Calculated mobility's are in agreement with experimental data points at high temperatures. However, we can conclude that the ionized impurity scattering dominate below 25 K, optical phonons scattering at high temperatures (>70 K) and the mobility is generated by the both mechanisms of scattering in the intermediate temperature regime. The average of Hall mobility is 2.7×10^5 cm^2/Vs at low temperature. This sample is a medium-infrared detector. We are trying to fit such mobility in p-type Hg$_{1-x}$Cd$_x$Te (x = 0.204) [15] and in the quasi two-dimensional (II–VI) p-type HgTe/CdTe superlattice [18]. The formalism used here predicts that the system is semiconductor, for our HgTe to CdTe thickness ratio $d_1/d_2 = 1.87$, when $d_2 < 14$ nm. In our case, $d_2 = 3$ nm and E_g (Γ, 4.2 K) = 111 meV. In spite of it, the sample exhibits the features typical for the semiconductor type p conduction mechanism. In the used temperature range, this simple is a medium-infrared detector ($7\,\mu$m $< \lambda < 11\,\mu$m), narrow gap and two-dimensional p-type semiconductor. The theoretical and magneto transport parameters are in good agreement. Note that we had observed a semimetallic conduction mechanism in the quasi 2D p type HgTe/CdTe superlattice [8]. In conclusion, the HgTe/CdTe superlattice is a stable alternative for application in infrared optoelectronic devices than the alloys Hg$_{1-x}$Cd$_x$Te.

References

1. J. L. Schmit, J. Appl. Phys. **41**, 2876 (1990).
2. L. Essaki and R. Tsu, IBM, J. Res. Develop., **14**,61 (1970).
3. G. Bastard, Phys. Rev. B **24** (10), 5693 (1981). Y. C Chang, et al., Phys. Rev. B **31**, 2557 (1985).
4. L. Hansen, J.L. Shmit, T.M. Casstleman, J. Appl. Phys. **53**, 7099 (1982).
5. J. Wenus, J. Rutkowski., A. Rogalski, IEEE Trans. On Electron Devices. **48**, 1326 (2001).
6. P. Kiérev in: Semiconductor physics, (Mir, Moscou, 1975), chap. 3, p. 213.
7. F. J. Blatt, in Solid state physics. edited by F. Seitz and D. Turnbull. vol. **4**, (Academic, New York, 1957), p. 332.
8. D. Chattopadhyay and B. R. Nag, J. Appl. Phys. **45**, Num 3 (1974).
9. K. Seeger in: Semiconductor physics, an introduction (Springer, 2002); chap. 6, p. 159.
10. W. Scott, J. Appl. Phys. **43**, No. 3, (1972).
11. J. G. Collins et al., J. Phys. C: Solid State Phys. **13**, 1649 (1980).
12. J. D. Wiley, Semiconductors and semimetals ,Vol. **10**, (Academic, New York, 1975).
13. R. Dornhaus, H. Happ, K. H. Müller, G. Nimtz, W. Schlabitz, P. Zaplinski, G. Bauer: in proc. XIIth Int. Conf. Phys. Semic., Stuttgart 1974, ed. by M. H. Pilkuhn (Teubner, Stuttgart 1974).

14. R. Dornhaus, G. Nimtz, W. Schlabitz, H. burkhard in : Solid State Commun. (1975).
15. A. El Abidi, A. Nafidi et al., Book of Abstracts (CA:235) 10th Moroccan Meeting On the chemistry of the Solid state (REMCES 10) Meknes, Morocco, April 27, 28 and 29 (2005).
16. Ab. Nafidi et al., the proceedings of the International Conference on theoretical Physics (HT 2002), Paris, France July 22–27, 2002, p. 274. G. Bastard, Phys. Rev. B **24** (10), 5693 (1981).
17. M. H. Weiler, in semiconductors and Semimetals, edited by R. K. Willardson and A. C. Beer (Academic, New York) vol. **16**, 119 (1981).
18. A. El Abidi, A. Nafidi and H. Chaib, submitted to Nato series Springer for the proceeding of the international conference Electron transport in nanosystems, 17–21 September, Yalta, Ukrania (2007).
19. A. Nafidi, A. El Abidi, and A. El Kaaouachi, the proceedings of the 27th International Conference on Physics of Semiconductors (ICPS27), AIP Conference Proceedings 772, New York: American Institute of Physics, 2005, pp. 1001–1002.
20. D. K. Arch, et al., J. Vac. Sci. Technol. **A4**, 2101 (1986).
21. A. El Abidi, A. Nafidi, and H. Chaib to be **published**.

AUTHOR INDEX

SUBJECT INDEX